DATA COMMUNICATIONS, NETWORKS, AND DISTRIBUTED PROCESSING

DATA COMMUNICATIONS, NETWORKS, AND DISTRIBUTED PROCESSING

Uyless D. Black

RESTON PUBLISHING COMPANY, INC.
A Prentice-Hall Company
Reston, Virginia

Library of Congress Cataloging in Publication Data

Black, Uyless D.
Data communications, networks, and distributed processing.

1. Data transmission systems. 2. Computer networks.
3. Electronic data processing—Distributed processing.
I. Title.
TK5105.B576 1983 001.64'404 82-23049
ISBN 0-8359-1230-2

10 9 8 7 6 5 4 3 2

Printed in the United States of America.

To David and Jim

CONTENTS

PREFACE

The idea for this book originated in data communications seminars I have conducted during the past three years for several U.S. and Canadian firms. Some of the seminar participants expressed difficulty in finding a reference on data communications that included presentations on network software and data bases. The dominant wish appeared to be for a book that described these concepts in practical, nontheoretical terms. I examined the offerings and came to the same conclusion. Many excellent references are available, but they are either beyond the grasp of the uninitiated, concentrate on one specific aspect of data communications, or are overly general. Therefore, my goal in writing this book is to provide a practical yet detailed explanation of data communications systems and networks, with highlights on software and data bases.

The task proved to be very challenging because the subject itself is not simple; an oversimplification would not serve any useful purpose. I decided to write the book using as few formulas and as little scientific jargon as possible, but I have also tried to provide sufficient technical detail to remove the mystery and confusion that often surrounds these topics.

I prepared and refined these approaches and ideas with many seminar participants, business associates, and clients. After concluding that I could meet my goal of practicality and technical clarity, I committed the project to prose and pen. I hope the result will be valuable to the reader.

I have many organizations and individuals to thank for their contributions. The IBM Corporation provided the background for the case study found in Chapter 11. My graduate students at The American University in Washington, D.C., provided useful input for several of the chapters. The seminar attendees provided valuable critiques (my thanks to Jim Calder at Control Data Corporation's Institute for Advanced Technology and Fred Karch and Lorraine Zimmer of the Institute for Professional Education for providing opportunities to conduct the seminars). My partners at the Center for Advanced Professional Education (CAPE) have provided tremendous help, and I am endebted to CAPE's Ken Sherman for his support and friendship during this effort. I especially wish to thank Karen Nold of Price Waterhouse, Inc., for her role in the completion of this book. I wish to acknowledge the encouragement and support I received from my friends and colleagues in the Federal Reserve System. John Denkler, Charles Hampton, and Bruce Beardsley at the Federal Reserve Board and Bob Boykin, Bill Wallace, and Neil Ryan at the Federal Reserve Bank of Dallas were instrumental in providing assignments that helped my efforts. Their support has been a great motivator. Bill Schleicher's friendship and perspective have been quite valuable to me. Also, my appreciation and thanks to Agnes Harris, Wanda Rolandson, Jean Stum, and Laura Celia for their tremendous administrative support.

Last and far from least, my special thanks to Miccio for his constant attention to my efforts and to my favorite son Tom for his continuous love and sense of humor.

1

INTRODUCTION

PURPOSE OF DATA COMMUNICATIONS SYSTEMS AND NETWORKS

During the past several years, there has been an increasing recognition within business and academic circles that certain nations have evolved into information societies. These countries now rely heavily on knowledge and information to spur economic growth. Many leaders today believe the generation of information has provided the foundation needed to increase the efficiency and productivity of a society. The use of information has lowered the costs of producing society's goods and has increased the quality of these products. The computer has been the catalyst for this information revolution.

To gain an appreciation of the statements of the business and academic leaders, one need only examine selected segments of society to understand what the computer has done to provide this vital information.

- The many benefits from space exploration would not be available if the computer did not exist to process the millions of data elements used to guide space vehicles in their probe of the solar system.

- Transportation reservations systems (airline, hotel) and the nation's highly efficient methods of distributing goods are actually made possible through the computer.
- Some industries have significantly lowered the costs and increased the quality of their manufactured goods through the use of automated processes and computerized robots.
- Office tasks requiring a handful of people would require hundreds of workers if office automation were not available.
- Oil exploration, energy conservation, defense measures, and medical research and development are all dependent upon the computer's ability to store and process information.

The computer has provided the foundation for the information society, but it is often a scapegoat for society's problems. One can read stories daily of the problems supposedly created by the computer and the reader has probably had similiar negative experiences. Yet, for all its perceived shortcomings, the computer and its contribution to our information society has made our lives immeasurably more pleasant and affluent. Like it or not, we have become dependent upon the computer.

The information society is also based on data—the raw resource that comprises information. The essence of automation and the information society is the processing, manipulation, and creation of data to provide something intelligible—information.

This powerful capability is strengthened by the use of data communications systems. The systems provide for the transport of data and information between the computers. Data communications provides the connections to computers located in the far reaches of a country and the world. The computer facilities are tied together by data communications components, forming a network of automated resources to support the many functions of a business or organization.

Data communications is a vital part of the information society because it provides the infrastructure allowing the computers to communicate with each other. An airline reservation system uses data communications to link the reservation offices to the computer. The Apollo space flights and the space shuttle use data communications systems to send data to and from the rockets and the command centers on earth. Office workers are dependent on networks to provide the use of data files and computers to different departments in an organization.

Thus, data communications systems are an integral part of our lives. They provide the foundation for our information society. Many people do not recognize how pervasive the systems are in our economy and social activity. Yet, the importance of data communications systems is so great that we should understand more about the field. The layman can benefit from knowing their characteristics and uses. The computer professional should understand data communications and networks in order to take advantage of their capabilities and avoid the pitfalls associated with their misuse. This book attempts to provide the necessary information for both the professional and the layman.

ORGANIZATIONS IN THE INDUSTRY

The United States communications industry is dominated by the American Telephone and Telegraph Company (AT&T). However, unlike many countries, the United States has opened communications to competitive offerings and, in recent years, the industry has become quite diverse. In other countries, the communications industry remains more highly regulated and is dominated by the countries' governments through an organization commonly known as the PTT (Postal, Telegraph, and Telephone Administration). It is quite likely that by the time this book is printed Congress and/or the Courts will divest AT&T, resulting in a communications structure that differs from the description presented herein. The Justice Department and AT&T have recently agreed to make some rather radical changes—more about this later.

The United States has one of the most reliable and efficient public communications systems in the world today. In terms of size, over 40% of the 300 million telephones in the world are in the United States. The future of communications in the United States will be greatly dependent on the action of the Congress, the Courts, and the regulatory agencies. Many people believe a sound communications infrastructure for the 1980s and 1990s is as important as the communications links (canals, railways, roads) were to the United States in previous times. Our future in this vital industry is greatly dependent upon the political and legal decisions that are being made today.

Common Carriers

The basic communication services in the United States, such as local connections to homes and offices and long-haul lines, are provided by common carriers. Common carriers were named in earlier days when certain industries (railroad, shipping) provided a common service for all people without discrimination and at a reasonable cost. These carriers provide the lines that link the computers in a network. Initially, the common carriers were given monopoly powers because Congress believed competition (with redundant services) made little sense. The wire communications industry grew up as a common carrier service. It was treated as a part of land transportation because the telegraph wires were constructed on railroad rights-of-way.

The best known communications common carriers are the telephone companies. They number well over 1,000 in the United States, although only 250 have more than 5,000 customers. With the exception of AT&T and GTE, the other carriers provide service for one-half the United States' geographical area but operate less than 20% of the telephones. Each telephone company serves a specific area under a franchise from a municipal or state government. Interstate communications are regulated by the Federal Communications Commission (FCC).

Specialized Common Carrier (SCC). In an effort to foster competition, the Federal Communications Commission has created an industry unique to the United States. Like the common carrier, the specialized common carrier provides the basic communications services such as long-haul lines, but unlike the common carrier, the specialized common carrier provides only a specific or specialized offering. An example is MCI. This organization was the first specialized common carrier, initially offering microwave facilities between Chicago and St. Louis. The specialized common carrier industry is now thriving, offering many services to the public and providing viable alternatives to the common carriers.

Value Added Carrier (VAC). The value added carrier industry is also a creation of the Federal Communications Commission and, like the specialized common carrier, it is an industry unique to the United States. In 1976, the FCC ordered all common carriers to eliminate restriction on resale and shared use of their services and thus provided for the genesis of the VAC. The value added carrier will lease the backbone communications facilities from a common carrier or a specialized common carrier. The VAC then augments the facilities with additional capabilities (i.e., adds a value) and offers these combined services to the public. Packet switching network vendors are an example of value added carriers.

Federal Communications Commission (FCC)

The Federal Communications Commission is responsible for the regulation of interstate radio, television, telegraph, and telephone communications and foreign-related communications. The FCC was created by the Communications Act of 1934 and is authorized to regulate the common carrier industry. Prior to 1934, the Interstate Commerce Commission regulated wire communications. All proposed interstate services must be reviewed and approved by the Federal Communications Commission. Intrastate services are under the auspices of state or municipal governments. Common carriers file an application (tariff) for the new service. The tariff describes the service, the price of the service, and the reason for offering the service. Tariffs, once filed and approved, can be used by other carriers. The approved tariff also acts as an agreement between the carrier and the user.

International Record Carriers (IRC)

These carriers provide for transborder data communications between countries through "gateway" facilities. The common carrier lines connect to the IRC lines at the gateway points. Profits are usually shared by a consortium of the companies in accordance with bilateral international agreements. The major United States IRCs are International Telephone and Telegraph (ITT), Western Union International (WUI), and RCA Globecon. The IRC role is changing; the companies are moving into domestic communications and other organizations are moving into the IRC arena.

National Telecommunications and Information Administration (NTIA)

NTIA is a part of the U.S. Department of Commerce and has replaced the Office of Telecommunications Policy (OTP). The old agency was established by President Nixon in 1970 to plan and coordinate communications programs in the government. NTIA provides similiar functions: it conducts analyses and prepares papers, promotes technology, and participates in managing the government's share of the radio spectrum.

The Communications Satellite Corporation (COMSAT)

COMSAT was created by Congress with the Communications Satellite Act of 1962 for the purpose of coordinating and promoting satellite communications in the United States and to act as the United States' satellite carrier. It is a private company with shareholders and distributed stock. COMSAT plays an important role for the United States in working with other countries to share worldwide satellite systems. COMSAT represents the United States in the INTELSAT organization.

International Organizations

INTELSAT. INTELSAT (the International Telecommunications Satellite Organization) is an organization of over 100 countries. Its purpose is to share in the development and use of satellite systems. INTELSAT was formed in 1964 and has been very successful in meeting its goals. The Early Bird system and the INTELSAT satellites are examples of INTELSAT activities. INTELSAT's first satellite was operational in 1965 and had 240 voice channels. Recent INTELSAT satellites have a capacity for 12,500 voice-type circuits plus two television channels.

CCITT. The CCITT (Comité Consultaif Internationale de Télégraphique et Téléphonique) is a standards body under the International Telecommunications Union, an agency of the United Nations. The CCITT is the primary organization for developing standards on telephone and data communications systems among participating governments. It is an influential body responsible for the development of standards such as X.21 and X.25. United States membership on CCITT comes from the State Department at one level (the only voting level); a second level of membership covers private carriers such as AT&T and GTE; a third level includes industrial and scientific organizations; a fourth level includes other international organizations; and a fifth level includes organizations in other fields that are interested in CCITT's work.

The United States has established several CCITT-related Study Groups. U.S. CCITT Study Group A addresses the regulatory aspects of international communications, including telegraph operations, telex service, tariffs, public networks, and many aspects of telephony. Study

Group B concentrates its efforts on telegraph transmission, equipment specifications, and telegraph signaling. Study Group C studies international telephone operations and Study Group D addresses problems relating to international data transmission service and international data networks. The CCITT Study Groups established for the period of 1980–1984 are listed in the Appendix.

International Standards Organization (ISO). The ISO is a voluntary organization consisting of national standards committees of each member country. The ISO coordinates its activities with CCITT on common issues. It is a fourth level or D class member of CCITT. ISO has produced several well-known standards such as High Level Data Link Control (HDLC). The organization has a number of subcommittees and groups working with CCITT and the American National Standards Institute (ANSI) to develop standards for encryption, data communications, public data networks, and the well-known Open Systems Interconnection (OSI) model.

American National Standards Institute (ANSI)

ANSI is a voluntary standards body in the United States and is a member of the ISO. It develops standards itself and accepts standards proposals from other organizations in the United States. ANSI activities are extensive. In addition to standards on programming languages, such as Cobol and Fortran, ANSI has several committees and groups working with communications standards as well. Its body, ANSC X3 (American National Standards Committee 3), and subcommittees provide the organization and mechanism for members to develop communications standards in conjunction with the ISO groups. The Appendix depicts the relationships of ANSI and ISO bodies. ANSI cooperates with CCITT through the State Department's Industry Advisory Group (IAG), which was created to allow ANSI and other standards groups more participation in CCITT issues.

These organizations are becoming more influential as more people recognize the value of standards. As one noted industry leader said, it is economically ridiculous and philosophically inane for each company to build it "own railroad gauge." We will see later how important the international standards have become in the United States and we will also learn how these standards work in a communications system and a network. Figure 1-1 summarizes the relationships of the international organizations.

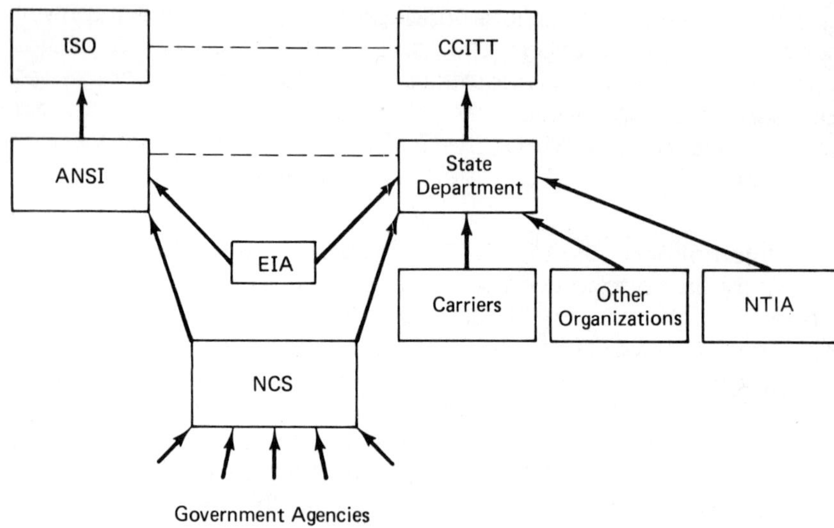

FIGURE 1-1. Data Communications Organizations.

Electronic Industries Association (EIA)

The EIA, a national trade association, is also very influential in developing standards in North American countries. The EIA work focuses primarily on electrical standards. Their more notable efforts include RS232-C and RS449. Subsequent chapters discuss the EIA standards.

National Communications System (NCS)

The NCS is responsible for U.S. government standards administered by the General Services Administration (GSA). Those government agencies having large telecommunications facilities belong to NCS. NCS works closely with these agencies as well as EIA, ANSI, ISO, and CCITT in fostering national and international standards. NCS is also responsible for preparing emergency plans for the nation's communications facilities.

National Bureau of Standards (NBS)

The NBS is responsible for making recommendations to the President concerning federal data processing and communications standards. NBS issues data processing guidelines for federal agencies called Federal Information Processing Standards (FIPS) and telecommunications standards called the Federal Telecommunications Standards Program (FTSP). The latter standards are developed in conjunction with NCS.

The Telephone System

The telephone system, still the most frequently used media for data communications systems and networks, is organized as depicted in Figure 1-2. The system is designed as a highly distributed multilevel switching hierarchy. At the bottom of the hierarchy is the user instrument, commonly the telephone, which is attached by two copper wires (local loops) to the phone company's local (central or end) office. Since each subscriber has a local loop connection, it can be seen that a considerable investment exists in this part of the

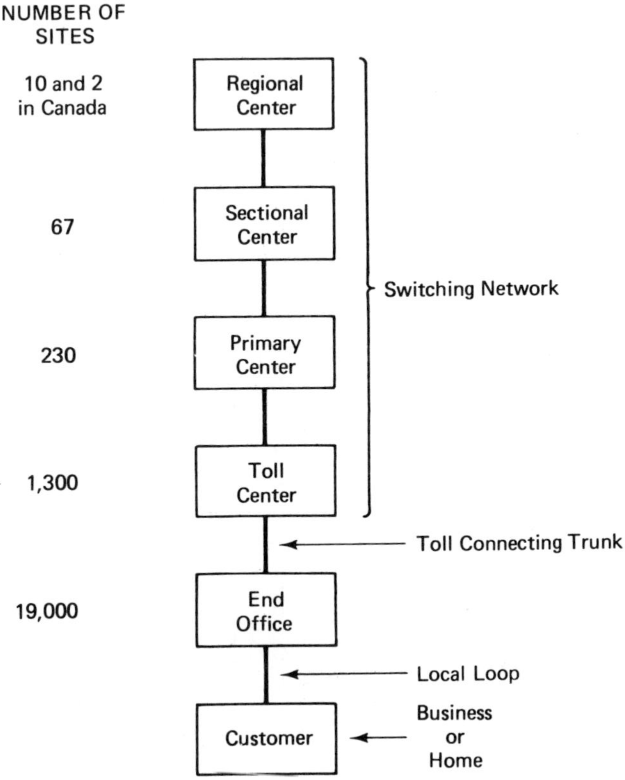

FIGURE 1-2. Telephone Company Hierarchy.

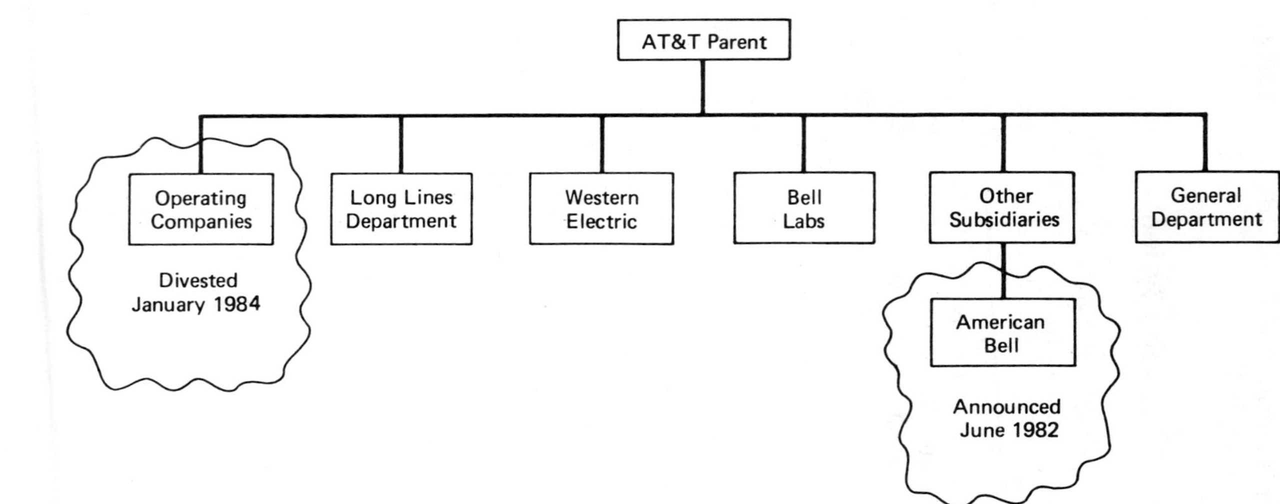

FIGURE 1-3. American Telephone and Telegraph Company.

network. The other offices in the telephone hierarchy are switching centers responsible for routing calls through the system based on the dialed number and the traffic load of other calls in the network. The switching is accomplished through computers and/or electromechanical devices.

The switching offices are connected by trunk circuits. Each local office also connects by toll connecting trunks to an associated toll office (or toll center) servicing a particular region of the country. The higher level offices form a network through intertoll trunks to provide for routing of calls between regions. See Chapter 2 for a more complete explanation of telephone switching.

The dialed number serves as an address to the called party. The local loop is identified by the last four digits of the 10-digit number; an end office by digits 4, 5, and 6; and a toll center by one of several possible area codes:

Toll	End office		Local loop
(XXX)	XXX	-	XXXX

American Telephone Telegraph (AT&T). AT&T (commonly known as the Bell system) is the dominant telephone company in this country. Until January 8, 1982, the organization was composed of 22 operating telephone companies (see Figure 1-3). These companies, to be divested from AT&T, provide for services within their respective areas in the country. Another major subsidiary, the Long Lines Department, is responsible for operating long distance interstate lines. The Western Electric Company is the manufacturing arm for AT&T. The other major subsidiary, Bell Labs, is responsible for research and development for the company. Bell Labs is one of the world's largest research organizations and has conducted some very innovative research in the past. (The transistor was invented at Bell Labs).

AT&T is the world's largest public company and one of the world's largest international corporations. It dwarfs most other organizations. Its assets are $134 billion and its 1980 revenues were greater than $51 billion—about 2% of this country's Gross National Product (GNP). Western Electric itself ranks very high in the Fortune 500. With over one million people in its employment, AT&T has more than 1% of the total U.S. labor force at its disposal. It is easy to understand why some other communications companies fear AT&T if the giant is loosened from its regulatory reins.

On January 8, 1982, the Justice Department and AT&T settled a seven-year-old antitrust suit. In the settlement, AT&T gives up its 22 operating companies (about $80 billion worth of assets). In turn, AT&T is allowed to enter into markets that were prohibited by a 1956 consent decree. These markets, such as computer data processing, information distribution, and cable television, are now open to AT&T, Bell Labs, and Western Electric. As of this writing, the long-range implications of this historic decision are not clear, but the impact will be quite significant and will affect the lives of all citizens who use the nation's telephone system. It is probable that the Courts and/or Congress will challenge the settlement.

General Telephone and Electric Corporation (GTE). GTE is the second largest telephone company in the United States. It has approximately 8% of the telephone business. GTE has moved aggressively into other areas of communications. It has two manufacturing subsidiaries and a nontelephone subsidiary. One of its more recent and well-known moves in the industry was the acquisition of the Telenet public network.

2

BASIC CONCEPTS OF DATA COMMUNICATIONS

HOW DATA ARE TRANSMITTED

Since the purpose of computers, data communications, and networks is to process data into information, a discussion of how data are represented on these media is in order. Data are stored inside a computer and transmitted on a communications system in the form of binary digits, or bits. The digits are either 1s or 0s and are coded in accordance with the binary (base 2) number system.

The binary bits inside a computer are represented by the level of polarity of electrical signals. A high-level signal within a storage element in the computer could represent a 1; a low-level signal, a 0. These elements are strung together to form numbers and characters, such as the number 6 or the letter A, in accordance with established codes.

Data are transmitted along the communications (usually the telephone network) path between computer-oriented devices using electrical signals and bit sequences to represent numbers and characters. In some instances, the data representation may be by light signals as in optic fibers. (More on this topic later.) The bit representations depict user data and control data. The control data are used to manage the communications network and the flow of the user data.

Figure 2-1 depicts how the data (the number 9 in base 2) moves from a sending computer device, through the communications

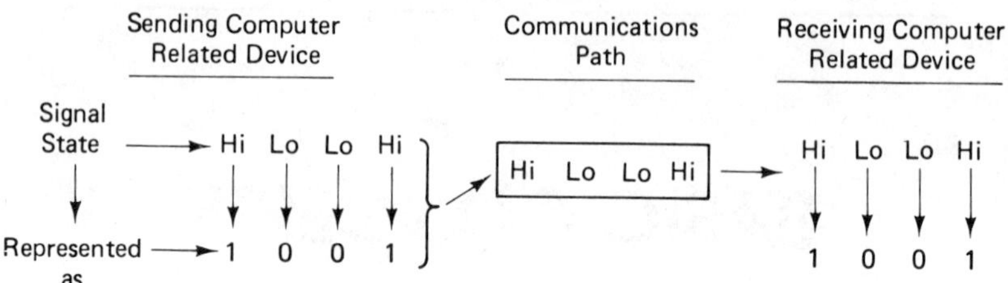

FIGURE 2-1. Transmitting Data.

medium, and into a receiving computer device. The reader should be aware that the binary data code is converted to base 10 for human consumption when it is displayed on terminals and printouts.

Throughout this book, the term *bits per second* (bit/s) is used. The term refers to the number of binary bits per second that are transferred through a communications path or component. If a 2400 bit/s line uses an eight-bit code to represent a number or character, then the character per second rate is 300 (2400/8 = 300). The majority of communications speeds are quoted in bit/s rates. The terms *bit/s*, *bs*, and *bps* are used interchangeably in this book.

It should be emphasized that a bit traveling down a communications path is actually a representation of the electrical or optical state of the line for a certain period of time. The bit 1 may be depicted by placing a stronger electrical signal on the line for a few fractions of a second, and a 0 would be represented by a low-level signal for the same period of time. This period is referred to as *bit time*, *bit length*, or *bit duration* on the line. Generally speaking, a 2400 bit/s speed requires the line to change its signal state 2400 times in one second. Stated another way, the bit time for each of the 2400 bit states is .000416 second (1 second/2400 changes = .000416 or 416 microseconds).[1]

Transmission Characteristics

A general knowledge of the characteristics of electrical transmission is essential if the reader is to gain an understanding of data communications. Line capacity, error control techniques, communications software, and many other network components

[1] The following notations are used in this book: .001 = one millisecond (ms); .000001 = one microsecond (μs); .000000001 = one nanosecond (ns). Also: 1,000 = kilo (k); 1,000,000 = mega (m). 1,000,000 = mega (m); 1,000,000,000 = giga (g).

are all analyzed and designed around the capabilities and limitations of electricity.

As stated earlier, data are transmitted by the alteration of an electrical signal to represent 1s and 0s. The electrical signal state manifests itself by either the signal level or some other property of the complex electrical signal. The movement of the signal over its transmission path is referred to as *signal propagation*. On a wire path, the signal propagation is a flow of electrical current. Radio transmission between computer sites without the use of wires is accomplished by emitting an electrical signal that propagates as an electromagnetic wave.

We know that all matter is composed of basic particles that may contain an electrical charge. Some of these particles, called *electrons* and *protons*, have negative and positive polarity, respectively. The particles group themselves in an orderly fashion to form atoms; the negative and positive charges attract each other to create the stability in the atom. An electrical current flow is generated by the introduction of an electric charge at one end of the communications path or conductor. For example, placing a negative charge at the transmitter end of a conductor would repel the negatively charged electrons in the path toward the other end, thus creating a current, i.e., a flow of electricity. In essence, electrical current (and a data communications signal) is the movement of these electrons down a conductor path.

Most signals consist of oscillating wave forms as shown in Figure 2-2(a). The oscillating signal has three characteristics that can be varied in order for it to convey computer-generated data (amplitude, frequency, and phase). The amplitude or voltage is determined by the amount of electrical charge inserted on the wire. This voltage can be set high or low depending on the binary state—that is, a 1 or a 0. Another characteristic of electricity is the power or strength as measured in watts. The signal power determines the distance the signal can travel or propagate on a wire communications circuit.

The signal is also distinguished by its frequency, or number of complete oscillations of the waveform in a given time. Frequency is measured in oscillations per second. The electrical industry has defined the unit of one hertz (Hz) to mean one oscillation per second. Other terms used to describe hertz are *baud* and *cycles per second*. The frequency of the wave has no relation to the amplitude. Signals can have many different combinations of the two. The *amplitude* indicates the signal level and the amount of negative or positive voltage, while the *frequency* indicates the time rate (in hertz) of the signal oscillation.

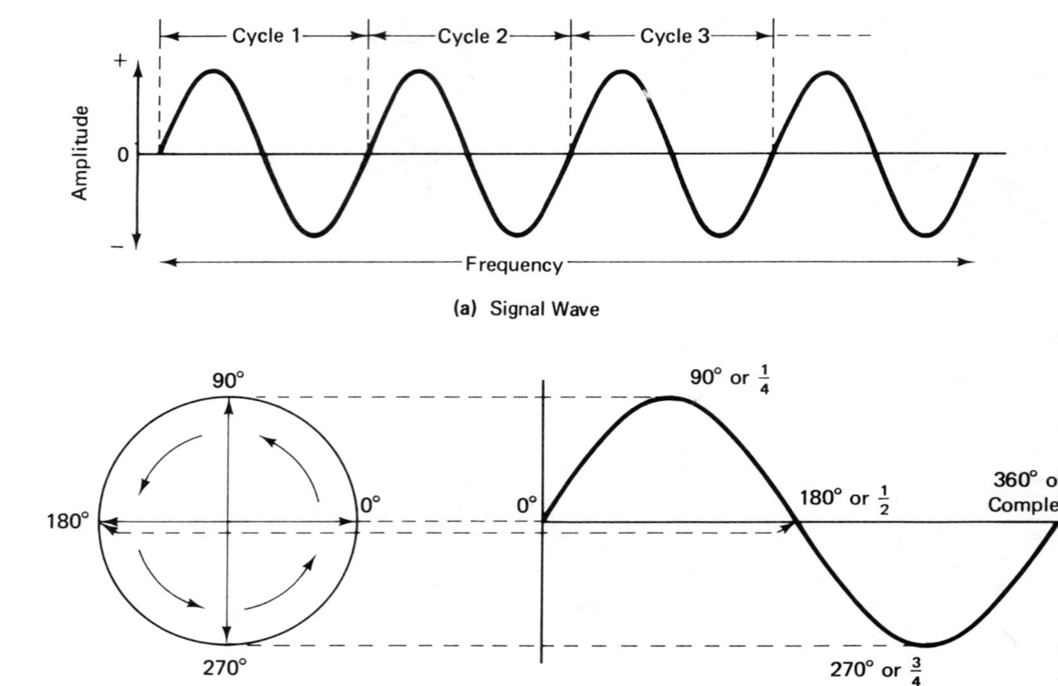

FIGURE 2-2. Oscillating Signal.

The *phase* of the signal describes the point to which the signal has advanced in its cycle. In Figure 2-2(b), the phases of the signal are identified as the beginning of the cycle, 1/4 of the cycle, 1/2 of the cycle, 3/4 of the cycle, and the completion of the cycle. The wave can also be labeled with degree markings like that of a sine wave or a 360° circle. The *sine wave* is so named because the wave varies in the same manner as the trigonometric sine function. The sine wave is derived from circular motion. The amplitude of the wave increases to a maximum at 90°, in the same manner that the sine of an angle of rotation increases to a maximum of 90°. Since a complete cycle represents a 360° rotation around the circle, 1/4 of a cycle point represents a phase value of 90°. The use of trigonometry to describe the electrical signal has proven very valuable for engineers.

The wave in Figure 2-2(b) depicts the signal at its maximum strength at two points in the cycle. One half of the cycle is represented by a positive voltage and the other one half by a negative voltage. The changing voltage state results in electrical charges changing their direction of flow down the wire circuit. The voltage continuously varies its amplitude and periodically reverses its polarity. The nature of the charge is the only difference. The strength, either positive or negative, is the same at the two peaks in the cycle.

The alternating nature of the signal voltage determines the direction of the current flow on a transmission wire. For example, a negative voltage at the transmitting end of the wire will repel the negative particles (electrons) of the conductor wire toward the receiving end—thus creating the direction of the current. On the other hand, the positive voltage will attract the charges back to the transmitter. The reader may recognize this form of signaling by the term *alternating current* (AC).

The information rate of a data signal on a path between computers is partially dependent upon the amplitude, frequency (or frequencies), and phase of the signal. As stated earlier, information rate in bits per second (bit/s) depends upon how often a signal changes its state. We will see later that the alteration of the amplitudes, frequencies, and phases of the signal will provide a change of state on the line, thus permitting a 1 to 0 or 0 to 1 representation with the change. The binary 1s and 0s are coded to represent the characters and numbers of user data messages flowing on the network between the computers.

Analog Transmission. The signal just described is called an analog signal because of its continuous, nondiscrete characteristics. This form of transmission was not designed to convey the discrete binary numbers that are found in computers. It is widely used, however, because analog facilities such as the telephone system were readily available when data communications networks were developing.

The telephone system is designed to carry voice, which is analog in nature. The human voice sends out analog waveforms of sound. The signals are actually oscillating patterns of changes in air pressure—in effect, air vibrations. These mechanical vibrations act upon the telephone microphone and are converted into electrical voltage patterns that reflect the characteristics of the speech pattern.

The analog voice signal is not one unique frequency, nor is the electrical signal. Rather, the voice and its signal on a telephone line consist of waveforms of many different frequencies. The particular mix of these frequencies is what determines the pitch and sound of a person's voice. Many phenomena manifest themselves as a combination of different frequencies. The colors in the rainbow, for instance, are combinations of many different lightwave frequencies; musical sounds consist of different acoustic frequencies that are interpreted as higher or lower pitch. These phenomena consist of a range or band of frequencies.

The human ear can detect sounds over a range of frequencies from around 40 Hz to 18,000 Hz. The telephone system does not transmit this full band of frequencies. The full range is not needed to interpret the voice signal at the receiver. Due to economics, only the frequency band of approximately 300 Hz to 3300 Hz is transmitted across the path. This explains why our voice conversations sound different on a telephone line.

Bandwidth. The range of transmission frequencies that can be carried on a communications line is referred to as the *bandwidth* of the line. Bandwidth is a very important ingredient in data communications because the capacity (stated in bits per second) of a communications path is dependent upon the bandwidth of the path. If the telephone channel were increased from a bandwidth of 3 KHz (300 – 3300 Hz) to 20 KHz, it could carry all of the characteristics of the voice. This also holds true for transmitting data; a better data transmission rate can be achieved with a greater bandwidth.

The effect of bandwidth was demonstrated by several individuals, notably Shannon, Fourier, and Nyquist. Fourier demonstrated that the sum of a minimum number of sine wave frequencies (whose frequencies (f) were integral multiples, e.g., f, 2f, 3f . . nf) were required to represent a signal. This is shown in Figure 2-3. The state of the line is changing 2000 times per second; in other words, the transmission rate is 2000 baud or Hz. A limited bandwidth of 500 Hz is insufficient to distinguish the signal accurately. As the bandwidth increases, the digital levels are more accurately portrayed.

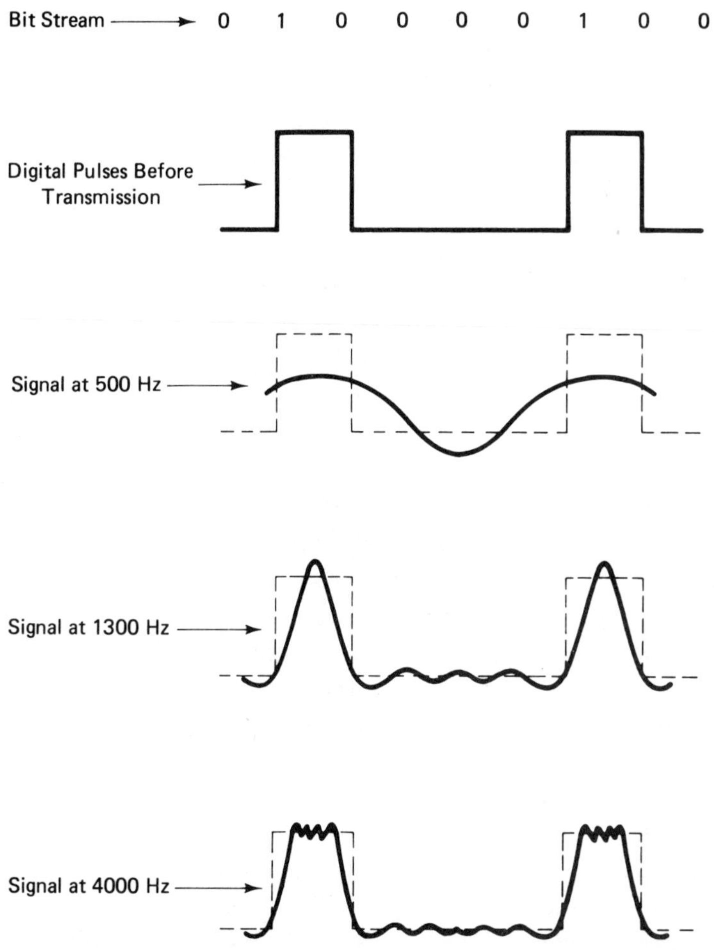

FIGURE 2-3. Effect of Bandwidth. (From *Telecommunications and the Computer* by James Martin. Englewood Cliffs NJ: Prentice-Hall, 1976, p. 149.

The greater the bandwidth the greater the capacity. This statement is explained by Table 2-1. The electronic frequency spectrum ranges from the relatively limited ranges of the audio frequencies through the radio ranges, the infrared (red light) ranges, the visible light frequencies, and up to X-ray and cosmic ray band. The importance of the higher frequencies can readily be seen by an examination of the bandwidth of the audio frequency spectrum and that of visible light (for example, optic fiber transmission). The bandwidth between 10^3 and 10^4 is 9000 Hz (10,000 − 1000 = 9000) which is roughly the equivalent to three 3000 Hz voice-grade lines.

TABLE 2-1. The Frequency Spectrum.

Approximate Frequency	Name	Use
10^3	—	Telephone Voice Frequencies
10^4	VLF	Telephone Voice Frequencies with Higher Speed Modems
10^5	LF	Coaxial Submarine Cables; Some High Speed Batch Data Transfer
10^6	MF	Land Coaxial Cables; AM Sound Broadcasting
10^7	HF	Land Coaxial Cables; Shortwave Broadcasting
10^8	VHF	Land Coaxial Cables; VHF Sound and TV Broadcasting
10^9	UHF	UHF TV Broadcasting
10^{10}	SHF	Short-Link Waveguides; Microwave Broadcasting
10^{11}	EHF	Helical Waveguides
10^{12}	—	Infrared Transmission
10^{13}	—	Infrared Transmission
10^{14}	—	Optic Fibers; Visible Light
10^{15}	—	Optic Fibers; Ultraviolet
10^{19}-10^{23}	—	X-rays and Gamma Rays

From *Telecommunications and the Computer* by James Martin. Englewood Cliffs NJ: Prentice-Hall, 1976, p. 149. Reprinted with permission.

The bandwidth between 10^7 and 10^8 (the HF and VHF spectrum) is 90,000,000 Hz, which is theoretically the equivalent of 30,000 voice-grade lines. While a somewhat extreme example, this does demonstrate that the telecommunications industry is moving to technologies utilizing the higher radio frequencies because of the greater bandwidth capabilities.

Period and Wavelength. The amount of time for a cycle duration is called the *period*. A signal of 2400 Hz has a cycle period of .000416 second (1 second/2400 = .000416). The period T is calculated as 1/F where F is the frequency in Hz. The important point to remember is the higher the frequency the shorter the period. This has implications for the use of certain types of communications paths and the selection of data rates for network equipment.

The wavelength and frequency are inversely proportional to each other. The higher the frequency, the shorter the wavelength, as proven by the formula: WL = S/F, where WL = wavelength, S = speed of signal propagation (usually quoted as 180,000 miles/second for radio propagation), and F = frequency of signal in Hz. The wavelength of the signal is an important factor in network equipment selection, protocol design, and response time analysis. Later we will learn how signal period and wavelength affect these important decisions in a data communications system, especially in satellite communications.

Other Waveforms. We will also see that the alteration of the analog signal will provide a means to transport the computer-oriented binary bits through the network. Another common approach is through the use of a symmetrical square wave. Figure 2-4 depicts this mode of transmission. The square wave represents a voltage that is switched *instantaneously* from positive polarity to negative polarity. The square wave is an excellent mode for the transmission of digital data since it can represent only the binary states of 0 and 1, with its positive and negative values.

DC Signals. Many communications systems do not use the analog (AC) from of transmission. A simpler approach is direct current (DC) transmission. DC signals resemble the symmetrical square wave in that they can take only the discrete values of 1 and 0. However, the DC transmitter does not use the oscillating waveform, but an on-off pulse of electrical energy. In addition, the DC signal is transmitted as it is, without being superimposed upon any other signal or frequency. An AC signal is often modified to "ride" other frequencies for purposes of efficiency, speed, and transmission distance. Many systems with limited distance requirements do not need the more powerful (and expensive) AC transmission scheme and, therefore, utilize DC signaling. The reader should remember that the sinusoid waveform, as in the symmetrical square wave, is the type

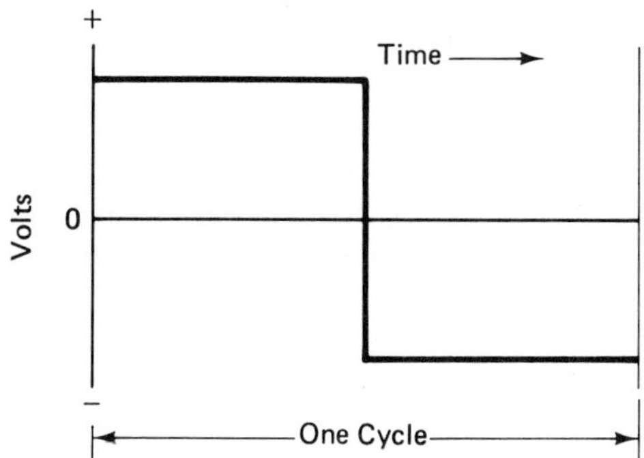

FIGURE 2-4. Square Wave.

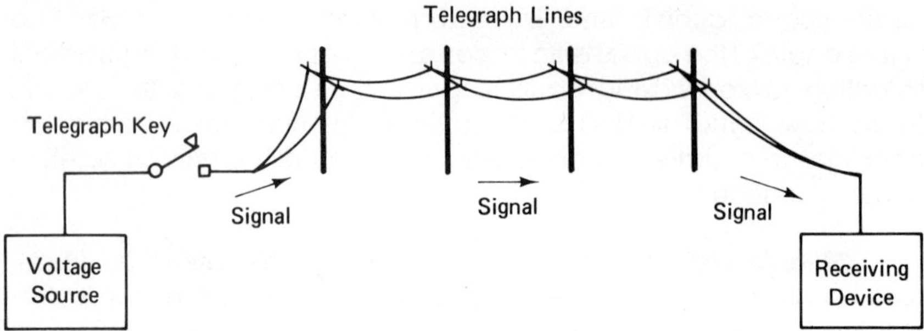

FIGURE 2-5. Telegraph Communications.

of transmission needed for long-distance data communications lines. While both DC and AC signals can be made to carry digital bit streams, the AC mode is used for long-haul transmission. Later discussions will explain this topic further.

The telegraph is an example of DC signaling. The key of the telegraph instrument is a switch that, when depressed by an operator, closes a circuit and places a voltage on the sending end of the line. The voltage produces a current that flows down the line and is detected at the receiver as a pulse. The receiver converts the pulse of current into a short audible tone. In earlier systems, the transmitted current activated at the receiving site an electromagnet powered by a battery. The electromagnet pulled or repelled the key (closed or opened the circuit) in accordance with the transmitted signal. The mechanical movement of the key produced audible clicks and the clicks established a pattern of code. The length of time the key is depressed provides the "dot," "dash" signals of the Morse code. Figure 2-5 shows an example of a DC signal on a telegraph line.

Transmission Capacity, Speed, and Delay. The transmission capacity, stated in bits per second (bps), of a communications system is quite important because response time and throughput for the user applications running on a computer depend upon the capacity of the system. For example, a 4800 bps line will provide twice the capacity of 2400 bps line and will result in increased throughput and better response time. From this statement, one might reasonably pose the following scenario: Let an engineer design a very fast transmitter that changes the signal state (baud rate or Hz) on the line very rapidly. Assuming a bit can be represented with each baud, one could then achieve a very high bps rate. Within certain limits, this can indeed be achieved. However, limits do exist and pose finite restrictions on the transmission rate.

The telephone network is designed to carry voice, which is a low bandwidth signal. Adequate voice fidelity requires a frequency spectrum from about 300 Hz to 3300 Hz. The frequency spectrum for voice-grade circuits does not allow for a high rate of bits per second to be transmitted.

The limiting factors on transmission capacity are the bandwidth, signal power, and noise on the conductor. Previous sections of this chapter explained the effect of bandwidth. We will cover signal power and noise in this section. An increased signal power can indeed increase the line capacity and also provide for greater distance for the propagation of the signal. However, excessive power may destroy components in the system and/or may not be economically feasible. Many years ago, a transatlantic communications line was rendered useless because excessively high voltage signals were used.

The noise on a line is a problem that is inherent to the line itself and cannot be eliminated. Noise (called thermal, Gaussian, white, or background noise) results from the constant, random movement of electrons on the conductor and provides a limit to the channel capacity. The hiss you hear on a telephone line is such a noise. Any electrical conductor is a source of noise. The power of the noise is proportional to the bandwidth; so, an increased bandwidth will also contain additional noise. An electronic technique known as *filtering* is used to reduce the added noise.

One of the fundamental laws in communications is Shannon's Law. Shannon demonstrated the finite limits of a transmission path with the following formula:

$$C = W \log_2 (1 + S/N)$$

Where C = Maximum capacity in bps
W = Bandwidth
S/N = Ratio of signal power (S) to noise power (N)

If the reader studies this formula, it will be evident that increasing bandwidth, increasing signal power, or decreasing noise levels will increase the allowable bps rate. However, changing these parameters may be physically or economically prohibitive. A voice-grade line with a 1000 to 1 signal to noise ratio yields a maximum allowable 25,900 bps rate.[2] The theoretical limit imposed by Shannon's Law is actually much lower in practice. Due to errors occurring in a

---

[2] James Martin, *Telecommunications and the Computer*. Englewood Cliffs NJ: Prentice-Hall, 1976, p. 304.

transmission, it is usually not desirable to push Shannon's Law to its limit. For example, the 25,900 bps rate might require such short bit times (e.g., 1 second/25,900 = .00004 bit time) that a small imperfection on the line could cause bit distortion. We will learn later that the signal state itself can be made to represent more than one bit and will provide some relief to the imposition of Shannon's Law.

One method to increase the signal to noise ratio is to place more signal amplifiers on the line. Amplifiers strengthen the signal periodically as it travels down the communications path. Since noise is constant throughout the line, the location of the amplifiers must not be spaced too far apart to allow the signal power to fall below a certain level. However, while frequent spacing of amplifiers improves the S/N ratio, it can also be quite costly. Moreover, the amplifiers must be carefully designed to minimize the amount of noise that is amplified along with the signal.

A circuit can actually carry a much greater signal rate than 25.9 Kbit/s through a technique called *digital transmission* (see Chapter 7). However, digital transmission entails the use of higher bandwidths and more frequent spacing of digital repeaters (the digital equivalent of an analog amplifier). Digital transmission need not have as high a signal to noise ratio since Shannon's Law shows that a relatively small increase in bandwidth offsets a much greater decrease in the signal to noise ratio.

Practically speaking, many people associated with data communications and networks are not interested in *why* a limited capacity exists, but *what* the limit is. The engineer's task is to determine the limits and convey this information to other people who then use the engineer's information to design and configure the network.

The range of options and costs for obtaining different speeds of lines varies greatly. The choice inevitably revolves around the user needs and the costs to meet those needs. Table 2-2 shows some ranges of transmission speeds available, as well as some typical user applications that utilize the transmission speeds. It can be seen that a wide range of options exist.

Transmission or propagation delay of the signal is yet another consideration for the engineer as well as the user. Propagation delay depends on several factors such as the type of circuit and the number and type of intermediate points between the transmitter and receiver. A good rule of thumb is that a transmission over a path with coaxial cables and microwave travels at approximately 130,000 miles per second. However, the velocity of the signal varies considerably with frequency. For instance, a typical telephone toll line (19 gauge) can operate at approximately 110,000 miles per second at 10 KHz and 125,000 miles per second at 50 KHz. The speeds are slower than the theoretical speed of 186,000 miles per second due to the frequencies and certain electrical characteristics in the cable. Additional and significant delays may be encountered as the message moves through a network into and out of intermediate stations. However, the primary transmission delays are experienced on the path itself; the intermediate components, such as switches and computers, while causing delays, usually operate at very high speeds (in nanoseconds: billionths of seconds). Of course, storing the message at these stations on disk or tape introduces considerably more delay.

Is a transmission speed of 130,000 miles per second sufficient? After all, a theoretical 3,000 mile line across the United States would require only .023 seconds (3,000 miles/130,000 miles per second = .023 seconds) for the signal to reach its destination. (Remember, additional delays may result from other intermediate components.) The answer to the question is that it depends upon the user

TABLE 2-2. Range of Transmission Requirements.

Type of Transmission	Typical Number of Bits	Transmission Time at 9.6 Kbit/s (sec)
A page or a full CRT screen of text (uncompressed)	$1–4 \times 10^4$	1–4
Facsimile image, black and white, two-tone (compressed)	$2–6 \times 10^5$	20–60
Full-page, three-color image, high quality (heavily compressed)	$2–10 \times 10^6$	200–1000
A 20-cm floppy disk, single-sided, double-density	5×10^6	500
A 720-m reel of computer tape (type 6250 BPI) or two medium-sized disk units (IBM 3310)	1×10^9	100 000 (29 hours)
One second of digitized telephone speech at pulse-code modulation	6.4×10^4	7
One second of digitized telephone speech (heavily compressed)	2.4×10^3	0.25
One second of Picturephone (moving video image)	6.3×10^6	660

From "Assessing the New Services" by Thomas Mandey. *IEEE Spectrum*, October 1979.

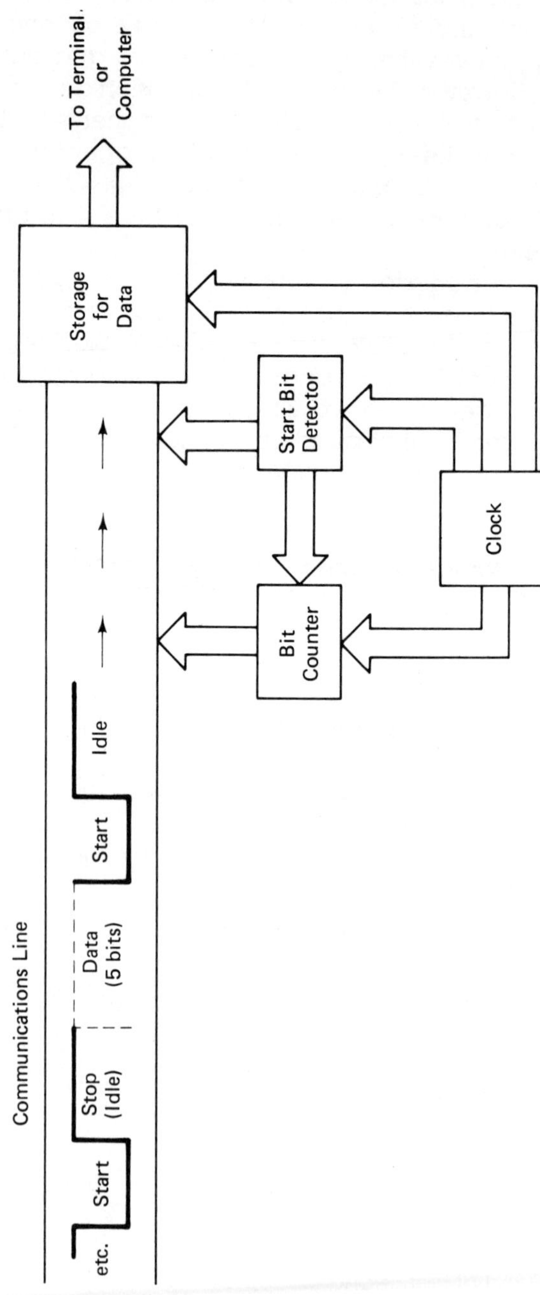

FIGURE 2-6. Asynchronous Transmission Process.

requirement and the user application. For example, a delay of .023 seconds to several seconds may be sufficient for message transfers between human operators yet insufficient in an environment where two computers are multiprocessing a distributed data base. The effect of a 23 ms wait time can be quite costly in terms of computer processor time and can create serious data base synchronization problems.

Asynchronous and Synchronous Transmission

Figure 2-6 provides an illustration of a simple transmission process. The transmitted bits are actually consecutive intervals of time represented by an analog signal, square waveform, or DC pulse, and are measured by a sensing and timing mechanism at the receiving site. The start bit precedes the data character and is used to notify the receiving site that data is on the path (start bit detection). The path or line is said to be idle prior to the arrival of the start bit and remains in the idle state until a start bit is transmitted. During the idle state, the signal voltage is held at the high level. The low signal state of start initiates mechanisms in the receiving device for sampling, counting, and receiving the bits of the data stream (bit counter). The data bits are represented as a high voltage level for the mark signal (binary 1) and a low voltage level for the space signal (binary 0). The low signal is either a zero voltage or a negative voltage. The user data bits are placed in a temporary storage area such as a register or buffer and later moved into a terminal or computer for further processing. Stop bits, consisting of one or more mark signals, provide for a time lapse in which the mechanism in the receiving site (of older equipment) can readjust for the next character. Following the stop bits, the signal returns to the high or idle level, thus guaranteeing that the next character will begin with a 1 to 0 transition. If the preceding character had been all 0s the start bit detection would become confused if a stop bit were not present to return the voltage to a high or idle level.

This method is called *asynchronous transmission* due to the absence of continuous synchronization between the sender and receiver. This allows a data character to be transmitted at any time without regard to any previous timing signal; the timing signal is a part of the data signal. Asynchronous transmission is commonly found in unbuffered machines or terminals such as teletypes or teleprinters and low-speed computer terminals. Its value lies in its simplicity.

An important component of any communications system is the clocking device. Its purpose is to continuously examine or sample the line for the presence or absence of the predefined signal levels. It is also used to synchronize all internal components. The device can be compared to any type of timing device in that its speed is dependent upon how fast it can change state from a "tick" to a "tock." Notice that the clock is connected to other components in order to maintain consistent timing of all parts.

The sampling clock actually samples the communication line at a much faster rate than the arriving data. For example, if the data were arriving at 2400 bps, the timing mechanism might sample at 19,200 times per second, or eight times the bps rate. The faster sample time enables the receiving device to detect the change of a 1 to 0 or 0 to 1 transition very quickly and, thus, keeps the sending and receiving devices more closely synchronized.

The importance of the sampling rate is evident in Figure 2-7. The bit time on a line of 2400 bps rate is .000416 seconds (1 sec/2400). A sampling rate of only 2400 times per second might sample at the beginning of the bit or at the end of the signal. In both cases, the bit was detected. However, it is not unusual for a signal to change slightly and remain on the line for a shorter or longer duration. A slow sampling rate may fail to sample the change of state of the line at the proper time and, as a signal "drifts," it is likely to lead to the end station not receiving the bits correctly.

A more efficient method, *synchronous transmission*, is distinguished by the existence of a separate clocking signal at the sending and receiving stations. The synchronous scheme is shown in Figure 2-8. The entire data field is now surrounded by control bits (called synchronization or SYN bytes; a byte is a character). The SYN bytes provide a similar function to the start bit in asynchronous

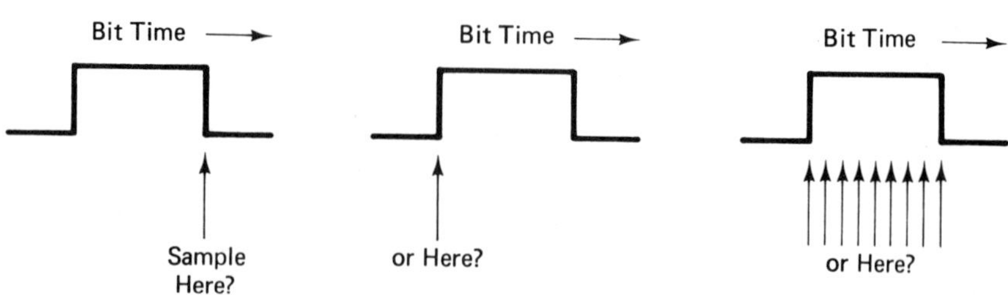

FIGURE 2-7. Bit Sampling.

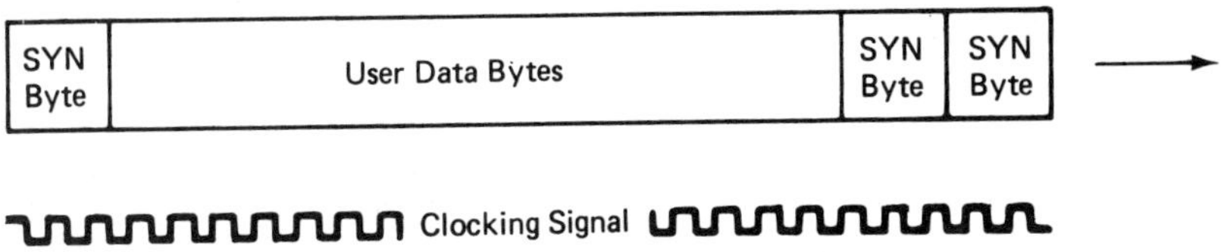

SYN Byte	User Data Bytes	SYN Byte	SYN Byte

Clocking Signal

FIGURE 2-8. Synchronous Transmission Process.

transmission by notifying the receiver that a message is arriving. As in the asynchronous mode, the receiving device searches for the unique bit pattern of the SYN character but uses a locally generated clocking signal to determine when and how often to sample the incoming signal. The SYN patterns provide the means to synchronize the clocking devices (oscillators) at the sending and receiving devices. Once the oscillators are synchronized, they are usually very stable. The clocks may drift some but a common oscillator clock is accurate to 1 part in 100,000, so an oscillator sampling at 2500 times per second would stay "in sync" several seconds. Most communications devices resynchronize against drift every one or two seconds by periodically inserting SYN bytes into the message. Others have devised special codes to represent the binary data streams that provide for periodic resynchronization. We discuss this technique in Chapter 7.

The SYN character or bytes must be distinguished from user data to allow the receiver to search for the SYN bit patterns. Vendors may use different bit streams to represent the control character. The American National Standards Institute's code (ASCII) uses a 10010110 byte. Many vendors now use the ASCII code standard.

Figure 2-8 shows that more than one SYN byte is placed in front of the user message. Multiple occurrences of SYN are necessary for several reasons. First, if the leading SYN character is distorted during transmission, subsequent control patterns are available to provide the initalization of the receiving site. Second, the receiving equipment may need additional time to "train" itself onto the incoming signal. For example, let us assume the receiving terminal requires 3.5 ms (.0035 seconds) to set up its internal circuitry to the incoming signal arriving at 2400 bps. The data transfer rate moves 2.4 bits per ms into the receiver (2400 bps/1000 ms per second = 2.4). If 3.5 ms are required to train onto the signal, 8.4 bits would be required to synchronize the equipment (3.5 ms × 2.4 bits per ms = 8.4 bits). Consequently, more than one 8-bit SYN byte is required.

BASIC TERMS AND CONCEPTS

Messages, Headers, and Codes

The electrical signals and bit sequences on the line are transmitted in the form of messages or frames (see Figure 2-9). The message is a logical unit of user data, control data, or both. The message usually contains the following elements (or fields):

- *User Data:* Consisting of one or many fields, user data is created by a terminal operator at a keyboard or is the output of a computer program. It is usually associated with an application such as accounts receivable or inventory control.
- *SYN Characters:* Previously explained in preceding section.
- *Address Field:* Contains unique numbers or letters that identify the transmitting and receiving stations in the network.
- *Control Characters:* Also control data, these data provide a means to keep the messages flowing in the proper order.
- *Error Check Data:* Used to verify a successful, error-free transmission.

SYN	Control	Error Check	User or Application Data	Sequence Number	Header (Address)	Control	SYN	SYN

FIGURE 2-9. Typical Message Format for Synchronous Transmission.

The message may also use headers, which contain addresses, sequence numbers, and other fields that control and/or further identify the message. The chapter on data link controls (Chapter 6) describes these fields in more detail.

The bit sequences for the characters depend upon the particular code set. A wide variety of codes are available today. The earliest codes used in data communications were designed for telegraphic transmission. For example, the Morse code (Figure 2-10) consists of "dots" and "dashes" in a particular sequence to represent characters, numbers, and special characters. The dots and dashes represent how long the telegraph operator presses the key on the transmitter, as illustrated in Figure 2-5. Telegraphy now commonly uses five bit codes.

A	• –	N	– •	1	• – – – –
B	– • • •	O	– – –	2	• • – – –
C	– • – •	P	• – – •	3	• • • – –
D	– • •	Q	– – • –	4	• • • • –
E	•	R	• – •	5	• • • • •
F	• • – •	S	• • •	6	– • • • •
G	– – •	T	–	7	– – • • •
H	• • • •	U	• • –	8	– – – • •
I	• •	V	• • • –	9	– – – – •
J	• – – –	W	• – –	0	– – – – –
K	– • –	X	– • • –	.	• – • – • –
L	• – • •	Y	– • – –	,	– – • • – –
M	– –	Z	– – • •	?	• • – – • •

FIGURE 2-10. Morse Code. (From *Data Communications: A User's Guide* by Kenneth Sherman. Reston VA: Reston Publishing Company, 1981, p. 87. Reprinted with permission.)

Figure 2-6 illustrates the use of an asynchronous transmission code, which is used in telegraphic communication. The mechanisms in the figure are more recent versions of the transmission process and are now available on microchips. Older versions (still widely used today) use electromechanical devices. The transmission begins when an operator presses down a key on a terminal keyboard. The keying closes specific contacts in the terminal circuitry, an armature sweeps around the contacts, and the closed contacts create current flow according to a specific bit sequence code across the communications line. A similar process decodes the signal at the receiving site. The code and the start and stop signals are all generated and detected through the closed circuits and the revolving armature around these circuits.

The industry has developed many five-bit codes. In the early 1960s, these codes were developed before standards were implemented. The Baudot code is widely used today (see Figure 2-11).

Many codes evolved from the Morse and Baudot codes. Today the more prevalent codes in use are the Extended Binary Coded Decimal Interchange Code (EBCDIC) and the American National Standard Code for Information Interchange (ASCII). EBCDIC is widely used in IBM architecture. It is an eight-bit binary code and thus provides for 256 possible characters in the code set (2^8). Figure 2-12 shows the EBCDIC code structure.

ASCII is probably the most extensively used code for synchronous data transmission (see Figure 2-13). It is a seven-bit code with one

Character Case Lower	Character Case Upper	Bit Pattern 5 4 3 2 1	Character Case Lower	Character Case Upper	Bit Pattern 5 4 3 2 1
A	–	0 0 0 1 1	Q	1	1 0 1 1 1
B	?	1 1 0 0 1	R	4	0 1 0 1 0
C	:	0 1 1 1 0	S	'	0 0 1 0 1
D	$	0 1 0 0 1	T	5	1 0 0 0 0
E	3	0 0 0 0 1	U	7	0 0 1 1 1
F	!	0 1 1 0 1	V	;	1 1 1 1 0
G	&	1 1 0 1 0	W	2	1 0 0 1 1
H	#	1 0 1 0 0	X	/	1 1 1 0 1
I	8	0 0 1 1 0	Y	6	1 0 1 0 1
J	Bell	0 1 0 1 1	Z	"	1 0 0 0 1
K	(	0 1 1 1 1	Letters (Shift)↓		1 1 1 1 1
L	)	1 0 0 1 0	Figures (Shift)↑		1 1 0 1 1
M	.	1 1 1 0 0	Space (SP) =		0 0 1 0 0
N	,	0 1 1 0 0	Carriage Return <		0 1 0 0 0
O	9	1 1 0 0 0	Line Feed ≡		0 0 0 1 0
P	0	1 0 1 1 0	Blank		0 0 0 0 0

1 = Mark = Punch Hole
0 = Space = No Punch Hole

FIGURE 2-11. Baudot Code. (From *Data Communications: A User's Guide* by Kenneth Sherman. Reston VA: Reston Publishing Company, 1981, p. 87. Reprinted with permission.)

additional bit added for error detection purposes. The code was first developed in 1963 and has become a standard.

The figures also show that certain bit configurations of the five-, seven-, or eight-bit codes are used to represent more than one character. In fact they represent, not specific characters, but control functions for the system. For example, the EBCDIC code of 00101110 represents the control function ACK, which is used to positively acknowledge the receipt of a message across a communications path. (Chapter 6 explores these control functions in more detail.) The use of different codes in a communications system obviously presents compatability problems. Code translation packages are available to bridge together communications devices that do not have the same code set.

Address Detection

It was stated previously that the bits comprising the bytes and fields of the message are electrical signals of high or low level. The binary aspect of the signals makes it ideal to check and manipulate the message contents with digital logic circuits. Boolean algebra is a mathematical tool, first devised by the Englishman George Boole in 1847, that is used to design such logic circuits.

The logic circuits provide a means to represent the binary states of 1 or 0, high or low, and true or false through symbols and algebraic expressions. Boolean algebra provides for procedures to determine whether a statement is true or false provided that the statement can be expressed in terms of two possible values. For example, the station address in a message header contains a binary value representing the unique identification of the station. The logic circuits, designed with Boolean algebra, provide the means for a station to examine the bits in the address and answer yes or no to the question, "Is this address mine?"

Bits 4	0	0	0	0	0	0	0	0	1	1	1	1	1	1	1	1
3	0	0	0	0	1	1	1	1	0	0	0	0	1	1	1	1
2	0	0	1	1	0	0	1	1	0	0	1	1	0	0	1	1
1	0	1	0	1	0	1	0	1	0	1	0	1	0	1	0	1
8 7 6 5																
0 0 0 0	NUL	SOH	STX	ETX	PF	HT	LC	DEL			SMM	VT	FF	CR	SO	SI
0 0 0 1	DLE	DC_1	DC_2	DC_3	RES	NL	BS	IL	CAN	EM	CC		IFS	IGS	IRS	IUS
0 0 1 0	DS	SOS	FS		BYP	LF	EOB	PRE			SM			ENQ	ACK	BEL
0 0 1 1			SYN		PN	RS	UC	EOT					DC_4	NAK		SUB
0 1 0 0	SP										¢	.	<	(	+	|
0 1 0 1	&										!	$	•	)	;	¬
0 1 1 0	–	/										,	%	–	>	?
0 1 1 1											:	#	@	'	=	"
1 0 0 0		a	b	c	d	e	f	g	h	i						
1 0 0 1		j	k	l	m	n	o	p	q	r						
1 0 1 0			s	t	u	v	w	x	y	z						
1 0 1 1																
1 1 0 0		A	B	C	D	E	F	G	H	I						
1 1 0 1		J	K	L	M	N	O	P	Q	R						
1 1 1 0			S	T	U	V	W	X	Y	Z						
1 1 1 1	0	1	2	3	4	5	6	7	8	9						¤

PF - Punch Off
HT - Horizontal Tab
LC - Lower Case
DEL - Delete
SP - Space
UC - Upper Case

RES - Restore
NL - New Line
BS - Backspace
IL - Idle
PN - Punch On
EOT - End of Transmission

BYP - Bypass
LF - Line Feed
EOB - End of Block
PRE - Prefix (ESC)
RS - Reader Stop
SM - Start Message
Others - Same as ASCII

FIGURE 2-12. EBCDIC Code. (From *Data Communications: A User's Guide* by Kenneth Sherman. Reston VA: Reston Publishing Company, 1981, p. 93. Reprinted with permission.)

American
National Standard Code
for
Information Interchange

Bits				7	0	0	0	0	1	1	1	1
				6	0	0	1	1	0	0	1	1
4	3	2	1	5	0	1	0	1	0	1	0	1
0	0	0	0		NUL	DLE	SP	0	@	P	\	p
0	0	0	1		SOH	DC1	!	1	A	Q	a	q
0	0	1	0		STX	DC2	"	2	B	R	b	r
0	0	1	1		ETX	DC3	#	3	C	S	c	s
0	1	0	0		EOT	DC4	$	4	D	T	d	t
0	1	0	1		ENQ	NAK	%	5	E	U	e	u
0	1	1	0		ACK	SYN	&	6	F	V	f	v
0	1	1	1		BEL	ETB	'	7	G	W	g	w
1	0	0	0		BS	CAN	(	8	H	X	h	x
1	0	0	1		HT	EM	)	9	I	Y	i	y
1	0	1	0		LF	SUB	*	:	J	Z	j	z
1	0	1	1		VT	ESC	+	;	K	[	k	{
1	1	0	0		FF	FS	'	<	L	\	l	¦
1	1	0	1		CR	GS	—	=	M	]	m	}
1	1	1	0		SO	RS	.	>	N	∧	n	~
1	1	1	1		SI	US	/	?	0	—	o	DEL

Example:

Bits: P*7 6 5 4 3 2 1

1 1 0 0 0 0 0 1 = letter "A" (Odd Parity)
0 0 1 1 1 0 0 0 = number "8" (Odd Parity)

P* = Parity Bit

FIGURE 2-13. ASCII Code. (From *Data Communications: A User's Guide* by Kenneth Sherman. Reston VA: Reston Publishing Company, 1981, p. 95. Reprinted with permission.)

The components in the sending and receiving communications devices are actually blocks of hardware that produce 1 or 0 signals as output when certain input signal requirements are satisfied. As early as 1938 it was explained that switching circuits could be built around logic gates and could be employed to manipulate binary bit streams (such as a station address) so that various logic functions are performed.

The logic gates have distinct graphic symbols and a simple algebraic function. The relationship of the input binary stream and the resulting output is usually described by an accompanying truth table. Figure 2-14 shows part of a Boolean structure. The Boolean name (AND, OR, NOT) describes the symbol and function.

The AND function states that inputs A and B must *both* be 1s (high voltage signals) in order to produce an output (x) of 1. If either A or both A and B are 0s, then the output gate is 0. We could add additional inputs, such as C, D, etc., but the AND rule still holds: all inputs must be 1s to produce an output of 1. The OR function states that *any* input of 1 will result in an output (x) of 1. The NOT gate (or inverter gate) changes a 1 to a 0, and a 0 to a 1.

As stated earlier, Boolean logic is quite useful in computers and communications. Our illustrations are message headers and station addresses. Let us assume a particular station on a line has an address

Gate Name | Symbol | Truth Table

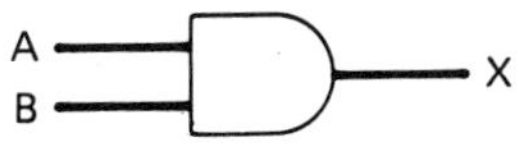

AND — A, B → X

A	B	X
0	0	0
0	1	0
1	0	0
1	1	1

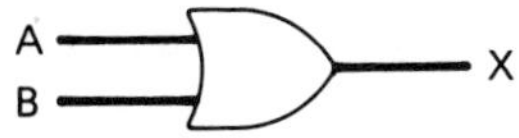

OR — A, B → X

A	B	X
0	0	0
0	1	1
1	0	1
1	1	1

NOT — A → X

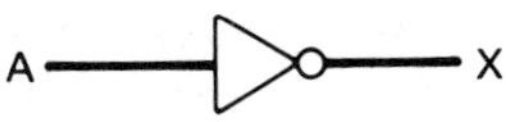

A	X
0	1
1	0

FIGURE 2-14. Logic Gates.

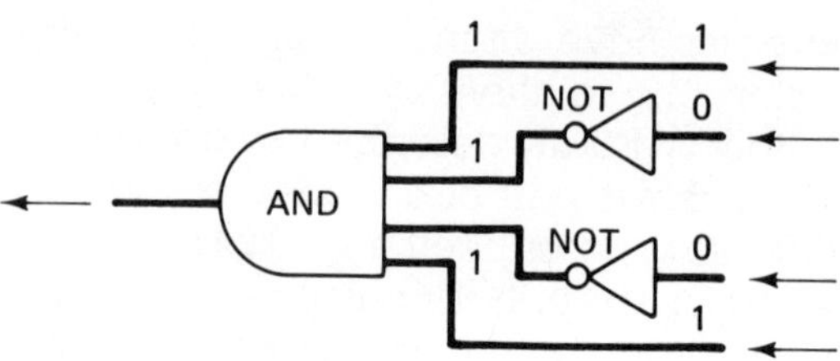

FIGURE 2-15. Address Detection.

of 9. The binary representation for a 9 is 1001 (Figure 2-15). The logic at the receiving site will check for the bit stream 1001 through the NOT and AND Boolean functions; this can be accomplished by hardware, software, or both. Notice that the two middle input lines into the receiving station have NOT gates attached in front of the AND gate. These gates reverse the 0s on those inputs, producing all 1s into the address detection logic. Thus, the input bit sequence 1001 causes the output to be high or a 1. Any other address with any other bit configuration would not produce an output of 1 through the logic. Since the address detection circuitry has produced a 1 from the incoming address, this station must then evoke other hardware and software to receive the full message and perform the needed functions on the message.

Digital logic circuits are used to perform many functions. For example, detection may not be restricted to a station address. The header might contain other identifiers, such as a message transaction type, a user department code, or an identification of a specific software application. These parameters could be examined to determine if the message belongs to that station. We see this practice increasingly used in networks today wherein detection is on functional identifiers and not specific physical addresses. This approach allows for more flexibility in configuring the network and moving components among the sites. It also isolates the user application from the physical network topology. For example, a user can insert an identifier of a data base in the message header. Network logic can detect the data base identifier and insert in another header an actual address of the location of the data base. If necessary, the data base can be relocated without affecting the user applications. One need only change the network logic tables and the second header. The physical network constraints become transparent to the user. This approach is discussed further in Chapter 8.

Sessions

The communications flow between two components in a network is called a *session*. The session can take several forms. For example, a session can exist between two operators at two terminals in the network; a session can exist between computers; a session can also exist between two software programs—either applications (payroll, accounts receivable, etc.) or network control programs. Other forms of sessions can exist as well. Whatever the form, sessions are ultimately established to serve the end user, such as a terminal operator or an applications program.

Several sessions are often established to provide for a user-to-user session. If a terminal operator at site A wishes to obtain data at site B, the following events and sessions are likely to occur:

1. Session establishment between local user and site A computer.
2. Session establishment between site A computer and site A communications control logic (perhaps at several levels).
3. Session establishment between site A and site B communications control logic.
4. Session establishment between site B communications control logic and site B computer.
5. Session establishment between site B computer and site B application (the retrieval program to the data base).
6. Finally, session establishment between the end users (the terminal operator at A and the applications software at B).

This process may appear to be cumbersome, but it is necessary in order to establish and maintain proper control of network users and resources. For instance, Session 3 is required to ensure that a path is available for messages to move between sites A and B. Each session provides for the establishment of resources to eventually service the end user. Session 4 might not be established if site B determined it was too busy to satisfy the request from the terminal operator at site A.

Session establishments utilize the information in the headers (or other parameters) of the messages. The data base retrieval request from the terminal operator at site A would provide a header in the message that identified an address of site B, the data base location, or some other kind of identifier, such as type of data. As the message is passed through the network, the header is examined in a manner similar to address detection (see page 36) and the proper resources (paths, data files, computer programs) are identified and allocated to service the request. The more advanced networks now provide for a layering structure of the sessions and a layering of the communications logic to provide the resources. Chapter 8 discusses the layering concept in more detail.

Line Characteristics

The transmission path or line provides the media for the data exchange between stations. This exchange includes session establishment, user messages, and session termination. In addition to the electrical attributes of the path, other characteristics play an important role in the performance and design of the communications system. Later chapters will cover the types of paths used in a network (e.g., microwave, satellite, etc.). This section discusses the following characteristics of those types of paths:

- Point-to-Point and Multipoint Configurations
- Simplex, Half-Duplex, Duplex Arrangements
- Switched and Leased Lines

Point-to-Point and Multipoint Configurations. A point-to-point line connects two stations (see Figure 2-16(a)); a multipoint line has more than two stations attached (see Figure 2-16(b)). The selection of one of these configurations is dependent upon several factors. First, a point-to-point arrangement may be the only viable choice if a prolonged dedicated user to user session is necessary. Second, the message traffic volume between two users may preclude the sharing of the line with other stations. Some computer-to-computer sessions require a point-to-point arrangement. Third, two users may be the maximum number involved in parts of a network. A multipoint arrangement is commonly used in situations where low-speed terminals communicate with each other or a computer. The line is shared by the stations, thereby providing for its more efficient use.

Multipoint lines require the use of more elaborate controls than do point-to-point lines. The stations on the multipoint path must be supervised to provide for the allocation and sharing of the line. Sessions must be interleaved and priorities must be established for the more important sessions. Data Link Controls (see Chapter 6) are used to control the flow of messages in these sessions.

Simplex, Half-Duplex, and Duplex Arrangements. These terms are often subject to more than one interpretation. They usually refer to

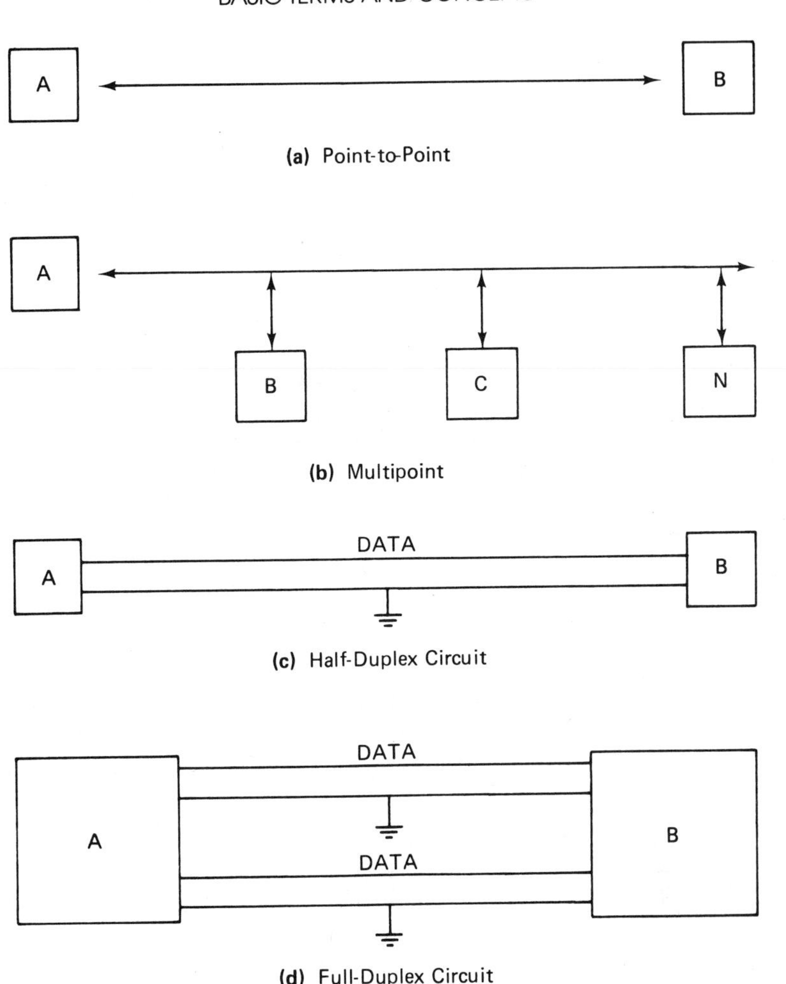

FIGURE 2-16. Line Configurations.

the manner in which message traffic flows across the line. Another common interpretation is the number of physical paths involved in the transmission. We will discuss both of these interpretations.

Message Flow. A simplex transmission provides for the movement of messages across the path in one direction only. The sender cannot receive and the receiver cannot send. A commercial radio broadcast is one example of a simplex transmission. The scheme is used in numerous applications. For instance, environmental filtering and sampling systems often use a simplex arrangement wherein the sampled data from water, air, etc., is transmitted in one direction only to a computer for analysis.

Half-duplex transmission provides for movement of data across the line in both directions, but in only one direction at a time. Human-operated keyboard terminals commonly use this approach. Typically, a message sent to the terminal requires the operator to decipher the message, enter an appropriate response, and send the reply. The terminal and the other station "take turns" using the line; the sending station waits for a response before sending another message.

Duplex transmission (also called full-duplex) provides for simultaneous, two-way transmission between the stations. Multipoint lines frequently use this approach. For example, Station A sends to the central computer at the same time the computer sends a message to Station B. Duplex transmission permits the interleaving of sessions and user data flow among several or many stations.

Physical Path. The physical lines or paths are sometimes described as half-duplex or duplex circuits. Figure 2-16(c) depicts a half-duplex configuration. This arrangement provides for two conductors but only one is used for message exchange. The second conductor is a return channel or common ground to complete the electrical circuit. The half-duplex circuit is also called a *two-wire circuit*.

A duplex (or full-duplex) circuit is shown in Figure 2-16(d). In this case, four conductors are used to provide two data transmission paths and two return channels. The duplex circuit is also called a *four-wire circuit*.

The reader should note that a half-duplex circuit does not necessarily mean that the message flow across the circuit is half-duplex. This can be confusing and, if you want an immediate explanation, refer to the section on modems and side channels in Chapter 3.

Switched and Leased Lines. The use of switched lines is well known to all of us because we use such facilities to make telephone calls. The dial-up telephone uses the public telephone exchange. The switched line is a temporary connection between two sites for the duration of the call. A later call to the same site might use different circuits and equipment in the telephone system. The leased line is a permanent connection between sites and does not require a dial-up to obtain the communications path. The path is configured and connected to alleviate dialing (addressing) the site in the network. The advantages, disadvantages of switched and leased lines are:

- A switched line requires several seconds to complete a call and obtain a connection. User needs may dictate the use of leased lines wherein no dial-up delay is encountered.
- Leased lines can provide for better performance and fewer errors. First, the leased lines can be monitored and "conditioned" to perform better because the connection does not change. Second, some switching systems introduce noise into the line and the noise sometimes distorts the data.
- Low-volume traffic users usually benefit from using switched lines due to the greater costs of leased facilities. Periodic use of a line favors the dial-up approach of leased lines.
- The switched lines are quite flexible. A failed circuit only requires that a user redial into the public network. A leased line failure requires more effort and entails additional delays to recover.

The choice between leased and switched facilities is a very important consideration for an organization. An analysis of traffic volume, peak loads, response time, and throughput performance is a prerequisite to making rational decisions about the use of leased or switched lines.

Use of the Telephone Network

The following section will give the reader an idea of how the telephone system provides the facilities for a call. Our illustrations are simplifications but sufficient to understand the process.

Figure 2-17 shows some of the components in the local telephone and in the end office. The telephone has switches (SH for switch hooks) that are kept open by the phone's weight in the telephone cradle. As long as the phone is "on hook," the open switches do not provide an electrical connection to the telephone office. When an individual picks up a phone, the SH switches are closed. This "off-hook" condition is detected in the central telephone office by the closed SH switches allowing a flow of DC current to the end office. A computer or other devices can "call" the office by activating off-hook through circuitry.

The end office has an electronic detector that scans the incoming local loop lines. At approximately every 100 ms, a line is scanned for the off-hook condition of a DC current flow.

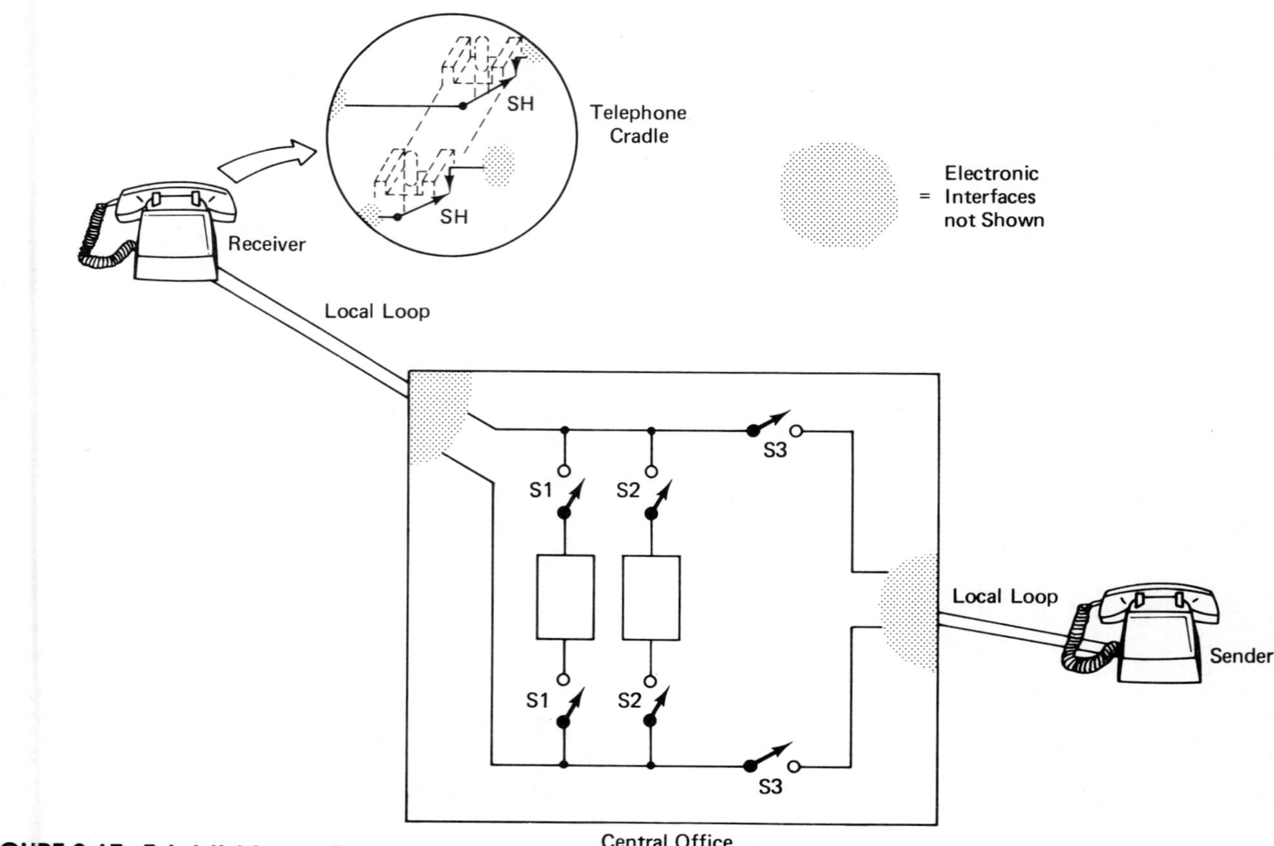

FIGURE 2-17. Establishing a Call.

The central office has several switches that it uses to establish the call. Upon detecting the DC current flow from the local subscriber, the office places a dial tone on the line by closing the S1 switch. This permits a tone of 480 hertz to be sent to the caller's telephone. Upon receiving the dial tone, the subscriber (or computer) is alerted to dial a number. The number is then entered by a rotating dial, the touch-tone push buttons, or by the computer itself.

The signals arrive at the local telephone office, and the call is passed to the local toll center. Increasingly, computers are being used to route the call through the telephone system. Upon receiving a dialed number, the computer examines routing tables to determine which path should be used. If the call is in another area of the country, the logic will determine the proper path to the remote toll center. As depicted in Figure 1-2, the call may be routed and switched through several levels of toll centers. The toll centers scan for free lines by checking for a 2600 hertz signal; upon finding the "free signal," the toll center replaces the 2600 hertz with the calling signal.

In a matter of seconds, the call arrives at the receiver's local central office. This office scans the appropriate local loop to determine if it is busy. The central office can so determine by checking for the absence or presence of DC current flow. The end office will close switch S2 to activate the ringing mechanism in the called party's telephone. S2 closure permits the sending of a 20 hertz signal to the telephone.

If the called telephone is taken off-hook to answer the ring (thus connecting the current flow to the end office), the ringing signal is removed by opening S2. The connection is then completed by closing S3. A switched call between cities typically passes through four to nine switching centers.

3

MAJOR COMPONENTS IN A DATA COMMUNICATIONS SYSTEM

THE TRANSMISSION PATH

The path for the movement of data between computer sites can take several different physical forms. This section describes the more prevalent methods in use today. The transmission characteristics of each path are explained. Emphasis is placed on the attributes of optic fibers, radio transmission, and satellite paths.

Wire Pairs

Wire transmission characteristics were covered in Chapter 2; however, some additional information might be useful to the reader. Wires are described by their size. The sizing system in the United States is called the American Wire Gauge (AWG) System. The AWG System specifies the size of the round wire in terms of its diameter. Higher gauge numbers indicate thinner wire sizes. The smaller the diameter of the wire the greater its resistance to the propagation of a signal. Increased resistance results in a decreased bit rate across the communications path. At higher transmission frequencies, the signal tends to travel on the outside surface of the wire. A smaller wire provides less total surface for the radiating signal, resulting in an increased signal loss. A larger wire with a greater cross-sectional area will allow for an increased signal intensity. The local subscriber loops (of the telephone system) are usually 22 to 26 gauge wire. Trunk and toll lines typically employ 19-gauge wires.

In the earlier part of the century, a pair of wires or 3 wires constituted the path for telegraphs and telephones. The wires typically provided for 12 circuits per pair. Wire pairs can still be seen in certain parts of the United States, primarily in the rural areas. The wires are openly suspended from telephone poles and connected to the poles with insulating points. The early facilities had 16 pairs on a pole, providing a capacity per route of about 200 circuits. The wire is usually made of copper and steel. Open wire pairs have been replaced by other technologies because of attenuation (signal loss) and cross talk (circuit interference) problems.

Cables

Wire cables have replaced most of the wire pairs. Several hundred of these wires are insulated and packaged into one cable. The wires are paired and twisted around each other to decrease certain electromagnetic problems. (Another term describing the technology is *twisted-wire pairs*). The cables can contain hundreds of pairs. They provide for better performance than the open pairs, but are still susceptible to cross talk. The cables are quite heavy and cumbersome; a foot section of the cable containing 400 pairs weighs several pounds. Wire pair cables are usually laid under the streets, suspended from poles, or installed inside the ducts of buildings.

Microwave

Electromagnetic radiation was predicted by James Clark Maxwell as early as 1873 and, in 1887, Heinrich Hertz actually produced radio waves. These waves are fields of force. They are similar to light, both are radiations of electromagnetic energy, but the radio waves are longer and more coherent in phase. The signals can be altered to carry information in the form of binary bits.

Electromagnetic radiation can be created by inducing a current of sufficient amplitude into an antenna whose dimensions are approximately the same as the wavelength of the generated signal. The signal can be generated uniformly (like a light bulb) or can be directed as a beam of energy (like a spotlight).

Microwave is a directed line of sight radio transmission (see Figure 3-1). It is used for radar and wideband communications

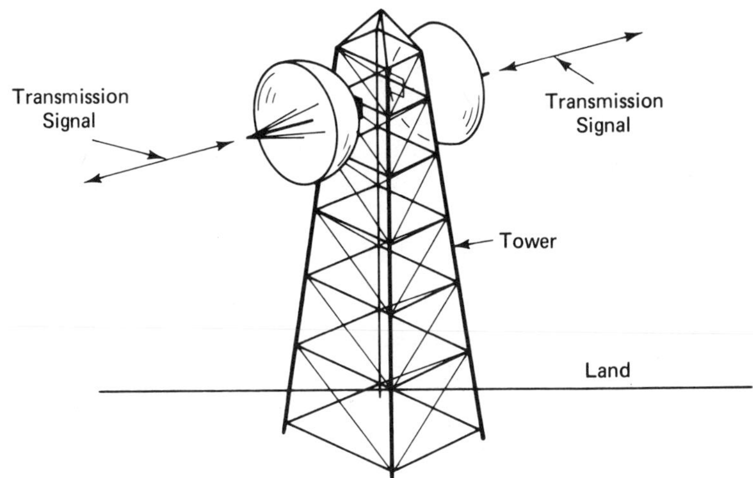

FIGURE 3-1. Microwave.

systems and is quite common in the telephone system. In fact, well over half of the toll and long-distance telephone trunks use microwave transmission. Television transmission also utilizes microwave transmission, because microwave transmission is above the one gigahertz frequency band and provides the capacity required for video transmission. The high bandwidth gives a small wavelength (See chapter 2 for a review of this point) and, the smaller the wavelength, the smaller one can design the microwave antenna. The antenna size has significant implications for distributed processing systems (see Chapter 10).

The transmitting towers are spaced 20 to 30 miles apart. The transmitted radio beam is focused narrowly to the receiving antenna. These towers can be seen throughout cities and the countryside. They have become a familiar landmark to the motorist. The antenna are also frequently placed on the rooftops of buildings.

Microwave is very effective for transmission to remote locations. Countries such as Canada and the Soviet Union have used this technology in the remote sections of their countries. Canada has one of the most extensive systems in the world and the Soviet Union has placed microwave systems in such remote areas as Siberia. Several U.S. specialized common carriers' primary product line is the offering of voice-grade channels on their microwave facilities.

Coaxial Cables

Coaxial cables have been in existence since the early 1940s and are another very popular media. They are used extensively in long-distance telephone toll trunks, urban areas, and local networks. The technology consists of an inner copper conductor held in position by circular spacers (see Figure 3-2). The inner wire is surrounded by insulation and covered by a protective sheath. The covering protects the conductor and prevents interference from signals of other coaxial cables.

Coaxial cables are designed to provide for greater bandwidth and faster bit rates than wire cables. Typical coaxial cable systems can carry from 3,600 to 10,800 voice-grade channels. As stated earlier, the use of higher frequencies on a wire pair is limited because higher frequencies tend to produce a current flow on the outside portion of the wire. This phenomenon (the "skin effect") creates significant attenuation and cross talk problems on uninsulated wires. The same holds true for coaxial cables: the current flows more on the outside of the wire at higher frequencies. However, the wire is encased in the shell so the current is actually flowing in the inside of the outer shell and the outside of the inner wire. This approach allows for a system above the 60 MHz bandwidth (and 10,800 voice-grade circuits). The technology is somewhat limited due to repeater design and signal loss at higher frequencies.

Coaxial cable systems continue to proliferate and will remain a popular technology for many years. AT&T has announced plans to install what is reportedly the world's largest high-speed coaxial system. It will run from Plano, Illinois, to Sacramento, California, and will have a capacity of 140 Mbit/s.

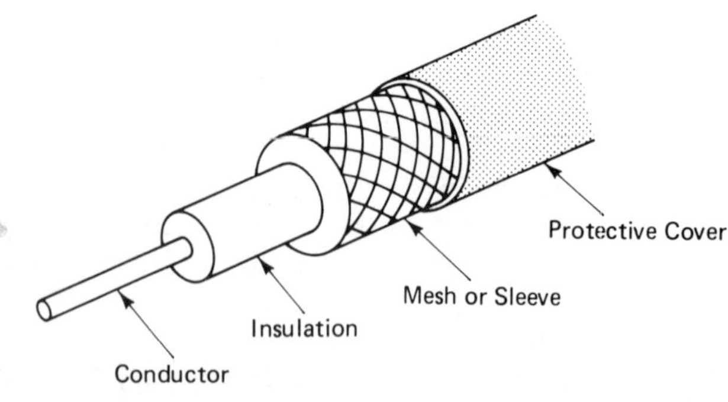

FIGURE 3-2. Coaxial Cable.

Satellite Communications

In 1945, Arthur C. Clarke, a noted science fiction writer, described in *Wireless World* the communications satellite technology as it exists today. Later, Clarke predicted satellite communications would create as profound an impact on the world as did the telephone. One may question the latter claim, but few people dispute the significance of satellite communications.

The technology is actually a radio relay in the sky. Satellite stations with transmitting and receiving antenna are launched into orbit by rockets where they receive signals from transmitting stations on earth and relay these signals back to the earth stations. Figure 3-3 illustrates this process.

FIGURE 3-3. Satellite Transmission.

Unique Aspects of Communications Satellites. Satellite communications are unique from other media for several reasons:

1. The technology provides for a large communications capacity. Through the use of the microwave frequency bands, several thousand voice-grade channels can be placed on a satellite station.
2. The satellite has the capacity for a broadcast transmission. The transmitting antenna can send signals to a wide geographical area. Applications such as electronic mail and distributed systems find the broadcast capability quite useful.
3. Transmission cost is independent of distance between the earth sites. For example, it is immaterial if two sites are 100 or 1,000 miles apart, as long as they are serviced by the same communications satellite. The signals transmitted from the satellite can be received by all stations, regardless of their distance from each other.
4. The stations experience a significant signal propagation delay. Since satellites are positioned 22,300 miles above earth, the transmission has to travel into space and return. A round trip transmission requires about 240 ms, and could be greater as the signal travels through other components. This may affect certain applications or software systems. This issue is discussed further in Chapter 6.
5. The broadcast aspect of satellite communications may present security problems, since all stations under the satellite antenna can receive the broadcasts.

History of the Technology. Although satellite communications concepts have existed for many years, the ideas could only be implemented after the advent of space age technology and solid state electronics. The United States and the Soviet Union lead in the development of the technology. The Soviet Union was first off the launching pad with its Sputnik I on October 4, 1957, and the United States followed shortly with Explorer I on January 1, 1958. Neither of these rockets carried communications satellites. The United States Army is credited with the first communications satellite (SCORE) which was launched December 18, 1958. The world's first commercial satellite, Early Bird, was orbited from Cape Kennedy on April 6, 1965. The WESTAR satellite launched by Western Union in 1974 was the first United States domestic satellite. COMSAT was credited with the first satellite communications system for ships at sea with the 1976 Marisat System.

Canada has also been a leader in communications satellite technology. As early as 1962, it launched the Alouette system and achieved the world's first geosynchronous domestic system with the Anik series orbits in 1972, 1973, and 1975. Canada has also pioneered satellites in the 14/12 gigahertz bandwidth—considerably greater in capacity than the 6/4 gigahertz that are commonly used today. Canada has benefited greatly from satellite communications—no other country has a population of 24 million people scattered over 5.9 million square miles. Due to Canada's extensive satellite system, fewer than one-fourth million people cannot be reached through the technology.

The earlier satellites were passive. The United States satellites Echo 1 (launched in 1960) and Echo 2 (launched in 1964) merely reflected the transmitted signal (up-link signal) back to the earth station (down- link signal). The improvement in rocket technology and the smaller weight of solid state electronics now provide for active satellites that receive a signal, amplify it, and retransmit it back to earth.

The earlier satellites were still not commercially viable. The limited rocket power could boost the satellite into orbits no greater than 6,000 miles above the earth. The low orbit resulted in the satellite moving faster than the earth's rotation and, as the satellite moved across the horizon, the earth station had to rotate its antenna. Eventually, the satellite would disappear and tracking would be passed to another earth station or satellite. The North Atlantic region would have required about 50 satellites for continuous coverage—a very expensive arrangement.[1]

The current satellites are in a geosynchronous orbit. They rotate around the earth at a speed to remain positioned over the same point above the equator. Thus, the earth stations' antenna can remain in one position since the satellite's motion relative to the earth's position is fixed. Further, a single geosynchronous satellite with nondirectional antenna can cover about 40% of the earth's surface. The geosynchronous orbit requires a rocket launch of 22,300 miles into space. Geosynchronous satellites can achieve worldwide coverage (some limited areas in the polar regions are not covered) with three satellites spaced at 120° intervals from each other (see Figure 3-4).

Today, the typical satellite operates in the 6 and 4 gigahertz (GHz) bands (the C Band) for the up-link and down-link, respectively. The bandwidth can be divided in a variety of ways. The Western Union satellites have 12 channels, each using a bandwidth of 36

---

[1] Pier L. Bargellini, "Commercial U.S. Satellites." *IEEE Spectrum*, October 1979, p. 30.

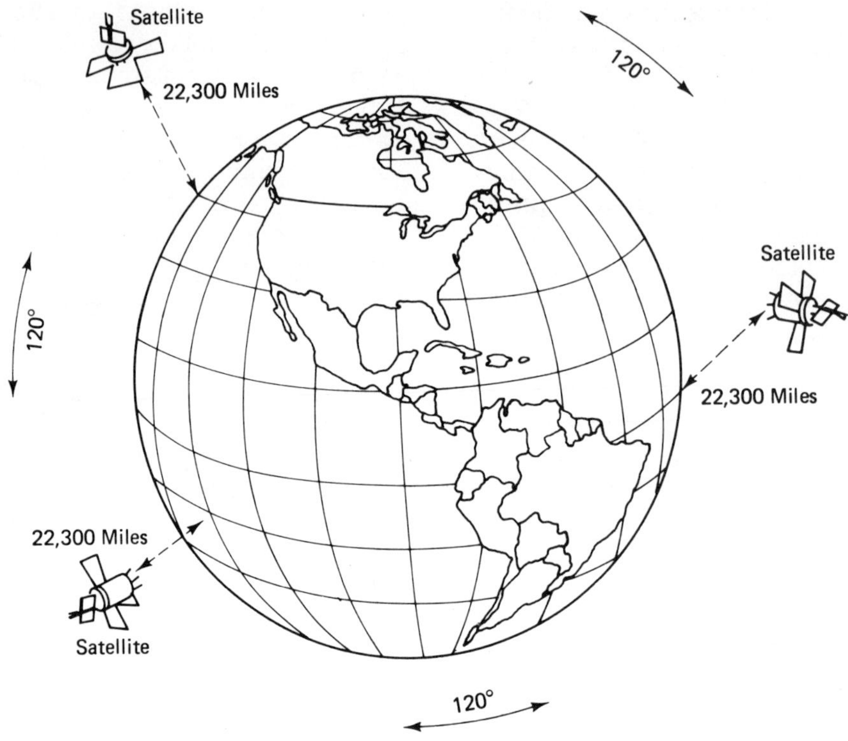

FIGURE 3-4. Geosynchronous Satellites.

megahertz, with a 4 megahertz spacing between the channels. These channels are further divided into lower speed circuits such as voice-grade channels. The voice-grade channels may be further divided to operate subchannels at speeds such as 1200, 2400, 4800, or 9600 bps.

Increasingly, satellites are using frequencies in the 11/12/14 gigahertz range (Ku Band). The 30/20 gigahertz (K Band) range will also be used. The 6/4 gigahertz bandwidths are also assigned to terrestrial microwave systems, and the use of the higher frequency spectrums will eliminate what is now serious interference and signal congestion problems between microwave and satellite systems. IBM's Satellite Business Systems (SBS) satellites use 14/12 gigahertz bandwidths. The new INSTELSAT satellites utilize the 14/12 gigahertz technology as well. The higher frequencies also provide for greater capacity. The new INTELSAT satellite can accommodate 12,000 telephone circuits and two television channels.

The 30 and 20 GHz satellites are also very attractive due to the spectrum/orbit problem. While ample space exists 22,300 miles above the earth for all the satellites, the satellites must be spaced apart in order to prevent radio interference from adjacent satellites. The 6/4 GHz requires about 4° (450 miles) of separation. The wavelengths of the 30/20 GHz satellites are more narrow and thus the satellites can be placed more closely together—about 1° of arc apart. This is an important consideration because the 6/4 slots over the United States and Puerto Rico are already over-committed. Satellites are now carrying antennas that operate both in the 6/4 GHz and 16/12 GHz bands, because these signals will not interfere with each other. The 14/12 and 30/20 bands also require smaller antenna because the wavelengths are shorter.

The 14/12 and 30/20 GHz bands are not without problems. The major obstacle to their use is their susceptability to rain. The water and oxygen molecules in the rain absorb the electromagnetic energy. Attenuation of the higher frequencies with the shorter wavelengths is particularly severe. In contrast, the 6/4 GHz bands are relatively immune to rain. The rain loss problem could be solved by boosting power in the satellites and in the earth stations, but this has not been cost-effective. Later, we will examine some methods of solving the rain problem (see discussions on multiplexing).

Currently, 14 satellites transmit audio, video, and data signals down to the United States and Canada. The satellites (see Figure 3-5), positioned west of South America and above the equator, are owned and operated by United States and Canadian firms. About 50 other satellites orbit in the geosynchronous plane; hundreds of others exist for military, weather, and other purposes.

The American communications satellites are launched and positioned by NASA at a cost of about $30 million. The cost to build the satellite itself is around $40 million. Recently, the space shuttle was used to launch satellites and this will lead to significantly decreased launching costs.

INTELSAT Satellites. The International Telecommunications Satellite Organization (INTELSAT) is the dominant carrier of international communications, handling more than 60% of all transoceanic traffic. INTELSAT furnishes more than 135 countries full-time channels for telephone, teletypewriter, data, facsimile, and television services. COMSAT uses the INTELSAT system and has about 25% ownership interest in the international organization.

INTELSAT has launched and operated five generations of satellites (see Table 3-1):

INTELSAT I (Early Bird) was placed in service in June 1965 and carried the first transoceanic TV broadcast. It was the first

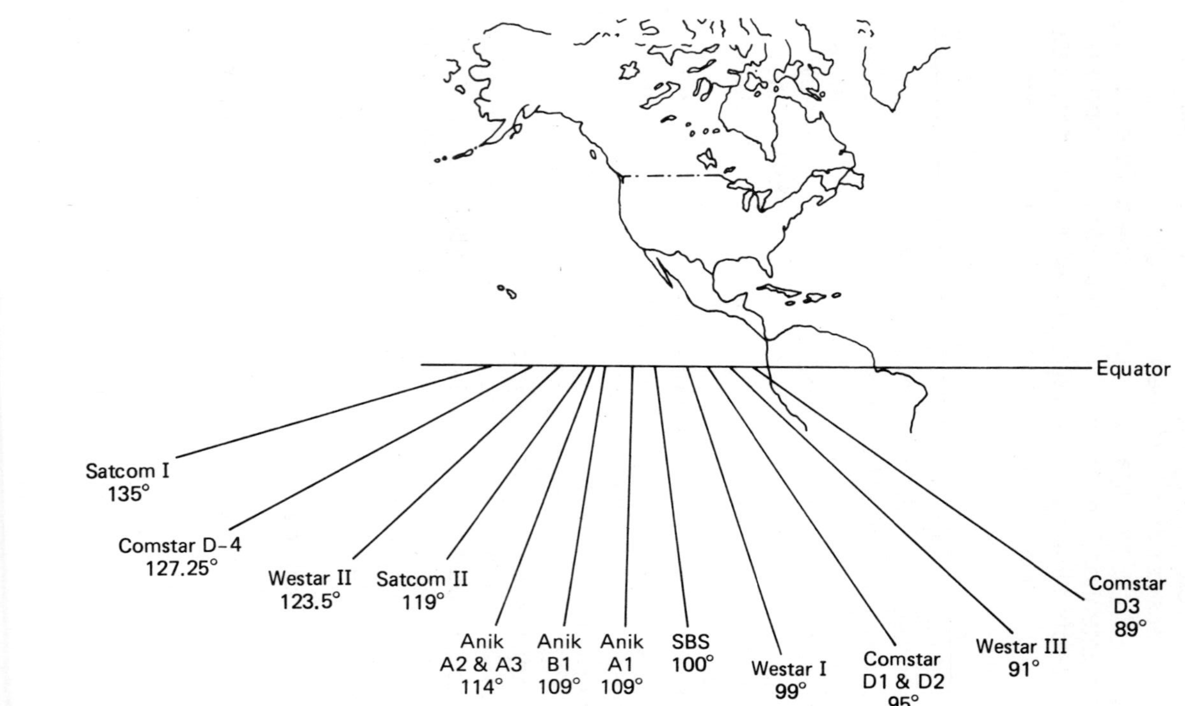

FIGURE 3-5. North American Coverage.

commercial communications satellite and established the first link between the United States and Europe.

INTELSAT II satellites were launched in 1967 over the Atlantic and Pacific oceans. The multiple satellites provided a multipoint capability and provided coverage to more than two-thirds of the world.

INTELSAT III provided full worldwide coverage. The system was placed in service betwen 1971 and 1975 and introduced simultaneous services in data, facsimile, television, telex, and telephone.

INTELSAT IV introduced the concept of spot beam transmission, a direct signal to specific areas on earth. These satellites were placed in service between 1971 and 1976.

INTELSAT V employs advanced concepts that allow for multiple use of the same frequencies. The system began operation in 1979 and employs the 14/11 gigahertz and 6/4 gigahertz bands.

Future Satellite Efforts. In the late 1960s, the Department of Defense initiated a program to develop a laser space communications system. The program has progressed to the point of a plan to launch the first laser communications equipped satellite. The system (LASERCOM) offers several advantages over conventional radio frequency methods. High data rates can be achieved with laser transmission. A rate of one giga bit per second (Gbit/s) has been achieved with a partial system, and an expanded system has the capability of supporting 16 Gbit/s. The narrow laser beamwidths provide for a more secure system that is less subject to detection and destruction.

Figure 3-6 illustrates the next likely generation of communications satellites—geostationary platforms. This approach will allow direct transmission and reception of signals between satellites. In effect, the satellites will switch signals among themselves. The platform could be designed to operate at the three frequency ranges of 6/4 GHz, 14/12 GHz, and 30/20 GHz. This approach has the advantage of transmitting message around the earth without incurring the delays of relaying signals to the earth stations.

Satellite communications certainly represent one of the more powerful technologies in data communications today. The advances made in the use of higher frequencies will result in smaller and cheaper earth station antennas. The potential applications for satellite use will increase dramatically as smaller antennas are made available. Within a few years, individual homeowners will be able to install small antennas (about three feet in diameter) for direct

TABLE 3-1. INTELSAT Satellites.

	I	II	III	IV	IV-A	V
Date Launched	1965	1967	1969	1971–1975	1976	1980
Size	28.4 inches in diameter 23.2 inches high	56 inches in diameter 26.5 inches high	56 inches in diameter 41 inches high	7.8 feet in diameter 17.3 feet high	7.8 feet in diameter 22.9 feet high	51-foot wing span 21 feet high
Weight	150 lbs. at launch 85 lbs. in orbit	357 lbs. at launch 190 lbs. in orbit	647 lbs. at launch 334 lbs. in orbit	3,120 lbs. at launch 1,610 lbs. in orbit	3,340 lbs. at launch 1,820 lbs. in orbit	4,110 lbs. at launch 2,231 lbs. in orbit
Attributes	240 circuits 18-month design life North American and European traffic	240 circuits 3-year design life Northern and southern hemisphere traffic	1,500 circuits and 4 TV channels 3-year design life Transmitted all forms of communications	4,000 circuits and 2 TV channels 7-year design life 12 transponders, each with 36 MHz bandwidth	6,000 circuits and 2 TV channels 7-year design life 20 transponders, each with 36 MHz bandwidth	12,000 circuits and 2 TV channels 7-year design life 6 antennas for 6/4 and 14/11 GHz

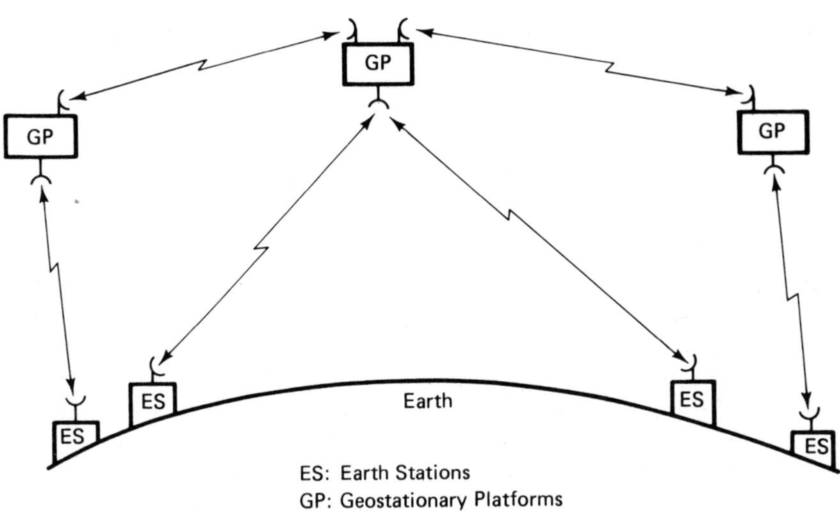

FIGURE 3-6. Geostationary Platforms.

television communications with the satellites. Satellite Television Corporation, a subsidiary of COMSAT, has already filed a request with the Federal Communications Commission for permission to construct and orbit several communications satellites of this nature. The channels can also be used for transmission of data.

The impact on the telephone companies, their local loops, and on distributed processing are enormous. Specialized common carriers will soon be able to bypass completely the telephone company facilities to service large corporations, small businesses, and even private citizens. The infrastructure of the nation's vital communications network will change dramatically in the 1980s as a result of satellite communications.

Optic Fibers

The use of light for sending messages has been in existence for hundreds of years. In earlier times, Greek armies transmitted coded light messages between their military units. In the 18th century, French scientists experimented with optical telegraph systems. These efforts had limited success because light signals attenuate rapidly in the atmosphere. Recently, the use of light for transmitting data has gained considerable attention and support in the industry. However, rather than use the atmosphere to transmit the light, newer techniques make use of cables (optic fibers) for the transmission.

Without question, fiber optics have a bright future. There are several reasons for this statement:

1. Optic transmission has a very large information capacity in terms of bandwidth. As Table 2-1 indicates, the frequencies encompassing light wave transmission are very high in the electromagnetic spectrum. The reader may recall that bandwidth is largely dependent on the frequency range. Optic fiber bandwidths in the range of 500 MHz are not unusual today; some researchers believe a fiber optic will carry 1,000 MHz; Bell Labs has successfully placed 30,000 simultaneous telephone calls on one optic fiber.
2. Optic fibers have electrically nonconducting photons instead of the electrons found in metallic cables such as wires or coaxial cables. This is attractive for applications in which the transmission path traverses environments that are subject to fire and gaseous combustion from electricity. Optic cables are not subject to electrical sparks or interference from electrical components in a building or computer machine room.
3. Fiber optics have less loss of signal strength than copper wire and coaxial cables. The strength of a light signal typically is reduced by only 50% after propagation through 9.6 miles of optic fiber cable. Repeaters can be spaced as far as 11.2 miles apart. In contrast, North American standards on existing copper cables stipulate repeaters every 2.8 miles (a T1 carrier operating at 1.544 Mbit/s).
4. Fiber optics are more secure than cable transmission methods. Transmission of light does not yield residual intelligence around the cable. Residual electromagnetic energy is found in electrical transmission. Moreover, it is quite difficult to tap an optic fiber cable. (In fact, this is one of the disadvantages at the present time, due to the limitation of multipoint optic fiber lines.)
5. Fiber optic cables are very small (roughly the size of a hair) and very light in weight (about 1/80 the weight of cable).
6. Optic fibers are easy to install and operate in high and low temperatures.
7. Due to the low signal loss, the error rate for optic fibers is very attractive. For example, a typical error rate on an optic fiber is 10^{-9} versus 10^{-6} in metallic cables.

8. Semiconductor technology has been refined to provide transmitting and receiving devices for the system. The rapidly decreasing costs of solid state chips have further spurred the optic fiber industry.

Methods of Transmission. The light signal is transmitted down the optic fiber in the form of on/off pulses representing a digital bit stream. The waves travel through the core of the cable, bouncing off a layer called *cladding* (see Figure 3-7). The refracting of the signal is carefully controlled through the design of the cable and the receivers and transmitters.

The light signal source is usually a laser or a light-emitting diode (LED). The lasers provide for greater bandwidth and yield significantly greater capacity than other methods. For example, a wire pair cable has a bandwidth distance parameter of 1 MHz/km; a coaxial cable has a 20 MHz/km parameter; and an optic fiber has a 400 MHz/km bandwidth distance parameter. The signal is emitted from microchips composed of semiconducting materials that transmit near infrared wavelength signals. Silicon photodetectors are used to receive the signals and convert the light ray to the original off/on electrical pulse for interface into the terminal, computer, or modem.

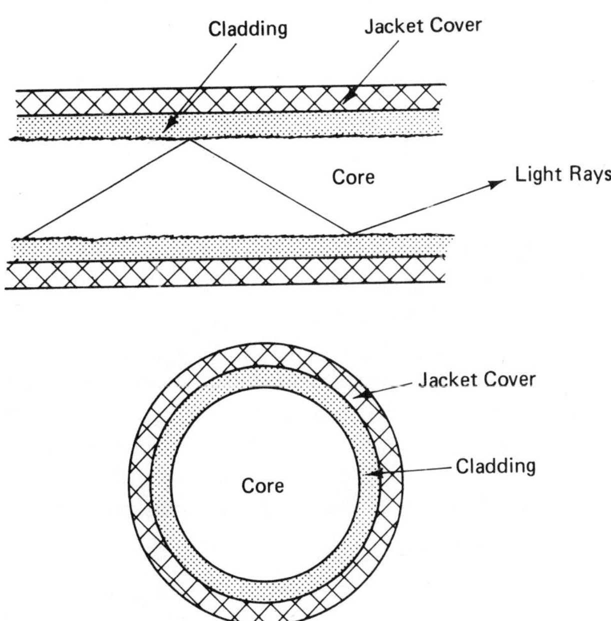

FIGURE 3-7. Optics.

Several methods are used to transmit the fiber light rays through the fiber.[2] In a *step index multimode* fiber (Figure 3-8(a)), the core and cladding interface is sharply defined. The light rays bounce off the interfaces into the core at different angles, resulting in different path lengths (modes) for the signal. This causes the signal to spread out along the fiber and limits the step index cable to approximately 35 MHz/km. This phenomenon is called *modal dispersion*.

A better approach, called *graded index multimode*, is to alter the cladding/core interface to provide different refractive indexes within the core and cladding [Figure 3-8(b)]. Light rays traveling down the axis of the cable encounter the greatest refraction and their velocity is the lowest in the transmitted signal. Rays traveling off axis encounter a lower refractive index and thus propagate faster. The aim is to have all modes of the signal attain the same net velocity through the fiber in order to reduce modual dispersion. The approach can result in bandwidths of 500 MHz/km.

A *step index single mode* fiber goes one step further (see Figure 3-8(c)). The core size and core/cladding index allow for only one mode to propagate down the fiber. This approach provides for very high bandwidth (as great as 2 GHz/km), but is subject to more attenuation as well as other problems.

Optic transmission is also subject to spectral or chromatic dispersion. The light passing through the fiber is made up of different frequencies and wavelengths. The refractive index differs for each wavelength and allows the waves to travel at different net speeds. LEDs, with a wide wavelength spread, are subject to considerable spectral dispersion. Lasers exhibit a nearly monochromatic light (limited number of wavelengths) and do not suffer any significant chromatic dispersion.

Future of Optic Fibers. It is estimated that optical systems will increase at an annual rate of 50% during the early 1980s.[3] Telephone companies in the United States are making increasing use of the new technology. AT&T is planning a lightwave system in the Northeast corridor from Cambridge, Massachusetts, to Mosely, Virginia. The total system will have a 270 Mbit/s capacity. Japan and Canada are conducting extensive research as well as implementing actual systems. Japan's Nippon Telephone and Telegraph is examining the use of 100 Mbit/s rates on cables with 83 mile repeater spacings.

---

[2] Dave King and Otto Szentes, "Fiber Optics: A New Technology for Communications." *Telesis*, February 1977, p. 5.

[3] Joel Fagenbaum, "Optical Systems: A Review." *IEEE Spectrum*, October 1979, p. 70.

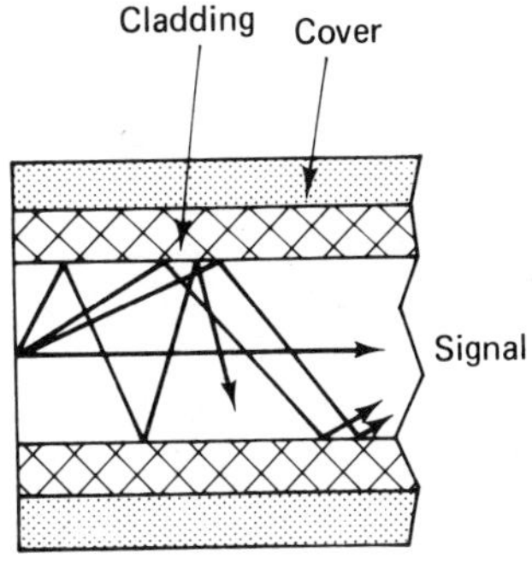

(a) Step Index Multimode

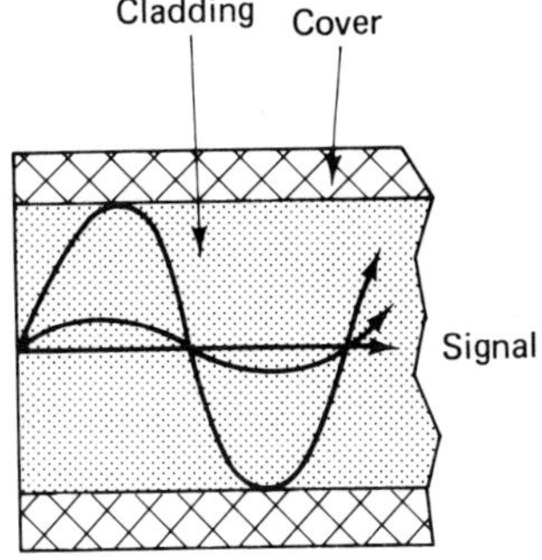

(b) Graded Index Multimode

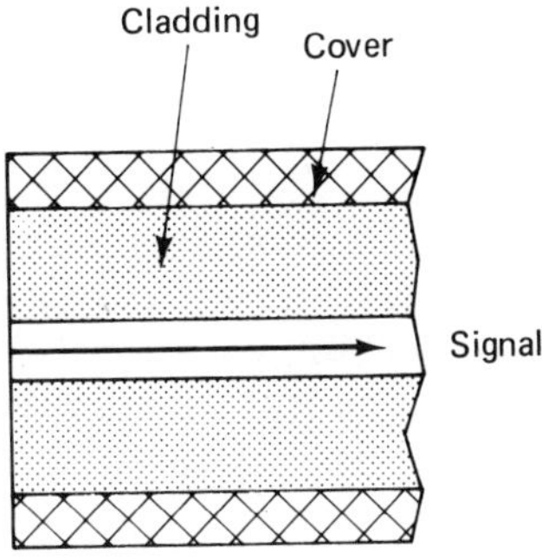

(c) Step Index Single Mode

FIGURE 3-8. Types of Fibers.

Canada is currently installing an extensive five thousand mile system in Saskatchewan to link 51 communities together. Britain is installing 280 miles of a lightwave network for a part of its telephone system.

This is not to say that the familiar coaxial and wire cables will be replaced rapidly. Optic technology is new and is still expensive. The capital investments in other technologies ensure their existence for many years. Nonetheless, the use of optic fibers in data communications and networks presents many opportunities to reduce costs and improve reliability.

Submarine Cable

Submarine coaxial cable was introduced for transmission over bodies of water to replace or augment high frequency radio schemes. The technology has been around since 1850, when a cable was laid across the English Channel. The first submarine cable in North America was laid in 1852 between New Brunswick and Prince Edward Island and in 1858 the first transatlantic cable was laid.

One might question the viability of underwater cable in view of the extraordinary satellite communications technology. However, many consider the path attractive due to its costs, security, and absence of a long signal propagation delay. Submarine cable is also designed to perform without repair for 20 years; the average life of a satellite is less than half that time. The technology has been reliable with the exception of fishing boats occasionally damaging the line. The recent installations have the shallow-water cables burrowed and this effort has provided much better reliability.

Waveguides

Prior to the advent of optic fibers, waveguides were considered to be the upcoming technology for short haul transmissions. The technology entails the transmission of radio waves through a metal tube. The tubes are designed to transmit waves of very high frequency and provide for very high data transfer rates. However, the tubes cannot be sharply bent and are expensive to build. Waveguide tubes are used in certain limited situations; for example, as a feeder between a microwave antenna and the equipment on the ground. The technology is sound, fast, and reliable but it cannot match the performance and flexibility of the optic fiber.

Several other transmission paths are available but are not used very much. The high frequency (HF) radio transmission was used extensively before satellite communications, although its error rate precluded any appreciable use for data transmission except for telegraph signals. Several organizations have experimented with meteor trail transmission. This technology relies on the reflection of radio waves off meteor trails. It is an experimental transmission path.

Summary

Today the prevalent technologies are microwave and coaxial transmission. Wire pair cables are also widely used. Usage of communications satellites is growing very rapidly and will continue to expand. Optic fibers will most likely replace much of the copper cable technology in the near future. The trend is toward integrated networks using a combination of technologies. For example, the FCC recently sponsored a system that links San Francisco and New York through a combination of microwave, coaxial, and satellite transmissions.

MODEMS

The digitally oriented computers and terminals often communicate with each other through the analog telephone facilities. Therefore, the digital messages must be translated into a form suitable for transmission across the analog network. The modem is responsible for providing the required translation and interface between the digital and analog worlds. The term *modem* is derived from (a) the process of accepting digital bits and changing them into a form suitable for analog transmission (modulation) and (b) receiving the signal at the other station and transforming it back to its original digital representation (demodulation). *Modem* is derived from the two words *modulator* and *demodulator*. Modems are designed around the use of a carrier frequency. The carrier signal has the digital data stream superimposed upon it at the transmitting end of the circuit. This carrier frequency has the characteristics of the sine wave discussed in Chapter 2.

Voice transmission is also subject to the modulation-demodulation process. Transmission media such as microwave and coaxial facilities have much greater bandwidth ranges than the three KHz used for voice transmission. Consequently, the 300–3300 Hz speech band is often changed through modulation to "ride" a carrier signal of much higher frequency. Other voice transmissions are spaced approximately four KHz apart to modulate different carrier frequencies on the same physical circuit. Typically, 12 voice transmissions can be modulated to share one two-wire circuit. The computer-generated bit stream can, therefore, undergo two separate modulation processes, one by the local modem and the other by the telephone company. The section in this chapter on multiplexing explains telephone company modulation in more detail.

Modulation Techniques

Three basic methods of digital bit modulation exist today. Some modems use more than one of the methods. Each method impresses the data message on the carrier signal, which is altered to carry the properties of the digital data stream in the message. The modulation methods are frequency, amplitude, and phase.

Frequency Modulation (FM). Figure 3-9(b) illustrates frequency modulation. This method changes the frequency of the carrier in accordance with the digital bit stream. The amplitude and phase are held constant. In its simplest form, a binary 1 is represented by a certain frequency and a binary 0 by another.

Several variations of FM modems are available. The most common is the frequency shift key (FSK) modem using four frequencies within the three KHz telephone line bandwidth (see Figure 3-10). The FSK modem transmits 1070 Hz signals and 1270 Hz signals to represent a binary 0 (space) and binary 1 (mark),

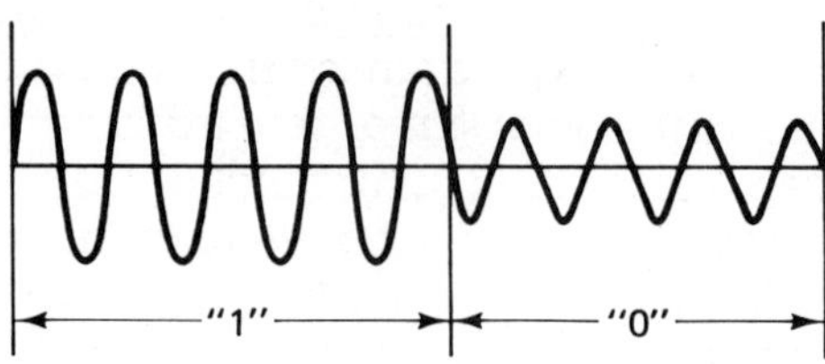

(a) Amplitude Modulation

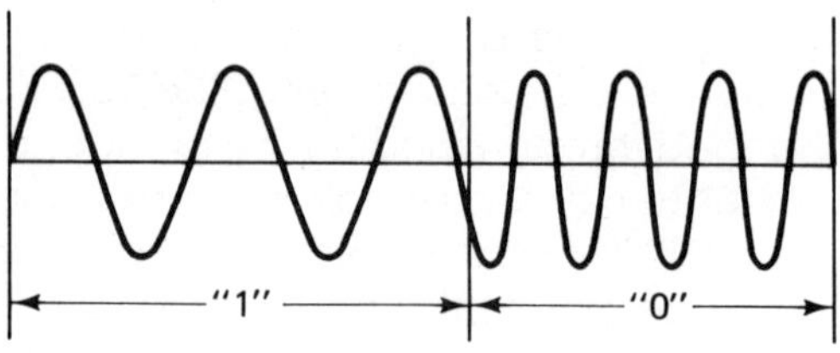

(b) Frequency Modulation

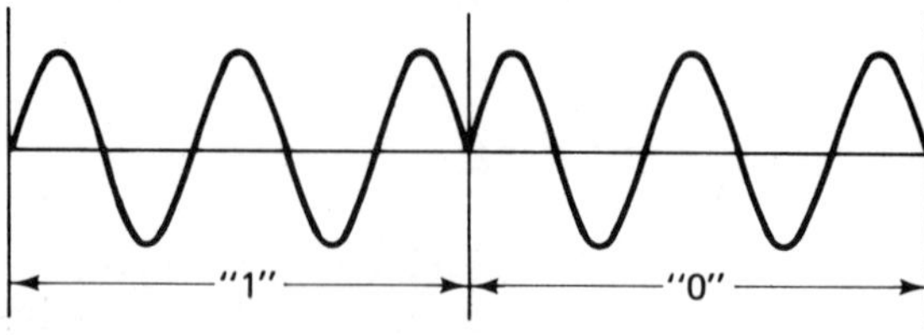

(c) Phase Modulation

FIGURE 3-9. Modulation Techniques.

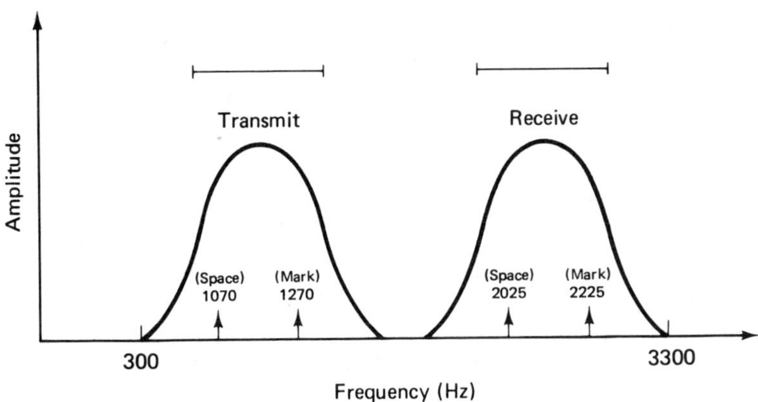

FIGURE 3-10. Frequency Shift Key Modulation.

respectively. It receives 2025 Hz and 2225 Hz signals as a binary 0 (space) and binary 1 (mark).

The FSK modem can operate with a full-duplex data flow on half-duplex (two-wire) or full-duplex (four-wire) facilities. (See Chapter 2 for an explanation of these terms.) The four frequencies permit the simultaneous sending and receiving of data within two channels. One channel is within the 300–1700 Hz bandwidth and the other is within 1700–3000 Hz. FSK modems are typically asynchronous low-speed machines. Speeds usually range up to 300 bps. In some instances, FSK modems operate at 600 bps and a few machines are available at bit rates greater than 600 bps. The low-speed constraint results from the need to increase separation of the mark and space frequencies at higher speeds. The three KHz bandwidth limits this separation.

FSK modems are also commonly found in the medium speed range of 1200 bps. These machines have a greater bit rate because they utilize two frequencies instead of four and thus allow greater separation of the carriers within the three KHz telephone bandwidth. A typical medium speed modem uses a mark frequency of 1200 Hz and a space frequency of 2200 Hz.

Amplitude Modulation (AM). AM modems alter the carrier signal amplitude in accordance with the modulating digital bit stream (See Figure 3-9(a)). The frequency and phase of the carrier are held constant and the amplitude is raised or lowered to represent a 0 or 1. In its simplest form, the carrier signal can be switched on or off to represent the binary state. AM modulation is not often used by itself due to transmission power problems and sensitivity to distortion. However, it is commonly used with phase modulation to yield a method superior to either FM or AM. This technique will be explained shortly.

Phase Modulation (PM). Previous discussions (in Chapter 2) of the sine wave described how a cycle is represented with degree markings to indicate the point to which the oscillating wave has advanced in its cycle. PM modems interrupt the continuous waveform and alter the phase of the signal to represent a 1 or 0 (see Figure 3-9(c)). The common approach today is to compare the phase of the cycle in a current time period to the phase in a previous time period. This approach is called *differential phase shift keying* (DPSK).

Multilevel Transmission

At this point in our discussion, the reader can surmise correctly that the bit rate (bps) of a modem is comparable to the baud rate. (Recall that baud rate is the same as hertz, i.e., the number of oscillations per second of the signal. Recall also that a baud rate is the number of times per second that a signal changes.) In low-speed modems, the baud rate is usually the same as the bit rate. In order to increase the bit rate across the limited bandwidth of the telephone channel, modem manufacturers frequently employ methods to represent more than one bit per baud. This technique is called *multilevel transmission*.

Multilevel transmission partially solves the problem of the short bit times that result from the high baud rate. A bit rate of 9600 bps requires a signal change every 104 μs (1 second/9600 = .000104). This is often insufficient time for the modem to detect and interpret the data stream. However, multilevel modems can use a rate of 1200 baud and, by representing two bits per baud, achieve a 2400 bit rate. Thus, multilevel modems achieve a higher bit rate per baud. The process is illustrated in Figure 3-11.

A common form of multilevel transmission is quadrature amplitude modulation (QAM). The QAM modem creates one of eight possible different bit conditions per baud by combining two AM levels with four PM changes. The eight level scheme allows three binary bits to be transmitted ($2^3 = 8$) with each baud.

Practically speaking, limitations exist on how many times the signal can be divided and remain detectable at the receiving modem. While the lower baud rate provides more time for detection

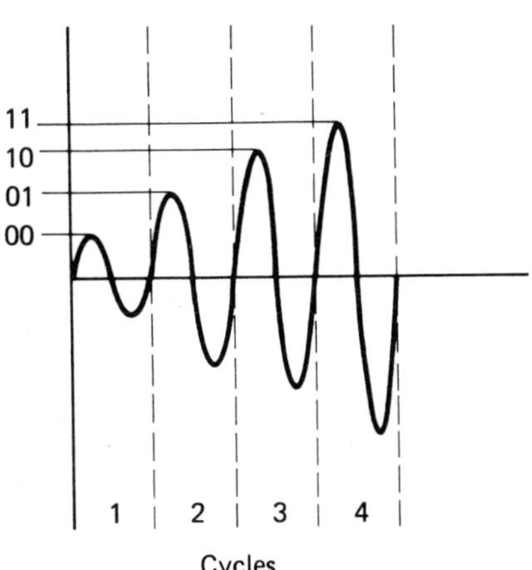

FIGURE 3-11. Multilevel Modulation.

and interpretation, smaller increments in the differences between signal states make accurate data recovery more difficult. Moreover, a noise spike (or other imperfection) on a line with multilevel transmission will distort more bits than with a single level modulation scheme. Multilevel transmission creates encoding and decoding delays and the schemes lead to more complex and costly modems. Nonetheless, the technique is effective and widely used.

Low Baud Modems

The conflict between the needs for a high data transmission rate and a low baud rate can be addressed by using multiple carrier frequencies in the modem. Each carrier is responsible for a portion of the total user bit stream. The carriers operate at a low baud rate, providing for better bit detection and interpretation.

A low baud modem generates 48 carrier frequencies, spaced between 500 and 2,800 Hz within the telephone bandwidth.[4] The data stream is divided into groups of five bits. The groups modulate each of the 48 carriers. The modem operates at 40 baud, permitting a 25 millisecond cycle duration ($1/40 = .025$). Since each carrier is modulated by five bits, the low baud modem yields a 9.6 Kbps (40 baud $\times$ 48 carriers $\times$ 5 bits per carrier = 9,600 bps).

---

[4]This example is from "Digital Supermodem: Why and How it was Developed" by G. Brian Hick. *Data Communications*, June 1980.

Side Channel Modems

The three KHz of the telephone circuit is often a greater bandwidth than is needed for the user application. Moreover, the allocation of the full band limits two-day (full-duplex) transmission and requires the modem to receive a transmission, reverse its mode from receiving to sending, and then transmit its reply back across the same three KHz circuit. Many modems are designed to divide the bandwidth into smaller size channels and provide for full-duplex transmission using one or two additional subchannels across a half-duplex (two-wire) circuit. Obviously, the smaller bandwidths necessitate a lower data transmission speed, but the side channels are often more than adequate for certain applications. The concept is quite similar to the usage of different frequencies in the FSK modem. The idea of side channel modems is illustrated in Figure 3-12.

The vendor offerings typically provide for one or two side channels, each operating at 5 to 150 bit/s. The main channel speed ranges from 1.2 to 9.6 Kbit/s. The combinations of channel speeds and number of side channels vary among the offerings; the reader should check several vendors before deciding upon one product.

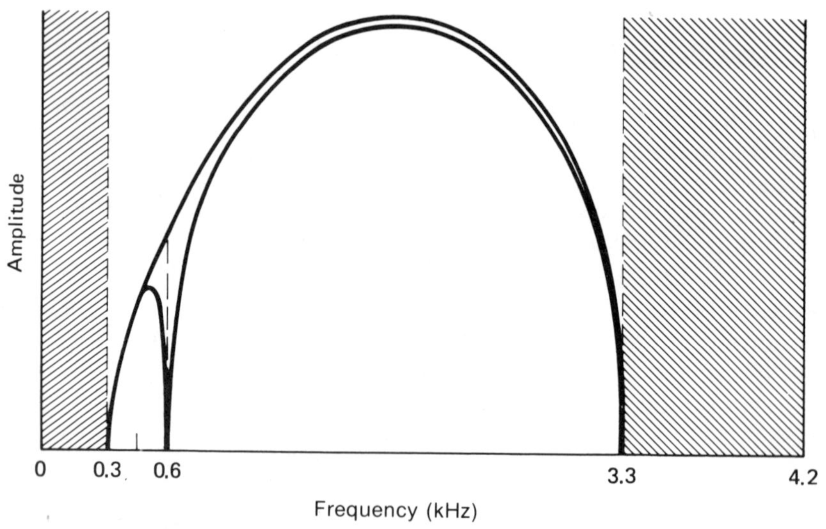

FIGURE 3-12. Side Channel Modems.

Other Modems

The short-haul or limited-distance modem (LDM) is used for transmissions of a few feet to approximately 20 miles. The distance is highly variable and depends upon the operating speed, type of transmission path, and configuration of the telephone company facilities. Typical LDMs use wire pairs or coaxial cables and some can operate at higher data rates (19.2 Kbit/s to 1 Mbit/s.). In some instances, these machines do not actually use AC modulation techniques but rather a DC baseband digital pulse stream. The use of the short-haul modem requires the removal of certain analog equipment on the telephone line. These machines are attractive due to their simplicity and low cost.

Line drivers are DC machines used to eliminate modems. Operating distances range from a few feet to several miles. The line driver is usually placed between two computer-related devices and replaces two modems (on short distances). A typical line driver operates at 2400 bps for four miles and 19.2 Kbps for two miles. Speeds are available up to 230.4 Kbps. These machines are attractive due to their low price.

Acoustic couplers are yet another alternative. These machines acoustically connect the terminal to the analog facilities. The telephone handset is placed into the coupler's transmittal receiver device for connection. Portable terminals often contain acoustic couplers. The technique is very simple and very effective for low data rates (up to 1200 bps; usually no greater than 450 bps).

Interface Standards

The Bell System, CCITT, the Electronics Industries Association (EIA), the computer industry, and other modem vendors have developed several standards defining how to interface the modems [the data circuit-terminating equipment (DCE)] with the terminals or computers [the data terminal equipment (DTE)]. Acceptance of the standards has been instrumental in the ability to use equipment from different vendors.

RS232-C. EIA sponsors the RS232-C standard which is used extensively in North America. (The CCITT has a comparable standard, V.24). Figure 3-13 defines the circuits of RS232-C. These circuits are actually 25 pin connections and sockets. The terminal pins plug into the modem sockets. All the circuits are rarely used; most modems utilize 12 or fewer pins.

Pin No.	CCITT Equiv.	Circuit	Direction	Description	Interface Type A	B	C	D	E	F	G	H	I	J	K	L	M	Z
1	101	AA	Both	Protective Ground	-	-	-	-	-	-	-	-	-	-	-	-	-	-
7	102	AB	Both	Signal Ground	x	x	x	x	x	x	x	x	x	x	x	x	x	x
2	103	BA	To-m	Transmitted Data	x	x		x	x	x		x		x		x	x	p
3	104	BB	To-T	Received Data			x	x	x		x		x		x	x	x	p
4	105	CA	To-m	Request to Send		x		x		x				x		x		p
5	106	CB	To-T	Clear to Send	x	x		x	x	x		x		x		x	x	p
6	107	CC	To-T	Data Set Ready	x	x	x	x	x	x	x	x	x	x	x	x	x	p
20	108.2	CD	To-m	Data Terminal Ready	d	d	d	d	d	d	d	d	d	d	d	d	d	p
22	125	CE	To-T	Ring Indicator	d	d	d	d	d	d	d	d	d	d	d	d	d	p
8	109	CF	To-T	Received Line Signal Detector			x	x	x		x		x	x	x			
21	110	CG	To-T	Signal Quality Detector														p
23	111	CH/CI	Either	Data Signaling Rate Selector/Indicator														p
24	113	DA	To-m	Transmitter Signal Element Timing (DT/DCE)	s	s		s	s	s		s		s	s	s	s	p
15	114	DB	To-T															
17	115	DD	To-T	Receiver Signal Element Timing (DCE)			s	s	s		s		s		s	s	s	p
14	118	SBA	To-m	Secondary Transmitted Data							x		x	x	x	x	x	p

16	119	SBB	To-T	Secondary Received Data	x		x		x	x	x	x	p
19	120	SCA	To-m	Secondary Request to Send		x			x	x	x		p
13	121	SCB	To-T	Secondary Clear to Send		x		x	x	x	x	x	p
12	122	SCF	To-T	Secondary Received Line Signal Detector	x		x		x	x	x	x	p

Legend:
- p To be specified by the supplier
- \- Optional
- d Additional Interchange Circuits required for Switched Service
- s Additional Interchange Circuits required for Synchronous Channel
- x Basic Interchange Circuits, All Systems
- m modem
- T term
- PIN Number 9 & 10 usually reserved for Modem testing
- PIN Numbers 11/18/25 unassigned

Key to Columns:
- A Transmit Only
- B Transmit Only*
- C Receive Only
- D Full-Duplex* Half-Duplex
- E Full-Duplex
- F Primary Channel Transmit Only*/Secondary Channel Receive Only
- G Primary Channel Receive Only/Secondary Channel Transmit Only*
- H Primary Channel Transmit Only/Secondary Channel Receive Only

Key to Columns (Cont'd):
- I Primary Channel Receive Only/Secondary Channel Transmit Only
- J Primary Channel Transmit Only*/Half-Duplex Secondary Channel
- K Primary Channel Receive Only/Half-Duplex Secondary Channel
- L Full-Duplex Primary Channel*/Full-Duplex Secondary Channel* Half-Duplex Primary Channel/Half-Duplex Secondary Channel
- M Full-Duplex Primary Channel/Full-Duplex Secondary Channel
- Z Special (Circuits specified by Supplier)

*Indicates the inclusion of Circuit CA (Request to Send) in a One-Way Only (Transmit) or Full-Duplex Configuration where it might ordinarily not be expected, but where it might be used to indicate a nontransmit mode to the data communication equipment to permit it to remove a line signal or to send synchronizing or training signals as required.

FIGURE 3-13. ***RS232-C***

The circuits perform one of four functions:

- Data transfer across the interface.
- Control of signals across interface.
- Clocking signals to synchronize data flow and regulate the bit rate.
- Electrical ground

The functional descriptions of the circuits are listed below. The reader should be aware that several options exist on how to use some of these circuits. Each vendor's offering should be examined carefully.

Circuit AA Protective Ground: Conductor is electrically bonded to equipment frame.

Circuit AB Signal Ground: Common ground for all circuits. This is the second wire of the two-wire electrical circuit, discussed in Chapter 2.

Circuit BA Transmit Data: Data signals transmitted from DTE to DCE. This represents the user data. Data cannot be transmitted unless circuits CA, CB, CC, and CD are all activated.

Circuit BB Receive Data: User data signals transmitted from DCE to DTE.

Circuits CA Request to Send: Signal from DTE to DCE. This circuit notifies DCE that the terminal or computer has data to transmit. Circuit CA is also used on half-duplex lines to control the direction of data transmission. The transition of OFF to ON notifies the DCE to take any necessary action to prepare for the transmission. For example, in a polled environment, the Request to Send signal would initiate the sending of a carrier signal to the remote modem. The transition from ON to OFF notifies the DCE that the DTE has completed its transmission.

Circuit CB Clear to Send: Signal from DCE indicating the DTE can transmit the data. The Clear to Send signal may be turned ON after receiving a carrier signal from the remote modem. The use of CB varies from modem to modem and is used differently on half-duplex and full-duplex circuits.

Circuit CC Data Set Ready: Signal from DCE indicating the machine is (a) OFF HOOK: connected to channel on a switched line, (b) DCE is in data transmit mode (not test, voice, etc.), (c) DCE has completed timing functions and answer tones.

Circuit CD Data Terminal Ready: Signal from DTE indicating terminal or computer is powered up, has no detectable malfunction, and is not in test mode. Generally, CD is ON if it is ready to transmit or receive data. In a switched arrangement, a ring from the remote site will normally activate CD.

Circuit CE Ring Indicator: Signal from DCE indicates that a ringing signal is being received on a switched channel.

Circuit CF Receive Line Signal Detector: Signal from DCE indicating the DCE has detected the remote modem's carrier signal.

Circuit CG Signal Quality Detector: Signal from DCE to indicate that the received signal is of sufficient quality to believe no error has occurred.

Circuits CH and CI Data Signal Rate Selector: Signals from DTE and DCE, respectively, to indicate the data signaling rate for dual rate machines. Some devices have the capability to transmit varying bit rates.

Circuit DA Transmitter Signal Element Timing: Signals from DTE to provide timing of the data signals being transmitted on circuit BA (Transmit Data) to the DCE. The mark and space elements are indicated by this circuit. (See Chapter 2 for a description of mark and space.) DTE provides this signal; if the DCE provides timing, then circuit DB is used.

Circuit DB Transmitter Signal Element Timing: Signals from DCE to provide timing of the data signals being transmitted on circuit BA (Transmit Data) to the DCE. Again, as in circuit DA, the mark and space elements are indicated by circuit DB. DCE provides this signal; if the DTE provides the timing then circuit DA is used.

Circuit DD Receiver Signal Element Timing: Signals from DCE to provide timing to DTE of the data signals being received on circuit BB (received data).

In addition to these circuits, RS232-C defines five other circuits designated as secondary channels: SCA, SCB, SCF, SBA, and SBB. These pins are used to implement the side channel modem transmission scheme that is described earlier in the chapter. The remaining circuits are used for testing and other vendor-specific functions.

The data flow across the RS232-C interface is illustrated in Figures 3-14 and 3-15. The use of half-duplex/full-duplex, leased line/switched line, or side channels determine the sequence and order of the activation of the circuits. These illustrations depict a half-duplex, private line configuration and a full-duplex, side channel arrangement.

RS449. Recently, the EIA announced RS449, a significant improvement over RS232-C. The latter standard has several electrical specifications that limit its effectiveness. For instance, RS232-C is

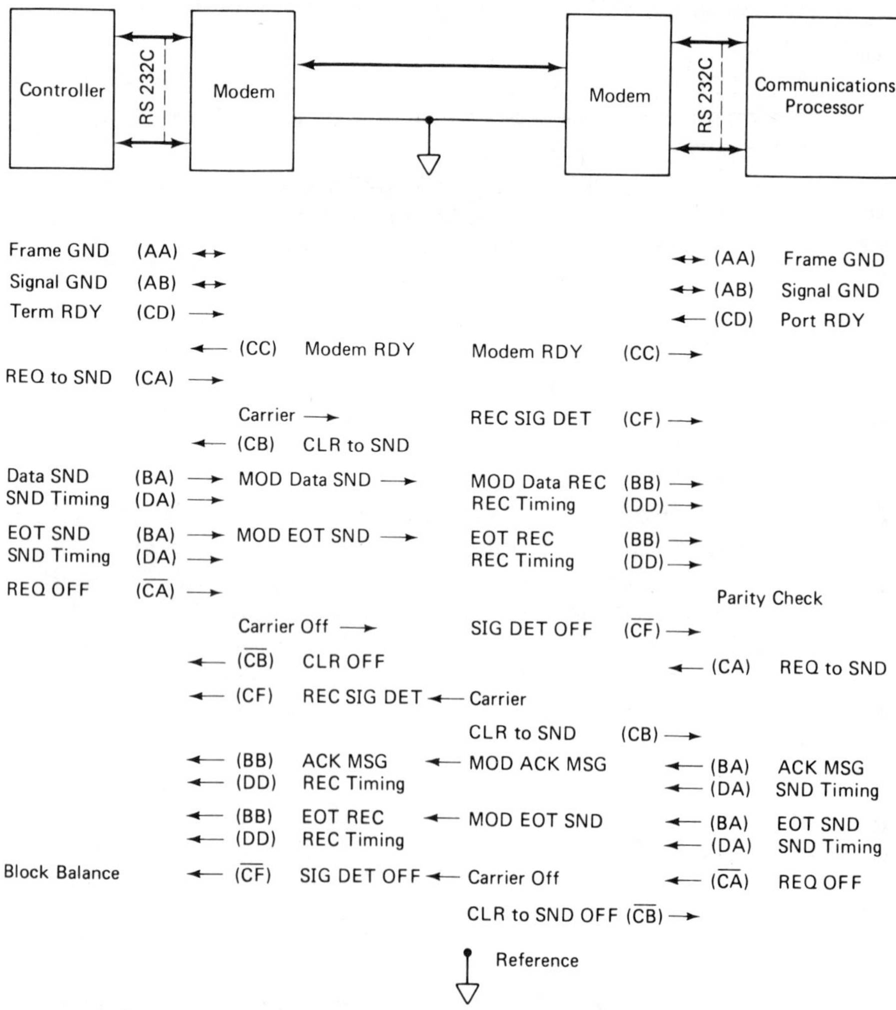

FIGURE 3-14. RS232-C Half-Duplex Private Line Data Flow (From *Data Communications: A User's Guide* by Kenneth Sherman. Reston VA: Reston Publishing Company, 1981, p. 119. Reprinted with permission.)

limited to a 20 Kbit/s and an allowable 50 foot spacing between the components. It also presents a noisy electrical signal. RS449 provides 37 basic circuits with 10 additional circuits and other testing maintenance loops. It is compatible with recent CCITT and ISO standards. In addition, the RS449 specification establishes a bit rate

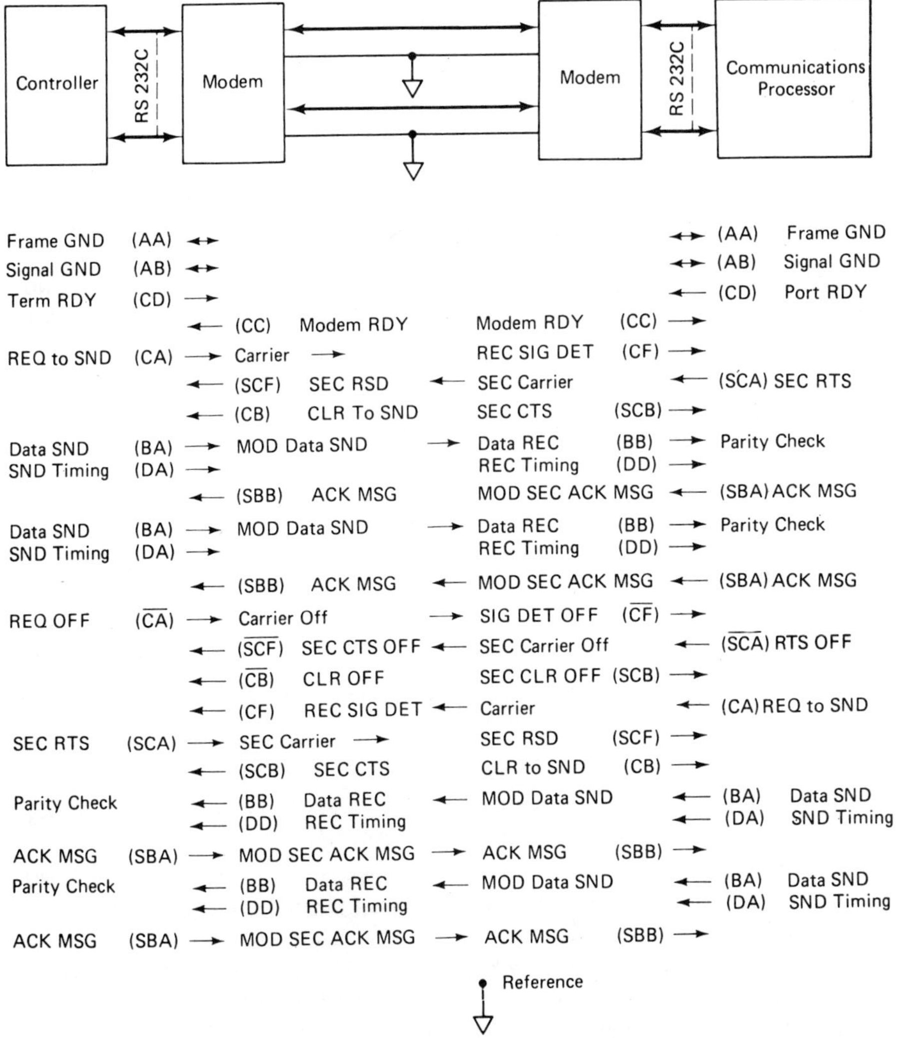

FIGURE 3-15. RS232-C Full-Duplex Side Channel Data Flow. (From *Data Communications: A User's Guide* by Kenneth Sherman. Reston VA: Reston Publishing Company, 1981, p. 121. Reprinted with permission.)

of 2 Mbit/s and an allowable distance of 200 feet between the machines (see Figure 3-16). EIA anticipates that RS449 will eventually replace RS232-C. A new federal standard required agencies to procure equipment that meets the provisions of RS449 by June 1, 1980, although waivers have been granted to many agencies.

The additional 10 circuits are defined as follows:

Circuit SC Send Common: Connected to DTE to serve as a reference voltage for unbalanced receivers.

Circuit RC Receive Common: Connected to DCE to serve as a reference voltage at DTE for unbalanced receivers.

Circuit IS Terminal in Service: Signal to indicate if DTA is available. This prevents an incoming call from being connected to the DCE when DTE is busy. If DTE is out of service, this signal can make a port busy during a rotary hunt to that machine.

Circuit NS New Signal: Signal to alert master stations in a multipoint network when a new signal is about to begin. Signal is from DTE to DCE. Use to improve response time in multipoint polling networks.

Circuit SF Select Frequency: Signal used to select the transmit and receive frequencies of the DCE. Used in a multipoint circuit where the stations have equal status.

Circuit LL Local Loopback: Signal to check the local DTE/DCE interface. The signal also checks the transmit and receive circuitry of the local DCE.

Circuit RC Remote Loopback: Signal to check both directions through the common carrier path and through the remote DCE up to the remote DCE/DTE interface.

Circuit TM Test Mode: Signal from DCE to DTE that DCE is in test condition (conditions LL or RC).

Circuit SS Select Standby: Signal used to increase reliability by switching to a backup channel or to allow switching between alternate applications' transmissions.

Circuit SI Standby Indicator: Signal indicates whether DCE is set up to operate with the SC (Select Standby) Circuit.

RS449 also provides standards for the transition away from RS232-C. EIA has specified adapter configurations between RS232-C and RS449. The adapters are meant to be temporary until the RS232-C functions are replaced. Figure 3-17 illustrates the adapter configurations.

The reader can obtain more detailed information on RS232-C and RS449 by writing to the Electronics Industry Association, Standards Sales Office, 2001 Eye Street N.W., Washington, D.C. 20006. The EIA has several documents that provide very useful information on the DTE and DCE interface.

The Modem Market

Modem suppliers number well over 80 and include over 500 different models and types. It is estimated that the United States has over three million modems installed, but that the installation base will decline in the 1980s as users migrate to end-to-end digital networks.[5] Perhaps, but the telephone company must also migrate to digital schemes at the local loop level for this decline to occur.

Bell modems are often cited as a benchmark for performance and feature comparisons. The more popular Bell modems are:

103/113: Low speed modems, 0 to 300 bps, FSK, Asynchronous, HDX or FDX, two- or four-wire line interface.

201: Medium speed modems, 2400 bps, PSK, four-phase, Synchronous, HDX or FDX, two- or four-wire line interface.

202: Medium speed modems, 1200 to 1800 bps, FSK, Asynchronous, HDX and/or FDX, two- and/or four-wire interface.

212: Medium speed modems, 1200 bps, FSK or PSK, Asynchronous or Synchronous, HDX or FDX, two-wire interface.

208: High speed modems, 4800 bps, PM, eight-phase, Synchronous, HDX, two- and/or four-wire interface.

209: High speed modem, 9600 bps, QAM, Synchronous, FDX, four-wire interface.

303: Wideband modems, up to 19.2, 50, or 230.4 kbs, operate on 5000 and 8000 tariffed channels.

The following list contains the major features available in modems today. The reader should be aware that the market is quite diverse; many variations of the features are available.

- Speed in bps (and multiple speeds)
- Leased line or dial-up

---

[5] "Modem Survey," Data Decisions, *DATAMATION*, October 1981, p. 159.

Circuit Name	MNEM	CAT	PIN No.	Circ. Class	Circuit Direct.	Usage Opt.	Nearest RS232 Equivalent
Signal Ground	SG	II	19	G	—	M	Signal Ground
*Send Common	SC	II	37	G	To DCE	M	—
*Receive Common	RC	II	20	G	From DCE	M	—
*Terminal in Service	IS	II	28	C	To DCE	O	—
Incoming Call	IC	II	15	C	From DCE	A	Ring Indication
Terminal Ready	TR	I	12,30	C	To DCE	S	Data Terminal Ready
Data Mode	DM	I	11,29	C	From DCE	M	Data Set Ready
Send Data	SD	I	4,22	D	To DCE	M	Transmitted Data
Receive Data	RD	I	6,24	D	From DCE	M	Received Data
Terminal Timing	TT	I	17,35	T	To DCE	O	Xmit Sig Element DCE
Send Timing	ST	I	5,23	T	From DCE	T	Xmit Sig. El Tim DTE
Receive Timing	RT	I	8,26	T	From DCE	T	Rec. Sig. El Timing
Request to Send	RS	I	7,25	C	To DCE	M	Request to sent
Clear to Send	CS	I	9,27	C	From DCE	M	Clear to send
Receiver Ready	RR	I	13,31	C	From DCE	M	Carrier Detect
Signal Quality	SQ	II	33	C	From DCE	O	Signal Quality Det.
*New Signal	NS	II	34	C	To DCE	O	—
*Select Frequency	SF	II	16	C	To DCE	O	—
Sig. Rate Selector	SR	II	16	C	To DCE	O	Data Sig Rate Select
Sig. Rate Indication	SI	II	2	C	From DCE	O	Data Sig Rate Select
*Local Loopback	LL	II	10	C	To DCE	O	—
*Remote Loopback	RL	II	14	C	To DCE	O	—
*Test Mode	TM	II	18	C	From DCE	M	—
*Select Standby	SS	II	32	C	To DCE	O	—
*Standby Indicator	SB	II	36	C	From DCE	O	—
Shield	—	II	1	G	—	O	—
Spares	—	II	3,21	—	—	O	—

(a) circuits on the 37-pin main connector

Signal Ground	SG	II	5	G	—	O	Signal Ground
Send Common	SC	II	9	G	To DCE	O	—
Receive Common	RC	II	6	G	From DCE	O	—
Sec. Send Data	SSD	II	3	D	To DCE	O	Secondary Transmit Data
Sec. Req. to Send	SRS	II	7	C	To DCE	O	Secondary Request to Send
Sec. Clear to Send	SCS	II	8	C	From DCE	O	Secondary Clear to Send
Sec. Rec. Ready	SRR	II	2	C	From DCE	O	Secondary Rec. Line Signal
Shieid	—	II	1	G	—	O	—

(b) circuits on the 9-pin optional connector

Legend and Notes:

Circuit Classifications
- G = Ground or Common
- D = Data
- C = Control
- T = Test

Circuit Category
- I =Category I 20,000BPS
- II =Category II 20,000BPS

Usage Options
- M = Mandatory for all two-way communications channels
- S = Additional circuits required for all switched channel
- A = Additional circuits required for all switched channel with answering signaled across the interface.
- T = Additional circuits required for synchronous primary channel
- O = Optional circuits.

Other:
- * = New circuits not contained in RS232 Standard

FIGURE 3-16. RS 449.

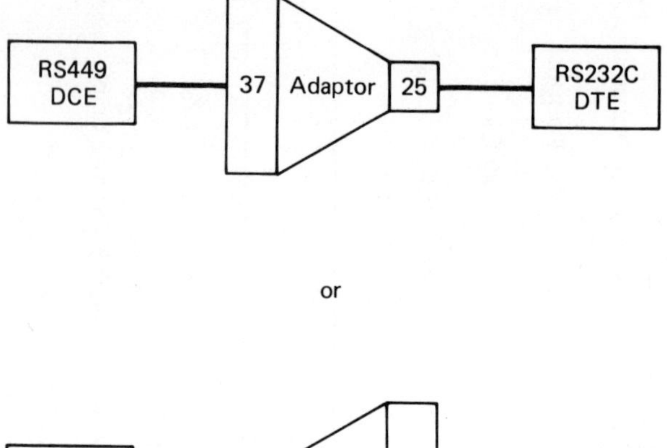

FIGURE 3-17. RS449 Adaptor.

- Diagnostic capabilities
- Electrical interface types (RS232-C, etc.)
- Type (line driver, coupler, etc.)
- Bell compatible
- Asynchronous or synchronous
- Voice/data
- Reverse channel
- Loopbacks
- Point to point, multipoint, or piggyback
- Automatic answer
- Modulation type (phase, frequency)
- Multilevel transmission
- Baud rate
- Half-duplex or duplex
- Line conditioning
- Distance of transmission

SWITCHING

Modern day networks consist of many components such as terminals and computers spread throughout buildings, cities, states, and nations. Many networks have hundreds of elements that, at any given time, must be able to establish a session and path with each other. It is obvious that a component cannot have a direct (point-to-point) connection to every other component. For example, a relatively small network of 500 components would require 124,750 individual interconnections [n(n—1)/2], where n is the number of components to be interconnected].

One solution to this problem is to place switches on the transmission path (as in Figure 3-18). The sites are not interconnected directly but effect the transmission first through a switch or switches and then to the receiving computer terminal, telephone, or some other component, such as a modem. In Figure 3-18, site C communicates with site L by sending its transmission to the switch; the transmission contains an address in a message header. In the case of a telephone network, the telephone number acts as the address. The switch uses the address to determine which path to use to deliver the transmission. The switch substantially reduces the number of interconnection paths.

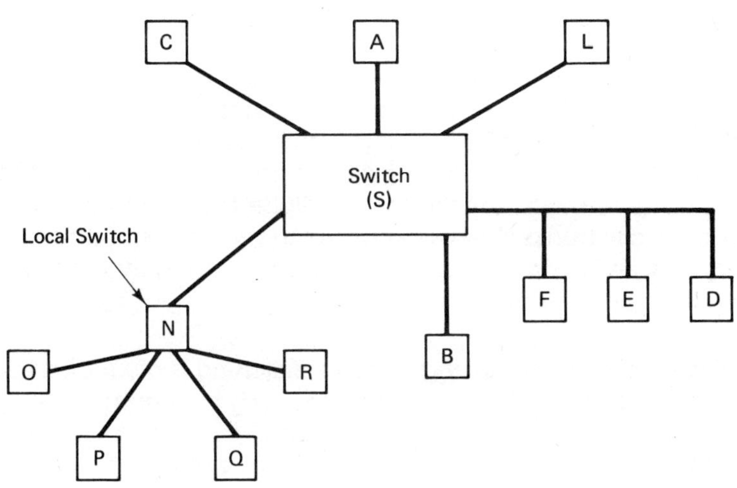

FIGURE 3-18. Switching.

Figure 3-18 also depicts a hierarchical switching arrangement. Site N provides switching facilities for sites O, P, Q, and R; N transmits to switch S only for traffic needs outside its switching domain. Most switches also provide for the sharing of the path by multiple sites. For example, sites D, E, and F share one of the paths through a multidrop arrangement.

Circuit Switching

Circuit switching is a direct electrical connection between two components. The telephone network is an example of circuit switching. (Chapters 1 and 2 contain introductory information on the telephone network.) The direct connection serves as an open "pipeline," permitting the two end users to utilize the facility as they see fit—within bandwidth and tariff limitations.

Figure 1-2 of Chapter 1 is reproduced as Figure 3-19 with more detail. The local subscriber initiates the call by going off-hook and activating the circuit on the local loop to the end office (central office); the dialed number is used to switch the call through the levels in the network to the proper destination. The four switching levels are:

Class 4: Toll Center

Class 3: Primary Center

Class 2: Sectional Center

Class 1: Regional Center

The telephone switching network is designed for calls to be connected at the lower levels in the hierarchy, if possible. Consequently, the lines in Figure 3-19 indicating first choice are used first; these facilities are designed for high usage. They will sometimes be unavailable, in which case the call is switched up the hierarchy to the lower capacity trunks. The approach is to provide economical and quality service by arranging the hierarchy into highly utilized, first-choice lines and less utilized, final-choice lines.

Crosspoints and Staging. Circuit switching is arranged in one or a combination of three architectures: (a) concentration—more input lines than output lines, (b) expansion—more output lines than input lines, and (c) connection—an equal number of input and output lines. In its simplest form, a circuit switch is an N X M array of lines that connect to each other at crosspoints (see Figure 3-20). In a large switching office, the N lines are input from the subscriber

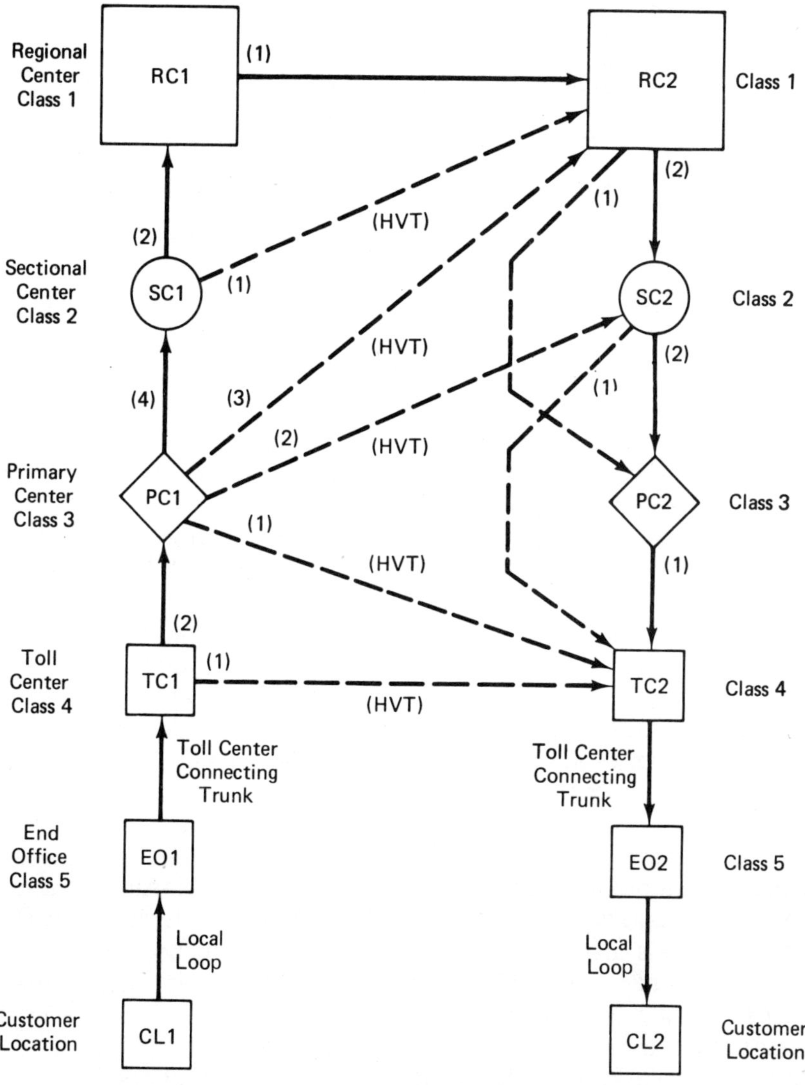

Dashed Lines Indicate High
Volume Paths (HVT = High Volume Trunk)

Numbers in Parentheses ( ) Show
Order of Route Choices at Each Center

FIGURE 3-19. Telephone Company Hierarchy.

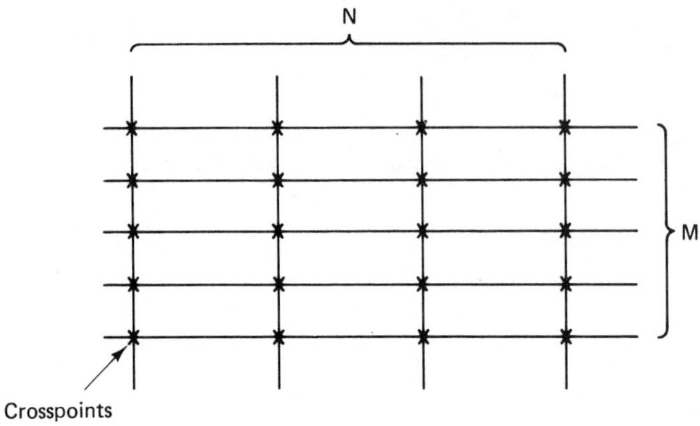

FIGURE 3-20. Crosspoints in a Circuit Switch.

(terminals, computers, etc.) and the M lines are output to other switching offices.

Circuit switching is usually performed in more than one array or stage in order to reduce the number of crosspoints. If N lines are to be connected as in Figure 3-20, N^2 crosspoints are required. Clos demonstrated that multistaging switches are economical for networks when $N > 16$.[6] Multistage networks are designed with fewer crosspoints, yielding a more economical arrangement. However, this approach usually allows for the blocking of a call. It is prohibitively expensive to implement a switch that handles *all* calls. The telephone office uses several stages and the path is selected through the switches to reduce the probability of a blocked call.

Circuit switching *only* provides a path for the sessions between data communications components. Error checking, session establishment, message flow control, message formating, selection of codes, and protocols are the responsibility of the users. Little or no care of message traffic is provided in the circuit switching arrangement. Consequently, the telephone network is often used as the basic foundation for a data communication network and additional facilities are added by the value added carrier, network vendor, or user organization. Subsequent examples of other switching technologies (e.g., message and packet) often use circuit switching to provide additional functions and facilities.

---

[6] Amos E. Joel, Jr. "Circuit Switching: Unique Architecture and Applications." *IEEE Computer*, June 1979, p. 11.

Early Switches. The late night movies provide a nostalgic example of an earlier form of circuit switching—the telephone switchboard operator. The operator, after requesting and receiving the number from the caller, plugged the proper cords and jacks into the switchboard to provide the direct connection. If the call were destined for a location outside the plugboard, the call would be routed to another switchboard operator. In this manner, the call would be switched through the network to the final destination.

The manual switchboard was largely replaced by the automatic exchange, a step-by-step switch called the Strowger switch. This technique operates on each dialed number by adjusting an electromechanical device to connect an incoming line to an outgoing line. The dialed number determines the physical position of one contact on a possible interface to one of 100 contacts or outgoing circuits. Each number moves the switch in a step-by-step fashion until the desired outgoing line is found. If the line is free, the caller is connected; if not, a busy signal is returned to the caller.

The Strowger switch is quite slow and subject to considerable electrical noise. The crossbar switch provided improvements over the Strowger switch and largely replaced the older technology. It also connects input lines to output lines but uses electromagnets to open or close vertical and horizontal bars to establish the physical connection.

The next major advance in circuit switching came with the introduction of common control. This technique avoids the time consuming and cumbersome serial step-by-step process by first storing the dialed number and then using it to set up the circuit. Common control allows for more efficient use of switching logic and provides increased flexibility in altering telephone numbers.

Computer Controlled Switching. More recently, the computer has been used extensively to control circuit switches. The Bell System is replacing its older systems with computer-controlled systems. Bell's network, the Electronic Switching System (ESS), uses the computer to scan the lines through sensing devices that detect the on-hook or off-hook condition (see Figure 3-21). The scanning information is read and stored by the computer, and the appropriate logic is executed to handle any changes in the status of the line.

Upon detecting that a line is off-hook, the computer sends a dial tone to the subscribing instrument through the signaling modules. (The signaling module is used for other control signals such as ringing and busy.) The dialed number is received; connection and path analysis determines which trunk should be used to establish the required connection. Thereafter, the call is switched through the

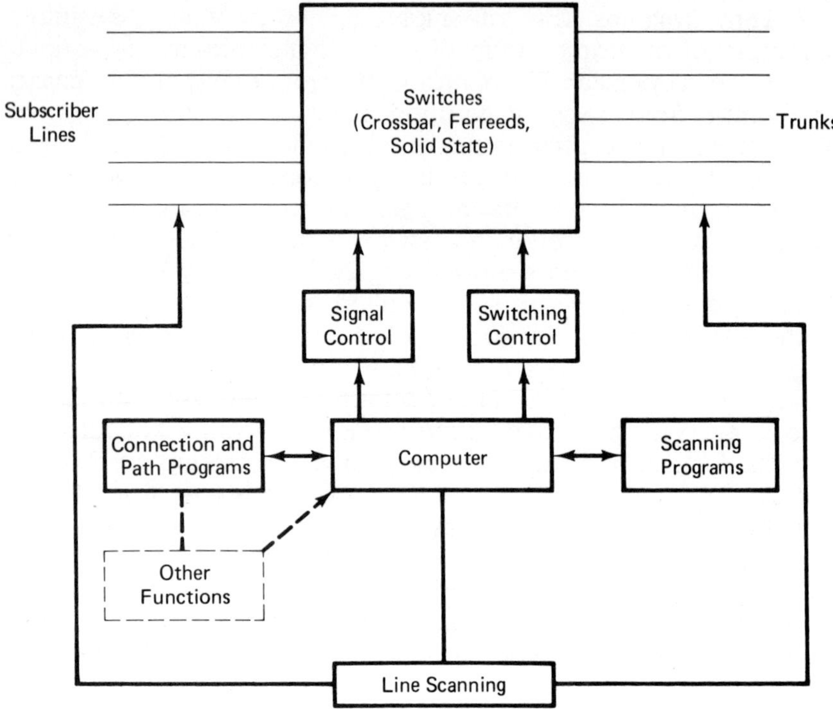

FIGURE 3-21. Computer-Controlled Switching.

network hierarchy, eventually resulting in the receiving end office placing a ringing signal onto the receiving subscriber's line.

Computer-controlled circuit switching is distinguished by the following characteristics:

- Real time data transmission is provided by the direct connection.
- Dial-up delay can be eliminated by the use of leased lines.
- User must provide all data communications functions such as message flow control and session establishments.
- Blockage can occur, in which case a busy signal is returned to sender.
- Transmissions are point-to-point.
- Once connection is established, any subsequent overload of the switch is invisible to the connected components.

Message Switching

Message switching is designed specifically for data traffic. As in circuit switching, the communications lines are connected to a switching facility, but the end users do not have a direct physical connection. Rather, the data message is transmitted to the switch, where it is stored on a queue for later delivery. The term *store-and-forward* is associated with message switching networks.

Early message switching networks, consisting of input messages to the "switch," were transcribed onto perforated paper tape. The tape messages were taken from one machine by an operator and manually carried to another machine for relay (switching) to an intermediate or final destination. In the 1960s, the author worked in a paper tape switching center aboard a U.S. Navy communications ship. During peak traffic periods, the backlog and queues of messages were almost overwhelming to the operators and managers. In one instance, the Gulf of Tonkin incident created a traffic backlog so immense that lower priority messages were stored for days. Moreover, the manual handling of the paper tapes created inordinate delays and many errors.

Computers are now used for message switching. The messages are stored on disks, drums, or magnetic tapes instead of paper tape. The computer programs read the incoming message header, decode the destination address, and route the message onto the proper communications line. Torn tape switches are still quite common in the world, but computer switches are becoming the prevalent mode for message switching.

Computer-controlled message switching is distinguished by the following characteristics:

- Connection is not a direct, physical interface as in circuit switching.
- Since the messages are stored onto a disk, tape, or drum device before transmission, real time processing is usually not feasible.
- Storage of messages allows for adjustments to peak traffic periods. Lower priority messages are queued for later transmission.
- The message switch usually provides for code and protocol conversions between different types of computers or terminals.
- Multiple line speeds are accommodated.

- Messages can be broadcast to all nodes in the network or a subset of nodes.
- The switch provides for care of message traffic; error checking, logging, and recovery procedures are part of the system capabilities.
- Priorities are allowed in the message traffic. The message switch can process higher priority messages before lower priority traffic.

Packet Switching

If one examines certain features of circuit and message switching, it becomes obvious that a combination of the two techniques provides a powerful tool for data communications: the real time mode of circuit switching and the routing power of message switching. Packet switching is intended to provide for real time processing and switching of data messages between computers or between terminals and computers. It was designed to provide optimum utilization of fixed long distance circuits.

Packet switching is so named because a user's logical message is separated (disassembled) and transmitted in multiple packets. The packet contains user data and control data, all of which are switched as a composite whole. Each packet of the user data contains its own control and identification. The addressed packets occupy a transmission channel (or line) for the duration of the transmission only; the line is then made available for other users, terminals, or computers. The idea of the packet is to limit the block size so that it does not occupy the line for extended periods.

The packet switched network (see Figure 3-22) has multiple routes available from each node in the network. The packets are routed across the paths in accordance with traffic congestion, error conditions, the shortest end-to-end path, and other criteria that are covered later in this discussion. In some networks, a user's individual packets are routed across different communications lines and later assembled as a logical message for presentment to the end user.

Packet switching has seen increasing use due to a number of factors:

- The alternate routing scheme gives the network considerable reliability. Failed switches or lines can be bypassed.
- Packet switching computers are designed for minimum delay between end users, usually in fractions of a second.

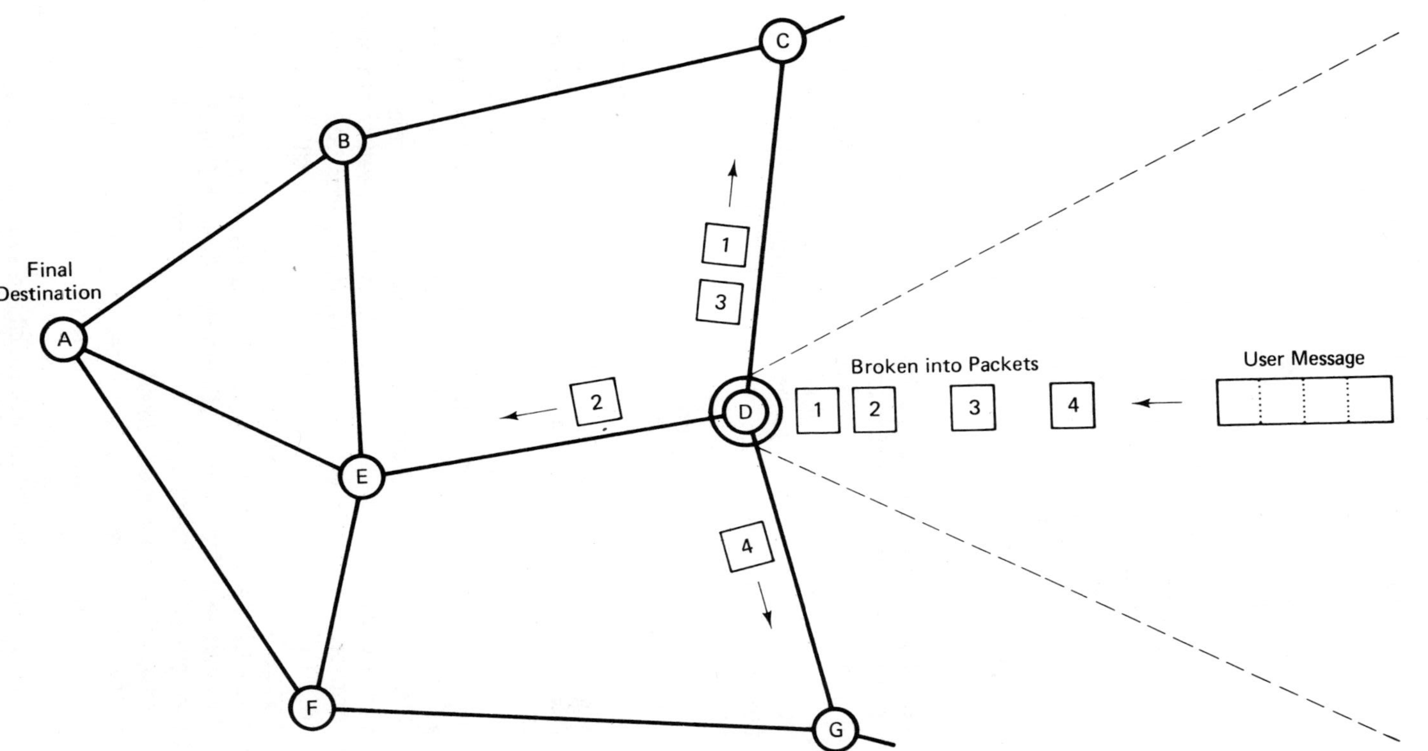

FIGURE 3-22. Packet Switching.

- The technology can use the existing telephone network; the packet switches thus provide a value added service for the end user.
- Packets containing a portion of user data are more secure than a message with a complete logical message.

Packet switching is a complex undertaking. The switches must be highly reliable and very efficient. If routing decisions change as conditions change, the routing algorithms must maintain a steady-state condition of all network components. The network logic must adapt quickly to dynamic conditions that would cause suboptimal routing. Adaptation must be done uniformly over all affected nodes. Control is required to prevent packet saturation at busy nodes in the network.

Packetizing to Reduce Delay. The size of a message, block, frame, or packet is a critical design factor in a network. The effect of packetizing and packet size can be seen in Figure 3-23. Referring to Figure 3-23(a), a message of length n bits is to be transported from node A, through nodes B and C, to the final destination, node D. The length of the message is such that, at time t + 5, all control and user data arrives at site B. At this node, it is checked for errors and relayed (assuming insignificant processing delay at the node) to node C. It finally reaches the end node D at t + 15.

In Figure 3-23(b), the message has been disassembled into four packets, each with its own control header. The packetizing results in an increased number of overhead bits in the headers, which also results in a longer transmission time to node B than in the example of Figure 3-23(a). Assuming no delay time between transmission of the four separate packets across one line, the four packets arrive at node B at time t + 8, which is 3 units of time slower than with the self-contained message block. However, notice that the first packet arrives at t + 2 and is checked for errors. Since it is a self-contained unit, it is immediately relayed to node C. In contrast, the longer message of Figure 3-23(a) cannot be relayed until all bits have arrived at node B and are checked for errors. As the four packets pass through node C, the delay is t + 10—the same delay as the self contained message in Figure 3-23(a). Finally, at node D, the overall delay of the four packets is t + 12, or 3 units of time faster than the message transported in Figure 3-23(a).

Of course, packetizing experiences a point of diminishing returns. Inordinately small packets will increase the delay beyond that of longer user data blocks due to the large ratio of overhead bits to user bits (as well as saturating queues at the switching sites).

There is no such thing as a free lunch. The network packet switches must be designed to handle more individual "pieces" and must have the speed and capacity to minimize queuing inside the machine. Most queueing systems cannot be loaded to over 70% of capacity and the packet switches must be loaded even lower. Packet networks seldom use more than 50% of the available band-

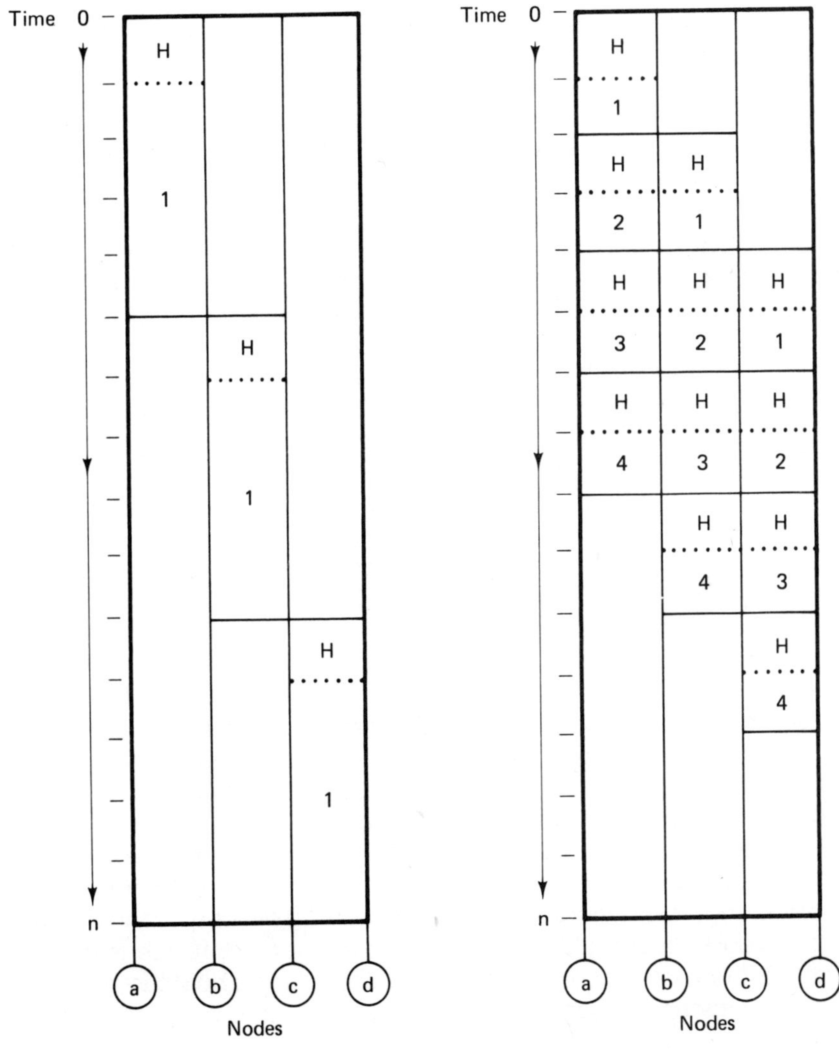

FIGURE 3-23. Packet Size.

width during peak periods. Packet networks have control mechanisms to warn against traffic congestion that is building up. Others control congestion by limiting the number of packets that a user can have outstanding at one time. Still others require a sending site to obtain a "reservation" from the receiving site before a session is established.

Packet Routing. Routing algorithms are implemented using the concepts of static or adaptive routing. The static scheme provides a fixed path through the network based on criteria such as user session or packet type. The rationale for static routing is that a session usually requires transmission of multiple packets between the same user end points and a static end-to-end path achieves simplicity and seriality.

Several variations of static routing exist. For example, alternate static routing provides for a first choice or second choice path based on line conditions, packet priority, type of user session, or other criteria. Once the path is established, it remains the same for the duration of the user session. Another popular approach is to provide for routing control from a network control center. The center changes routing tables as conditions in the network change. Some implementations allow network management to establish routing the packets based on minimum cost, the shortest route, or minimum delay considerations. The decisions are input as parameters into the logic of the packet switching computers.

Adaptive routing algorithms alter the route of packets in a dynamic manner, based on network conditions at any given instant. The routing decisions are typically based on obtaining performance information of adjacent switches and determining the readiness of the adjacent nodes to receive and relay the packets. The adjacency readiness information is obtained by sites exchanging tables with each other, usually every 400 to 700 ms. These tables provide data to determine the optimum route out of the sending switch. The tables are updated by each node or by a central network control center. The former method is called *distributed route maintenance* and the latter is *centralized route maintenance*.

Another method of obtaining adjacency readiness information is for the sending switch to maintain data on the delays encountered in obtaining acknowledgments from receiving switches in the receipt of previously transmitted packets. Still another approach is for the packets themselves to contain delay information about the switches through which they pass. Lengthy delays alerts the system to transmit the packet across another channel to other nodes.

Adaptive routing is a powerful, sophisticated approach and is used by several networks such as the ARPA network and GTE's Telenet. However, the approach can lead to nonsteady-state conditions and must be studied in detail before implementation. Figure 3-24(a) shows why. Let us assume node D wishes to transmit a packet to node A, the final destination (FD). The arrows indicate the time unit delays experienced between adjacent nodes. The optimum path for the packet is D→C→B→A (NN means next node); the overall delay (OD) is 7 time units (D→C=2, C→B=2, B→A=3). All other routes entail a longer delay. (Some routing algorithms determine the routing path based on the number of hops or intermediate nodes that must be traversed; this example does not use the hop method.)

Suppose, after the packet reaches the intermediate node C, the network conditions have changed such that in Figure 3-24(b), the optimum route to node A is C→D→E→A. The packet is returned to its original site D without making any progress toward reaching its final destination. Adaptive routing techniques are refined enough to handle this kind of problem, but the techniques are complex and use resources at the packet switch. To solve our specific problem of the "lost" packet in this example, several vendors increment a count in the control header of the packet. Upon the count exceeding a threshold, the packet is given special handling by the network control center or by error handling logic at the packet switch.

Packet Switching Standards. The international standards organizations ISO and CCITT have developed standards for a packet switching network. In the United States, the Department of State, the common carriers, vendors, and national standards organizations have participated in defining these standards. One such standard, called X.25, defines the interface between data terminal equipment (DTE) and data circuit-terminating equipment (DCE) for terminals operating in the packet mode on data networks. The reader may wish to refer to Chapter 8 for further information on X.25.

Other Switching Techniques

Hybrid switching is a combination of packet and circuit switching. It is receiving considerable attention as a method for integrating voice and data. Hybrid facilities are dynamically shared by voice and data traffic. Circuit switching is used for voice to accommodate traffic that has relatively long waiting periods; packet switching is used for data traffic to accommodate burst-like traffic that has short data blocks with longer pauses between the data blocks. Both voice and data traffic are converted into digital bit streams (see Chapter 7 on digitizing voice) and multiplexed into time slots on the communications path.

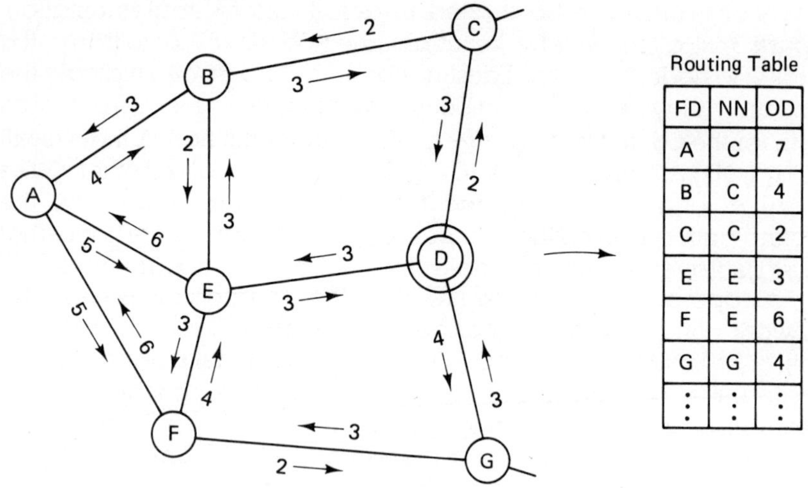

FD	NN	OD
A	C	7
B	C	4
C	C	2
E	E	3
F	E	6
G	G	4
⋮	⋮	⋮

(a) Packet Starts at Node D

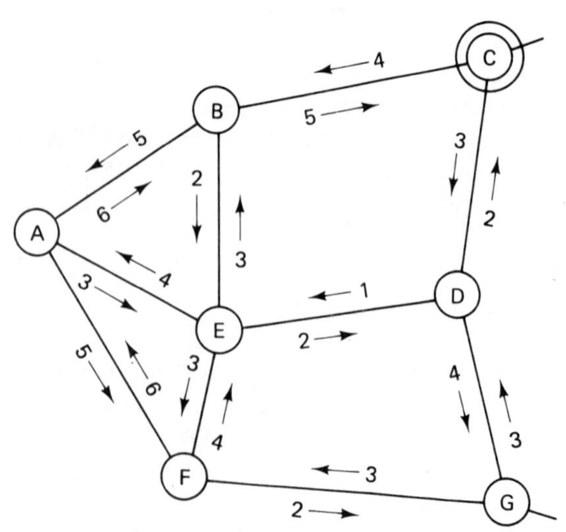

(b) Packet Arrives at Node C

FIGURE 3-24. Adaptive Routing.

Fast circuit switching has been in existence for several years. This technique provides a circuit for a data transmission and disconnects the user upon the completion of the transmission. The channel is not dedicated to an individual user, but shared as the need arises. The concept depends on the need for fast circuit setup and disconnection times for interactive applications, usually no longer than a 200 to 300 ms delay.

Private Branch Exchanges

The private branch exchange (PBX) switch is used extensively in offices. The PBX serves an individual private organization, provides telephone switching capabilities within the building, and also connects outside to the common carrier facilities. The PBX user base has grown dramatically as computers have been placed in the PBX to provide for sophisticated functions such as call forwarding, automatic redial, and abbreviated dialing for speed calling. The computerized PBX (CBX) now provides more functions than most users know how to utilize. It will be used in the future to integrate voice/data transmission and to provide for local networking within a building. Later chapters discuss the PBX in more detail.

MULTIPLEXING, LINE SHARING, AND COMPRESSION

A voice-grade communications line is capable of a 9.6 kbit/s rate using high speed, multilevel modems. Theoretically, this type of line can transmit 17,280,000 bits in 30 minutes (1800 seconds × 9600 bits = 17,280,000). Yet, many devices use only a small fraction of this line capacity. A keyboard terminal operated by a human typically sends and receives a few thousand bits during a 30-minute session with the computer. Assuming 4,000 eight-bit characters were exchanged during this period (a very fast typist!), the efficiency ratio of the total capacity of the 9.6 kbit/s line would be .0018 [(4,000 bytes × 8 bits per byte)/17,280,000 bit capacity = .0018]. This is a very poor use of an expensive component in the data communications system. Moreover, with the increasing use of faster lines operating at 56 kbit/s and 1.5 mbit/s, the ratio is worse.

The solution is to provide more traffic (in user data streams) for the path. Since many applications use low-speed keyboard terminals as input into the network, the high-speed lines are given more work by placing more than one terminal device on the line. Increased work provides for better line utilization.

Multiplexing

Multiplexers (MUXs) accept low-speed data streams from terminals and combine them into one high-speed data stream for transmission to the central site. At this site, a multiplexer demultiplexes and converts the combined data stream into the original multiple low-speed terminal data streams. Since several separate transmissions are sent over the same line, the efficiency ratio of the path is improved (see Figure 3-25).

Frequency Division Multiplexing. Until the last few years, the most widely used multiplexing technique was frequency division multiplexing (FDM). This approach divides the transmission frequency range (the bandwidth) into narrower bands (called subchannels). Figure 3-26(a) depicts the FDM scheme. FDM decreases the total bandwidth available to each terminal but, since the devices are low speed, the narrower bandwidth is sufficient for each device. The data transmissions from the attached devices are sent simultaneously across the path. Each device is allocated a fixed portion of the frequency spectrum. For example, an ASR-33 teletype terminal usually requires a bandwidth of 170 Hz. Given this requirement, a voice-grade channel can accommodate 12 to 15 110 bit/s ASR-33 terminals.

The subchannels must be separated by unused frequencies (guard bands) to prevent overlapping signal interference. Therefore, the full 3 KHz telephone voice-grade bandwidth cannot be fully utilized. The guard bands exist to compensate for signal filters that do not sharply cut off the outside frequencies of the subchannel. Even with these guardbands, the contiguous channels will experience some cross interference (cross talk). Figure 3-27 shows several alternatives for dividing the 3 KHz spectrum.

FDM is often used in short-distance, multidrop arrangements. It is code transparent and any terminal of the same speed can use the same subchannel after the subchannel is established. Multiplexer vendors have different requirements for connecting to their equipment, however, and the user should check for vendor variations. Most connections to terminals are accomplished through the RS232-C connection. An FDM provides the modulation/demodulation within its own circuitry.

The common carriers' communications channels are grouped together in packages to take advantage of the greater bandwidths of coaxial cable, microwave, and satellite transmission schemes. The channels are subdivided by using frequency modulation techniques. This is accomplished by a carrier frequency being generated for

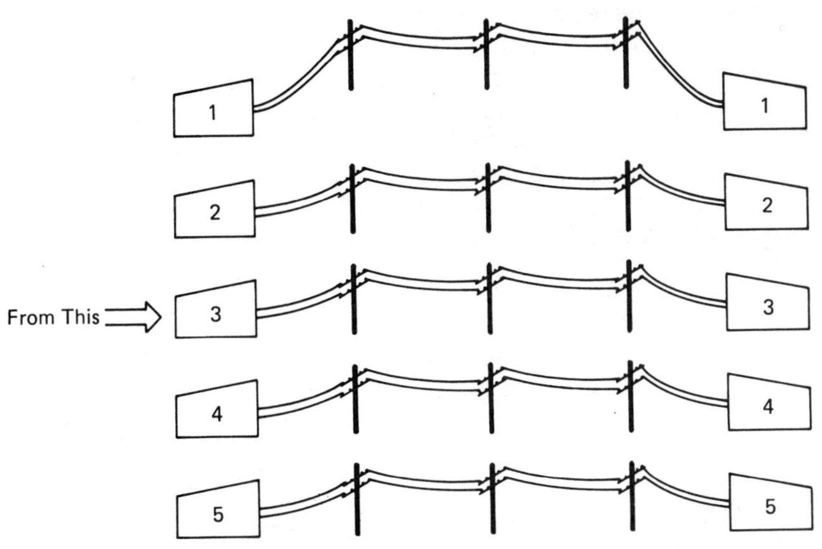

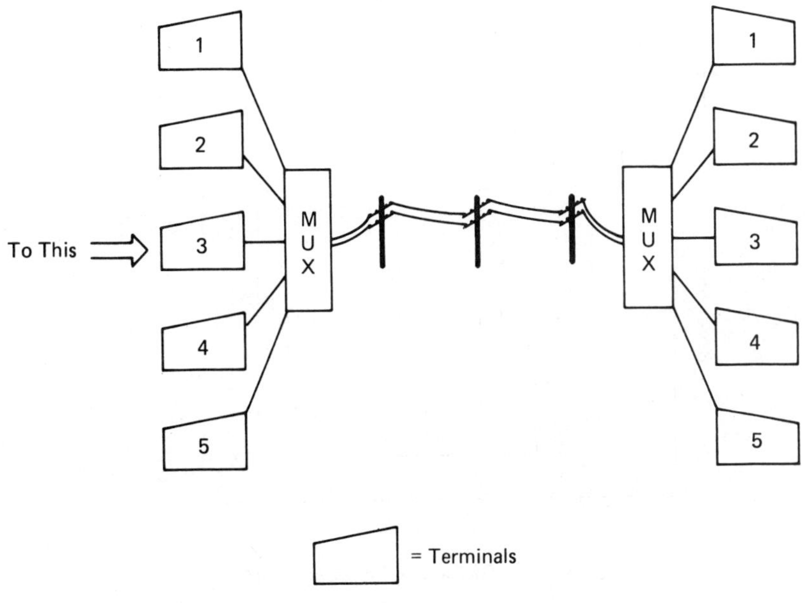

FIGURE 3-25. Multiplexing.

each channel; the separate carrier is required to translate and "carry" each device's signal. Each carrier channel is assigned a different frequency to prevent interference from other channels.

The process is illustrated in Figure 3-28. Each signal is placed onto (modulated) a carrier frequency. The carriers are generated by a central oscillating device, which also generates a pilot or reference signal. The pilot signal is used at the receiving end to provide a reference for demodulation and demultiplexing. The multiplexed

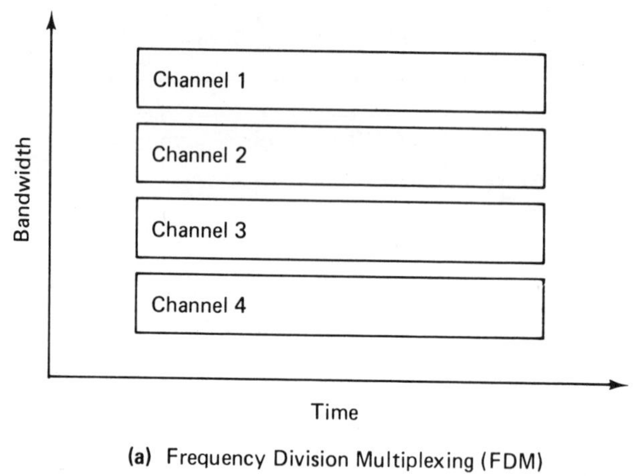

(a) Frequency Division Multiplexing (FDM)

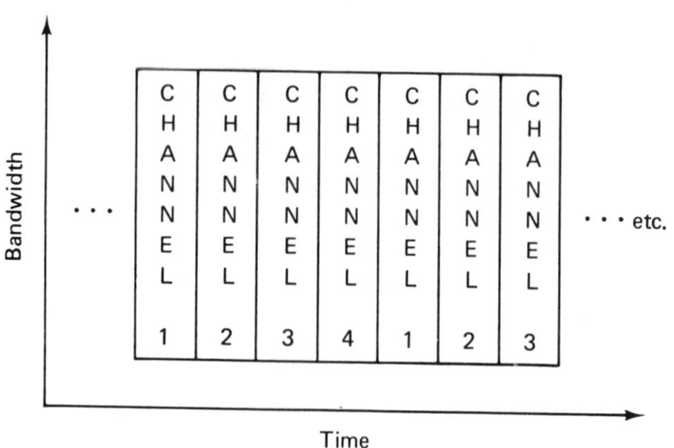

(b) Time Division Multiplexing (TDM)

FIGURE 3-26. FDM and TDM.

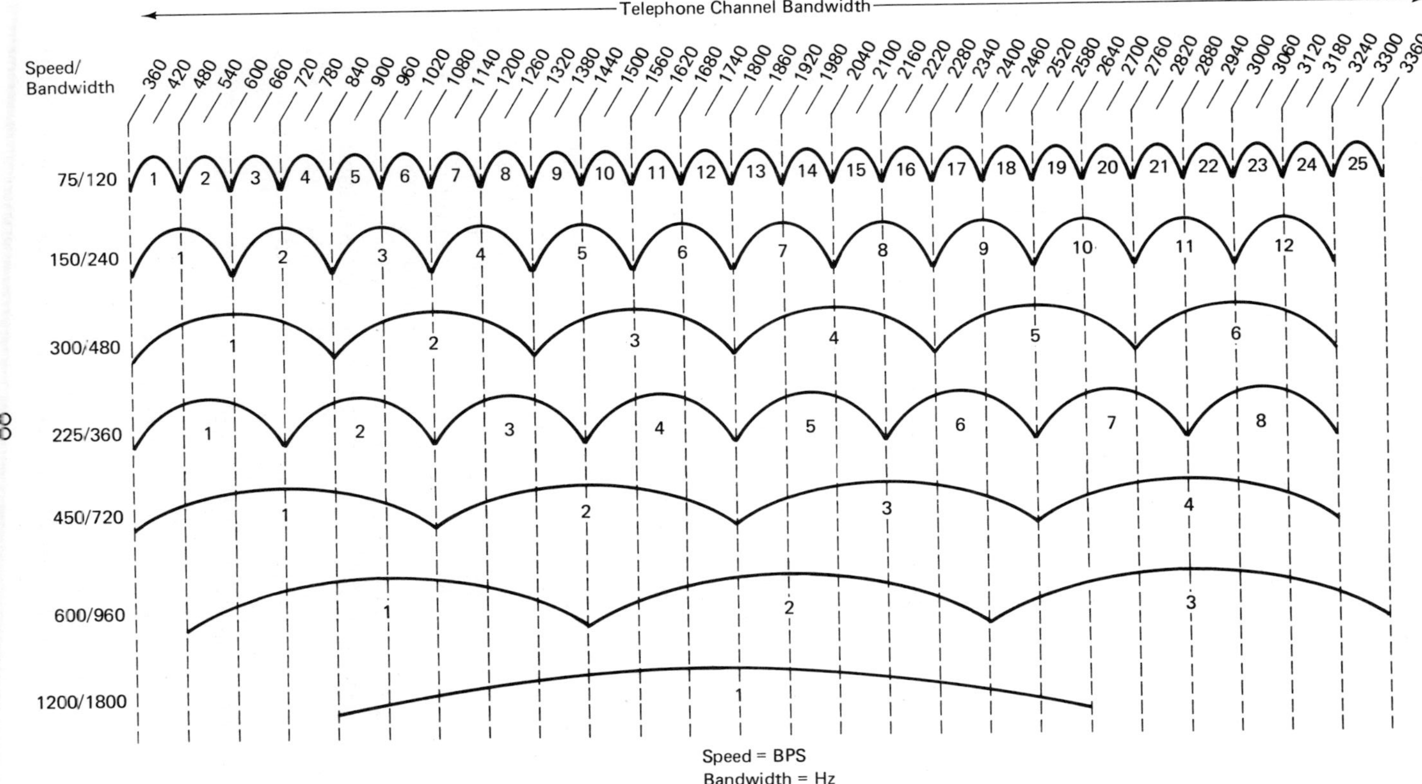

FIGURE 3-27. Dividing the Spectrum.

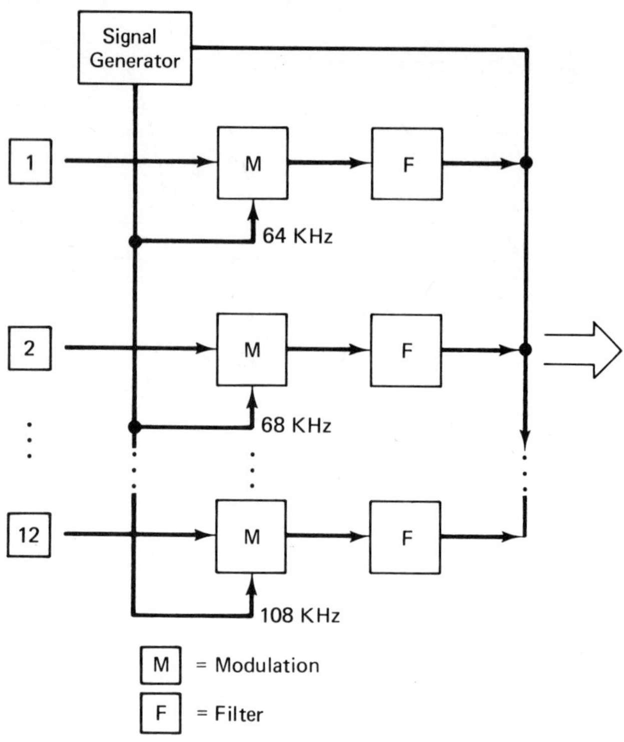

FIGURE 3-28. FDM Carrier and Filters.

signals travel across the high capacity communications link, arrive at the end site, and are then converted back to the original signals.

Figure 3-28 shows that signal filters are involved in the process. As stated earlier, filters suppress or eliminate certain unwanted frequencies by increasing the signal loss of those frequencies. Filters are necessary due to the nature of the output of modulation. For example, let us assume a voice-grade channel with a bandwidth of 300 to 3300 Hz is to be modulated onto a carrier frequency of 60 KHz. The resultant output contains the basic carrier frequency and two modulated frequencies, an upper sideband, and a lower sideband:

Upper sideband: 60,300 63,300 Hz

Carrier frequency: 60,000 Hz

Lower sideband: 56,700 Hz 59,700 Hz

Since it is necessary to carry only one of the sidebands, the filters are used to remove the unwanted band and perhaps the carrier signal as well. At the receiving end, the filters are also required since the demodulation process also creates the sidebands.

Time Division Multiplexing. Time division multiplexing (TDM) provides a user the full channel capacity but divides the channel usage into time slots. Each user is given a slot and the slots are rotated among the attached user devices [see Figure 3-26(b)]. The time division multiplexer cyclically scans the input signals (incoming data) from the multiple incoming points. Bits or bytes are peeled off and interleaved together into frames on a single higher speed communications line. TDMs are discrete signal devices and will not accept analog data.

TDMs operate in either a bit or byte fashion. A bit TDM peels off one bit at a time. Each frame contains a bit from each sampled device. A typical bit interleaved TDM can handle 18 terminals on a voice-grade line. The character (or byte) interleaved TDM assembles an entire character or byte into the frame. The character MUX is generally more efficient since it permits fewer overhead bits than a bit MUX. A character interleaved TDM can handle 29 subchannels within a voice-grade line. TDMs process binary, digital data; consequently, modems are required for interface into the common carrier analog network. Chapter 7 explains digital transmission techniques.

Isochronous Multiplexing. A TDM generally sends out signal pulses to the attached input devices to keep the data streams synchronized. The pulses are provided by a master clock in the multiplexer. Some communications lines have a large number of input devices operating with different physical transmission times and signal propagation delays. The isochronous MUX does not "sync" through the master clock, but provides internal buffers for the time-independent data streams to store data. The buffers allow the data to arrive at random intervals where it is stored and later multiplexed out onto the high-speed side of the MUX.

Statistical Multiplexing. The conventional TDM wastes the bandwidth of the communications line because the time slots in the frames are often unused. Vacant slots occur when an idle terminal has nothing to transmit in its slot. Statistical TDM multiplexers (STDMs) dynamically allocate the time slots (i.e., the bandwidth) among active terminals; thus, dedicated subchannels (FDMs) and dedicated time slots (TDMs) are not provided for each port and idle terminal time does not waste the line's capacity. It is not unusual for two to five times as much traffic to be accommodated on lines using statistical TDMs.

In most statistical multiplexers, the length of the frames varies in accordance with the input data streams. These streams usually come from asynchronous start-stop terminals. The majority of statistical multiplexers support synchronous devices as well. Isochronous transmission is accepted on some vendor's models. The frames must contain information on which channels have transmitted data within the frame. Typically, each frame provides "mapping" (MAP) that tells which devices have data and the number of data bits or bytes in each frame. The frames also have headers, sequence numbers, and error checking fields for purposes of identification and control.

Statistical multiplexers have evolved in a short time to become very powerful and flexible machines. Today, vendors sell machines that overlap the functions found in PBXs, message/packet switches, front ends, concentrators, and even satellite delay compensation units. STDMs have virtually taken over the FDM market and now offer serious competition to the TDMs. Statistical multiplexers also provide extensive error-checking techniques and buffer management, as well as data flow control. Some STDMs provide modulation circuitry for interfacing into analog networks. Otherwise, separate modems are required. Flow control is used to prevent the transmitting devices from sending data too fast into the multiplexer's buffers.

Statistical multiplexers are seldom beneficial in networks using full-duplex data flow protocols, or in local networks where the organization owns the communications path. Notwithstanding, these components should be given serious consideration by organizations using long-haul leased lines.

Multiplexed Common Carrier Systems

As stated earlier, the telephone companies and other carriers use frequency multiplexing extensively. For example, the Bell System provides a multiplexing hierarchy that multiplexes 12 subvoice-grade channels into one voice-grade channel at the low capacity end and up to 10,800 voice channels into a large capacity link at the high capacity end.

The Bell hierarchy uses multiplexed groups (Figure 3-29). The channel group contains 12 voice-grade channels. The voice channels are spaced 4 KHz apart and occupy the 60 to 180 KHz range. In Bell's system, only the lower sidebands are used and the upper sidebands and related carriers are suppressed (filtered out). Five channel groups are then multiplexed onto a supergroup. The supergroup

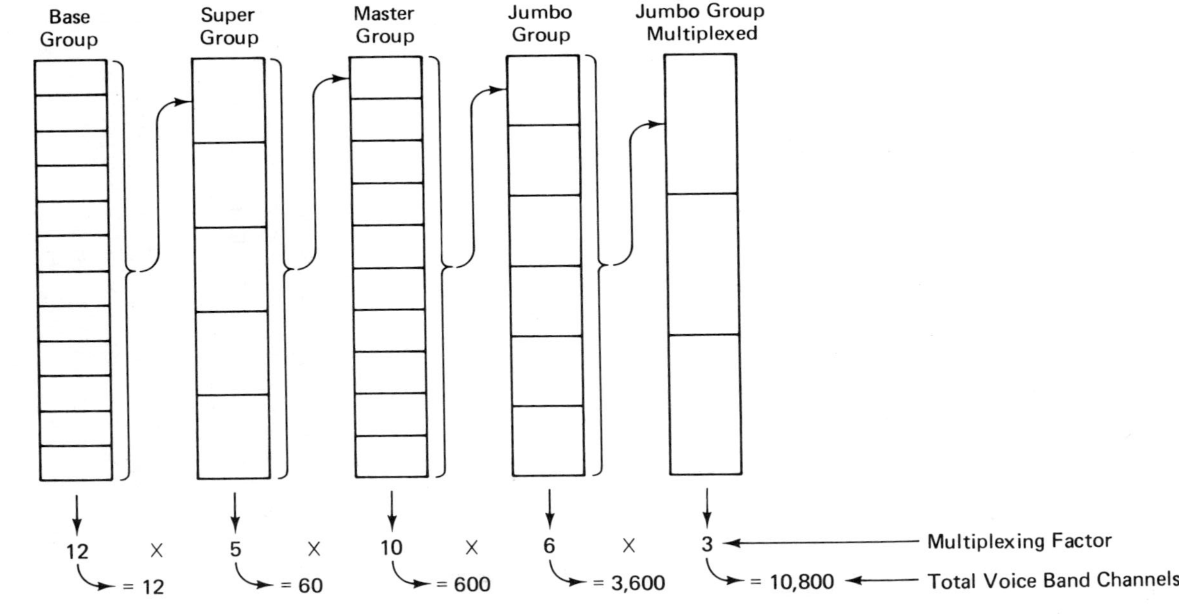

FIGURE 3-29. Carrier Multiplexing Hierarchy.

contains 60 voice-grade channels (12 × 5 = 60). Ten supergroups are in turn multiplexed onto a master group yielding a capacity for 600 voice-grade channels (12 × 5 × 10 = 600). Six mastergroups make up a jumbogroup with 3,600 voice-grade channels (12 × 5 × 10 × 6 = 3,600). At the upper level in the scheme is the jumbogroup multiplexed, which is composed of three jumbogroups and 10,800 voice-grade channels (12 × 5 × 10 × 6 × 3 = 10,800), occupying a frequency range of 3,124 KHz to 60,556 KHz. The common carriers also have facilities for digital multiplexing using the concepts of time division multiplexing. This area is discussed in Chapter 7.

The original high capacity carrier system is the Bell/Western Electric L1. Installed in 1941, L1 provides for 600 voice-grade channels utilizing the 60 KHz to 2,540 KHz frequency band on coaxial cable pairs. As the nation's communications requirements continued to grow, the multiplexing hierarchy grew to include supergroups and the L3 system, supergroups and mastergroups of the L4 system, and the jumbogroup multiplexed of the L5 system. The voice frequency systems have the following major characteristics:

Name	Implemented	Circuits per Cable Pair	Total Circuits
L1	1941	600	1,800
L3	1953	1,860	9,300
L4	1967	3,600	32,400
L5	1973	10,800	108,000

Multiplexing Satellite Signals

Satellite signals experience considerable loss from rain and clouds at frequencies above approximately 10 GHz. The lack of adequate frequency spectrum space at the 6/4 GHz bands has provided impetus to develop methods to overcome this problem. Time division multiple access (TDMA) transmits signals as digital bursts at the higher frequencies (for example, the 14/12 GHz bands). The signals are time division multiplexed into time slots but the same signal is transmitted several times within the allocated slot. The ground station receives the signals, compares the redundant transmissions, and corrects any errors by comparing the multiple "copies" of the data. The redundant transmissions consume extra bandwidth but the higher transmission bands of 14/12 GHz provide for the needed extra capacity.

The TDMA approach uses statistical analysis to examine the multiple-bit streams. Programs examine and compare the messages. Research has shown that repeating the bits of a data stream increases the probability of detecting and correcting errors that occur during the transmission. This approach has opened the way for using the 14 and 12 GHz bands. The 30/20 GHz band will likely see increasing use in the future.

Concentrators

The term *concentrator* is often confused with *statistical multiplexer*. This is certainly understandable since the functions of the two components often overlap. Strictly speaking, a concentrator has n input lines which, if all input devices are active, would exceed the capacity of the m output line. Consequently, in the event excessive input traffic is beginning to saturate the concentrator, some terminals are ordered to reduce transmission or are not allowed to transmit at all. We also find this kind of function in STDMs and in communications front ends. The overlapping functions of multiplexers, concentrators, and front ends often make it difficult to apply any clear-cut definitions to the components.

Statistical multiplexers are used as a combination concentrator/front end to enable a computer connection (port) to communicate with multiple channels (see Figure 3-30). The stat MUX in this environment is called a *port concentrator*. The port concentrator is responsible for control of the line. It provides buffering, error detection, and line synchronization with the remote components.

The port concentrator often provides for dial-up access to the port(s). This allows a large number of terminals to contend for a limited number of ports at the computer. The more sophisticated concentrators (called *port selectors*) integrate dedicated and dial-up terminals. The port selector switches the devices to the available ports as required.

Data Compression

The transport of data across the communications path is an expensive process. Multiplexers, concentrators, and multipoint techniques allow for more efficient utilization of the expensive medium. Data compression provides yet another option by reducing the number of characters in a transmission.

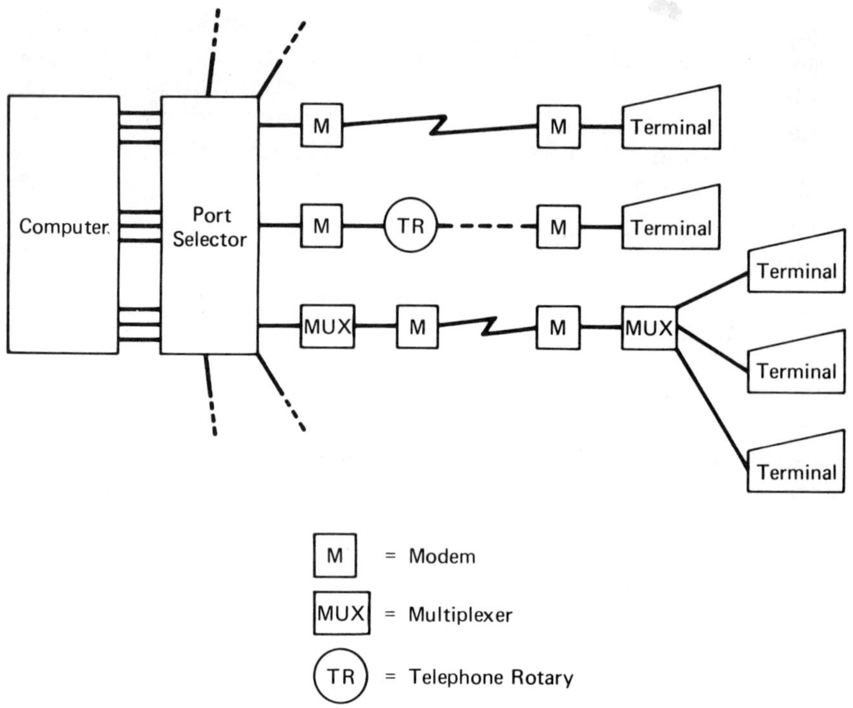

FIGURE 3-30. Port Concentrators and Selectors.

Messages are comprised of a fixed number of bits coded to represent a character. The codes have been designed as fixed length because most computers require a fixed number of bits in a code to efficiently process data. The fixed-length format means that all characters are of equal length, even though all characters are not transmitted with equal frequency (Figure 3-31 [a]). For example, characters such as vowels, spaces, and numbers are used more often than consonants and unique characters such as a question mark.

One widely used solution is to establish a variable-length code for purposes of transmission. The most frequently transmitted characters (bytes) are "compressed" and represented by a unique bit set smaller than the conventional bit code [Figure 3-31(b)]. This data compression technique can result in substantial savings in communications costs.

Data compression techniques are classified as reversible and nonreversible; reversible techniques are further divided into semantic independent and semantic dependent procedures.[7] Nonreversible techniques (often called *data compaction*) permanently eliminate the irrelevant portions of the data. Obviously, what is relevant or irrelevant depends upon the data itself. For example, the figure representing the national debt is described in billions of dollars and is often calculated at an aggregate level with the tens, hundreds, thousands, etc., positions compacted out of the number. While data compaction might be appropriate for the national debt, one would

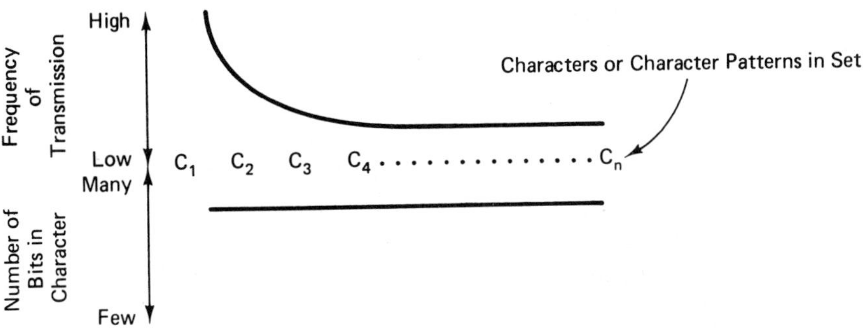

(a) Fixed Code with Variable Transmission Patterns

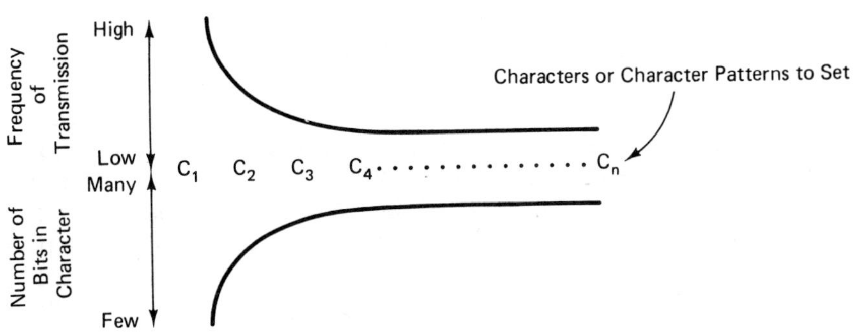

(b) Variable Codes with Variable Transmission Patterns

FIGURE 3-31. Fixed-Length Format Problems.

---

[7] Classification scheme source is "An Overview of Data Compression Techniques" by H. K. Reghbati. *Computer*, (an IEEE publication), April 1981, pps. 71–75.

be hesitant to do it with a customer's checking account balance. Typically, data items with leading blanks or trailing zeros are eliminated with data compaction.

Reversible techniques provide for temporary elimination of portions of the data. Reversible techniques using semantic dependent procedures depend on the content and context of the data for data reduction; semantic independent procedures do not depend on either the content or context of the data. Some common data compression techniques are described below:

Huffman Code. The Huffman code is a widely used compression method. With this code, the most commonly used characters contain the fewest bits and the less commonly used characters contain the most bits. For example, assume the characters A, B, C, and D comprise the complete character set. Analysis reveals that, on an average, A accounts for 50% of the total transmitted traffic, B accounts for 25%, C for 15%, and D for 10%. A code to take advantage of the skewed character distribution might appear as in Figure 3-32(a). The following transmitted message could then be interpreted by the decision chart [Figure 3-32(b)]. The data is examined and decoded by reading the bit stream left to right:

111	10	0	0	110	10
↓	↓	↓	↓	↓	↓
D	B	A	A	C	B

Huffman codes have been used to achieve compression rates of 50% or better. The rate depends on individual user data streams. However, as an illustration, typical text transmissions using an eight-bit EBCDIC code can be reduced to an average character length of 4.8 bits. The overhead of computer resources to accomplish the compression at the transmitting site and decompression at the receiving site is offset by the increased throughput across the communications line.

Adaptive Scanning. This technique uses a dictionary to store frequently occurring strings of characters. The common character string is substituted by a shorter code. For example, program source code is often transmitted across a network. Thus, the character string PERFORM might appear often in a program written for a Cobol compiler. Adaptive scanning would examine the text and substitute a code (for example, @) in place of each PERFORM. The result is fewer bits transmitted.

Facsimile Compression. One of the most useful applications for data compression is the transmission of documents or graphics, commonly known as facsimile transmission (FAC). Documents lend themselves to compression because their contents have many recurring redundant patterns of space (white patterns) or print (black patterns). Facsimile compression treats each facsimile line as a series of white and black runs. The runs are coded based on their

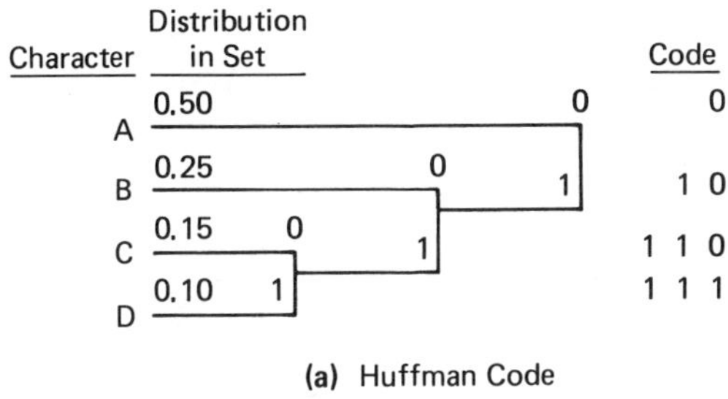

(a) Huffman Code

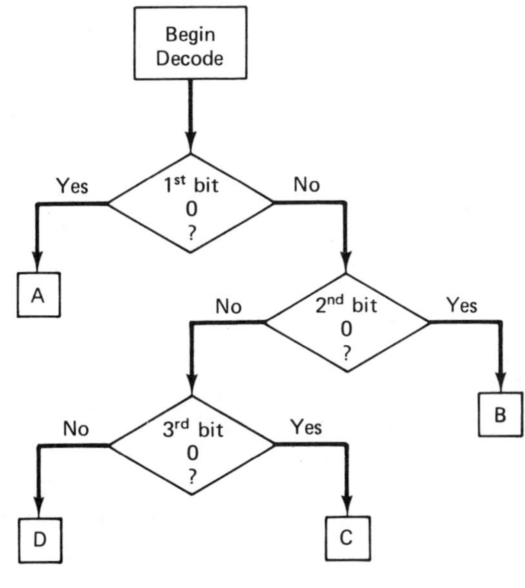

(b) Decision Chart

FIGURE 3-32. Huffman Data Compression.

length and the code is transmitted instead of the full "bit picture" of the document.

The compression is established through preestablished codes representing a picture element (PEL). The PEL defines the length of recurring white and black images. Each PEL is given a unique bit code. Figure 3-33 shows the codes for the modified Huffman code technique. Figure 3-33(a) contains codes for PELs with white or black run lengths up to 63; Figure 3-33(b) provides for encoding of PELs greater than 63. The two tables are used together to establish the exact code for longer PELs.

To further understand facsimile compression, let us assume that the following PELs have resulted from a document line scan through a facsimile device. The PELs are decoded as:

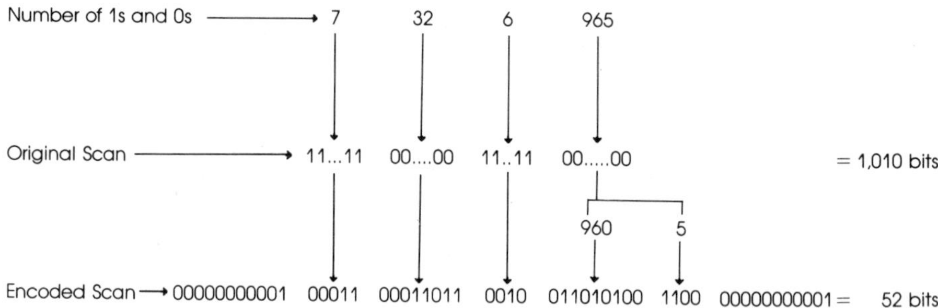

The control bits are used to delineate the beginning and ending of a line. Additional control bits may also be inserted for purposes of synchronization. The original scan established for overall run length of 1,010 bits and the compression resulted in an encoded line of 52 bits. This results in a 19:1 compression ratio, which is a conservative achievement. Far better ratios are possible.

Just a few years ago, a standard 8½ × 11 inch document took over six minutes to be transmitted over a 48 Kbit/s line. The modified Huffman technique cut this time significantly to approximately one minute. Recent improvements in compression techniques and modern speeds have further improved the process. The author recently shared a taxi at La Guardia airport with a representative from a facsimile vendor. The individual was carrying a portable FAC machine that his company was prototyping. The machine was purported to be able to transmit a standard 8½ × 11 inch document across a dial-up line in 20 seconds.

WHITE RUN LENGTH	CODE WORD	BASE 64 REP	BLACK RUN LENGTH	CODE WORD
0	00110101	0	0	0000110111
1	000111	1	1	010
2	0111	2	2	11
3	1000	3	3	10
4	1011	4	4	011
5	1100	5	5	0011
6	1110	6	6	0010
7	1111	7	7	00011
8	10011	8	8	000101
9	10100	9	9	000100
10	00111	a	10	0000100
11	01000	b	11	0000101
12	001000	c	12	0000111
13	000011	d	13	00000100
14	110100	e	14	00000111
15	110101	f	15	000011000
16	101010	g	16	0000010111
17	101011	h	17	0000011000
18	0100111	i	18	0000001000
19	0001100	j	19	00001100111
20	0001000	k	20	00001101000
21	0010111	l	21	00001101100
22	0000011	m	22	00000110111
23	0000100	n	23	00000101000
24	0101000	ø	24	00000010111
25	0101011	p	25	00000011000
26	0010011	q	26	000011001010
27	0100100	r	27	000011001011
28	0011000	s	28	000011001100
29	00000010	t	29	000011001101
30	00000011	u	30	000001101000
31	00011010	v	31	000001101001
32	00011011	w	32	000001101010
33	00010010	x	33	000001101011
34	00010011	y	34	000011010010
35	00010100	z	35	000011010011
36	00010101	A	36	000011010100
37	00010110	B	37	000011010101
38	00010111	C	38	000011010110
39	00101000	D	39	000011010111
40	00101001	E	40	000001101100
41	00101010	F	41	000001101101
42	00101011	G	42	000011011010
43	00101100	H	43	000011011011
44	00101101	I	44	000001010100
45	00000100	J	45	000001010101
46	00000101	K	46	000001010110
47	00001010	L	47	000001010111
48	00001011	M	48	000001100100
49	01010010	N	49	000001100101
50	01010011	Ø	50	000001010010
51	01010100	P	51	000001010011
52	01010101	Q	52	000000100100
53	00100100	R	53	000000110111
54	00100101	S	54	000000111000
55	01011000	T	55	000000100111
56	01011001	U	56	000000101000
57	01011010	V	57	000001011000
58	01011011	W	58	000001011001
59	01001010	X	59	000000101011
60	01001011	Y	60	000000101100
61	00110010	Z	61	000001011010
62	00110011	•	62	000001100110
63	00110100	≠	63	000001100111

(a)

WHITE RUN LENGTH	CODE WORD	BASE 64 REP	BLACK RUN LENGTH	CODE WORD
64	11011	1	64	0000001111
128	10010	2	128	000011001000
192	010111	3	192	000011001001
256	0110111	4	256	000001011011
320	00110110	5	320	000000110011
384	00110111	6	384	000000110100
448	01100100	7	448	000000110101
512	01100101	8	512	0000001101100
576	01101000	9	576	0000001101101
640	01100111	a	640	0000001001010
704	011001100	b	704	0000001001011
768	011001101	c	768	0000001001100
832	011010010	d	832	0000001001101
896	011010011	e	896	0000001110010
960	011010100	f	960	0000001110011
1024	011010101	g	1024	0000001110100
1088	011010110	h	1088	0000001110101
1152	011010111	i	1152	0000001110110
1216	011011000	j	1216	0000001110111
1280	011011001	k	1280	0000001010010
1344	011011010	l	1344	0000001010011
1408	011011011	m	1408	0000001010100
1472	010011000	n	1472	0000001010101
1536	010011001	ø	1536	0000001011010
1600	010011010	p	1600	0000001011011
1664	011000	q	1664	0000001100100
1728	010011011	r	1728	0000001100101
EOL	00000000001		EOL	00000000001

(b)

FIGURE 3-33. Modified Huffman Code. (From "Data Compression in High-Speed Digital Facsimile" by Timothy McCullough. *Telecommunications*, July 1977, pp. 41 and 43. Reprinted with permission.)

TERMINALS

The number and diversity of terminals is truly bewildering. Over 600 models are available from more than 150 vendors. The terminal market is one of the fastest growing segments in the communications industry and over four million terminals are already installed in the United States. Terminal selection is a difficult process, if for no other reason than the range of choices that exist. Moreover, the word *terminal* is open to many definitions since some terminals today have the capabilities of computers.

Perhaps the best method to describe terminals is to begin with the simplest machines and move up to the more powerful and elaborate devices. In so doing, a classification scheme should prove useful for the initial discussions.[8] Most terminals are broadly defined as teleprinters and/or cathode ray tube (CRT) devices (see Figure 3-34). The teleprinter provides output through a printer device; the CRT uses a television-like tube. Some terminals have both facilities. These two type terminals are further classified as:

Dumb: Terminal has very limited functions and operates under complete control of other devices.

Smart: Terminal has hard-wired logic to give some capabilities such as selectable line speeds, special function keys, and screen cursor movement and scrolling.

Local storage: Terminal contains auxiliary storage such as diskette or floppy disk.

Programmable: Software programs can be coded and executed in the terminal.

Teleprinters are usually serial printers in that they print one character at a time. Serial printers are further classified as impact printers (they mechanically hit or impact the page) and nonimpact (they use nonmechanical means to produce the image). The impact printers produce the character by using a solid character, like that of a typewriter, or by using a matrix of dots. Nonimpact teleprinters use the dot-matrix aproach with either electrothermal or ink-jet mechanisms. Figure 3-35 provides a classification scheme for teleprinters.

---

[8] Donald H. Haback, "As Terminal Types Multiply, So Do Problems of Selection." *Data Communications*, July 1978, p. 54.

	Teleprinter			CRT
	Impact		Nonimpact	
	Solid	Matrix		
Dumb	1	2	3	4
Smart	5	6	7	8
Local Storage	9	10	11	12
Programmable	13	14	15	16

FIGURE 3-34. Classification of Terminals. (From "As Terminal Types Multiply, so Do Problems of Selection" by Donald H. Haback. *Data Communications*, July 1978, p. 54. Reprinted with permission.)

Dumb Teleprinters and CRTs

The basic dumb terminal or teleprinter consists of a keyboard for data entry and a display station, usually a printer. Some terminals of this type are read-only (RO) and do not contain a keyboard. These terminals are best exemplified by the ubiquitous Teletype 33 and Teletype 35. The Teletype Corporation has shipped over 700,000 model 33 and 35 terminals since 1962. These units are rather slow, using impact solid character printing at 10 characters per second (CPS). Although they require frequent maintenance due to the large number of electromechanical parts, they have served the industry

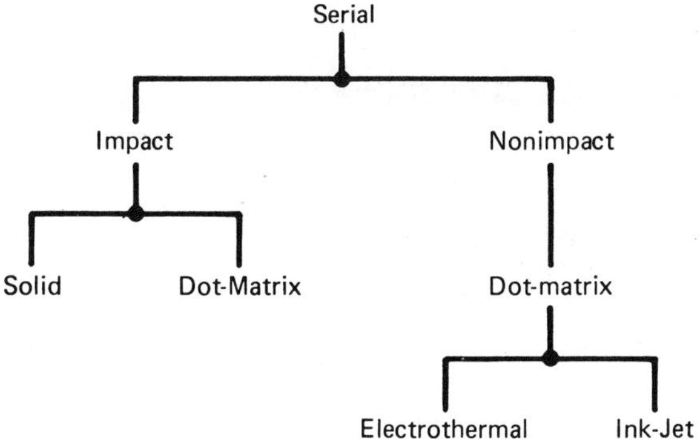

FIGURE 3-35. Classification of Teleprinters.

well. The IBM 2740 is another well-known terminal of this type. (Chapter 6 describes how the 2740 operates in a communications system.)

In 1974, Digital Equipment Corporation (DEC) introduced a formidable competitor to the Teletype and IBM machines with the DECWRITER II. This terminal prints (dot-matrix) at 30 CPS, operates in half- or full-duplex on 110/150/300 bit/s lines. DEC's success with this terminal was extraordinary and it soon became an industry standard. The DECWRITER II (and its later improvement DECWRITER II LA36) is very reliable [meantime between failure (MTBF) is 2,000 hours] and comes with a variety of options. Teletype has marketed the model 42 and 43 to counter the DEC machines.

The dumb CRT, also called an alphanumeric display terminal, provides basic output display functions such as screen cursors, line insertions/deletions, screen paging, and screen erasing. The keyboards are typewriter style. These terminals usually operate at half- or full-duplex using ASCII code. Typical line speed is 9.6 Kbit/s with some vendors offering 19.2 Kbit/s.

Smart Teleprinters and CRTs

The smart terminals provide quality output and more features than the dumb devices. They can often be used as word processors as well as data communications terminals. Smart terminals may be controlled by a microprocessor. They operate with a wide variety of line speeds, typically at 1.2 Kbit/s. Some of these devices also provide for graphics support with the controlling software resident in the host. The smart CRTs also have additional facilities. For example, some devices have limited graphics and terminal bypass printing. The following features are usually standard: scrolling, paging, cursor positioning, cursor blinking, protected screen (keying into a form on the screen), tabulation, transmitting a partial screen (what is keyed in), line delete, screen erase, and character repeat.

Local Storage Terminals

This type of terminal provides considerable overlap in the classification scheme in Figure 3-34 because storage is frequently an add-on option. In fact, many of the dumb and smart terminals provide storage as a standard feature [usually in 5-16K byte random access memory (RAM)]. The local storage feature is an important consideration for certain applications and should be examined as part of the terminal package. Some vendors provide the capability with little extra cost to the customer. The following storage options are available: (a) memory (RAM), (b) paper tape, (c) tape, (d) cartridges, (e) floppy disks, (f) voice-data unit (VDU), and (g) bubble memory.

Programmable (Intelligent) Terminals

These devices have had a major impact on the industry due to their effect on distributed processing (see Chapter 10). They are quite powerful and often include a computer, multiple workstations, and disk and software packages (compilers and utilities). The typical systems consist of one to four workstations with 64K to 128K of main memory, low speed printers, and disk or tape storage. Their programmability has provided considerable task offloading from the mainframe host. Table 3-2 describes the typical characteristics of these devices.

Organizations should establish plans for integrating these systems into their data processing environment. The intelligent terminal market is growing at the astounding rate of over 30%

TABLE 3-2. Characteristics of Typical Intelligent/Programmable "Terminals."

Attributes	Description
Memory Size	4K to 512K
Communications Capabilities	• Full- or half-duplex • Synchronous or asynchronous • Up to 9.6 Kbit/s typically, others up to 50 Kbit/s • ASCII or EBCDIC code • Typical connection is RS232-C • Use Bisync, SDLC, others (see Chapter 6)
Auxiliary Storage	• Diskettes, floppy disks, tapes • Disk storage ranges up to 300 megabytes of data
Software	• Operating systems on larger devices • Languages such as Cobol, Fortran, ALC, and BASIC • Support utilities such as SORTS
I/O Devices	• Support 1 to 100 work stations • CRTs, teleprinters, or various classes (see Figure 3-34)
Prices	Range from $3,000 to $35,000; some very powerful systems up to $80,000

annually. In some companies, the systems are being acquired by users and individual line managers. The prices ($3,000 to $35,000) often enable the user or line manager to purchase/lease the system from their own budget without coordination with data processing management. Obviously, this approach can lead to incompatibility among the systems, as well as duplicate efforts. The issue is becoming more critical as the industry moves toward local networks and automated offices, and personal computers. Chapter 10 addresses this issue in more detail.

COMMUNICATIONS FRONT ENDS

During the 1960s and early 1970s, data communications tasks were processed in the mainframe computer and terminals were controlled by software residing in the host. In many instances, the software was written as part of the application code. This environment presented two rather serious problems.

First, as communications systems grow in size and complexity, the mainframe computer used more of its resources to manage the network. In turn, fewer resources were available for processing the user applications programs. Second, the applications programmers/analysts spent an increasing amount of their time and effort developing code to interface their system with the components in the network. Yet, many of the communications tasks were common to all applications. For example, even though common code would allow the polling of a terminal to send messages for payroll, sales orders, and other applications, the code was written for each application and imbedded within each unique system. These problems dictated a solution—the communications front end processor.

This communications component is used to offload the many communication tasks from the host. It is actually a computer with software, a logic unit, registers, and memory. In contrast to a generalized mainframe, the machine is designed to perform specialized tasks.

With the front end, the host is relieved of many communications-oriented functions, thereby allowing itself more resources for running the application's programs (see Figure 3-36). The front end also provides a very valuable service to the application programmer. It has powerful software for network and line management. As stated previously, earlier applications contained this code in each application. Today's application code need only contain a relatively simple I/O instruction to the communications interface in the host. The host passes the instruction to the front end software for interpretation and execution.

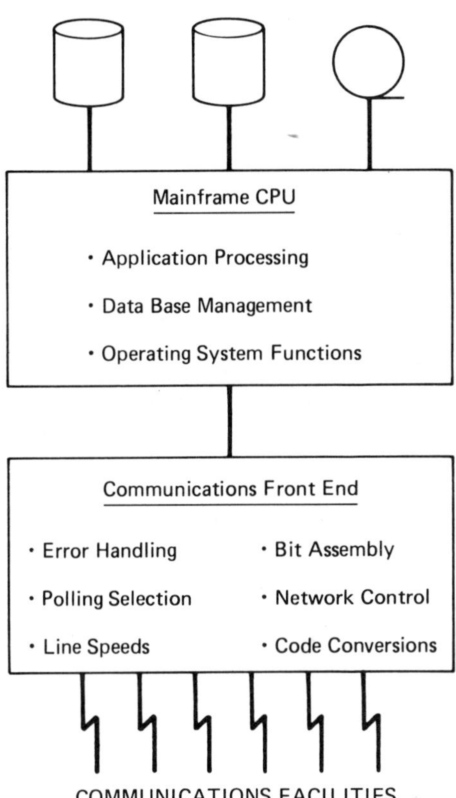

FIGURE 3-36. The Front End.

Figure 3-37 shows a front end in more detail. The control unit is quite similar to the CPU in the host computer. It is capable of accessing memory and executing program instructions. These instructions come from the host communications software and the operating system within the front end. Next, memory is available for buffer areas to store messages coming into and out of the front end. The memory also stores the operating system. Moving up in Figure 3-37 we see a channel interface. This component provides an interface between the front end and the host computer. At the bottom of the figure, notice the line interface devices. These components provide the interfaces to the network.

Figure 3-37 also depicts the communications scanner. The maximum data rate of a line is determined by the frequency at which the specific line interface port is scanned (i.e., sampled) by the communications scanner. The front end can service different line speeds by increasing or decreasing the scanning frequency.

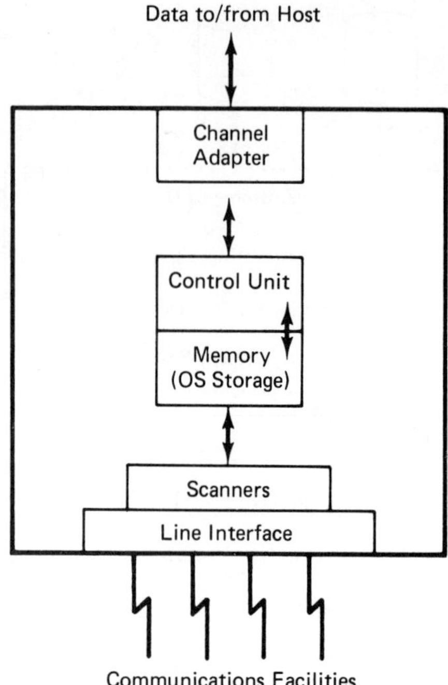

FIGURE 3-37. The Front End's Components.

The communications scanner provides an example of the value of off loading functions from the best. Obviously, each attached line must be examined at periodic intervals for the bit streams of messages flowing into and out of the user applications. Assuming a front end has five lines attached at 1200 bit/s speed, the scanner must examine each port every 8.3 μsecs (1 sec/1200 = .00083 sec.). The five lines require bit-sensing operations to occur 6,000 times a second (1200 bit/s × 5 lines = 6,000). Moreover, the data bits must be moved through the port and into a temporary buffer register to allow for the next arrival of a bit. Eventually, the received bits are assembled into bytes and then into messages for transfer to the host.

All these activities require some kind of control and the use of machine instruction cycles. Also required are interrupts to notify the system that the registers are full or empty. In our example, the five active lines feed 750 bytes into the node every second (1200 bit/s × 5 lines ÷ 8 bits per byte = 750 bytes). Assuming an interrupt is required to transfer a byte out of the temporary register, 750 interrupts per second must occur for the system to operate correctly. If we assume further that an interrupt requires an average of 30 μs, then 22.5 ms of every second is expended to service these lines (750 × .00003 = .0225). While this example represents only 2¼% of a processor's capacity, a typical network with many lines—some of much higher speeds—would consume substantial resources to do only one communications function: service the lines for bit/byte assembly/disassembly.

The front end performs the scanning and the byte assembly/disassembly process without borrowing valuable machine cycles from the host. Periodically, the host is interrupted to transfer blocks of data to it. In this manner, the front end relieves the host for other processing.

Line servicing and bit/byte assembly/disassembly is only one function provided by front end processors. Listed below are other services typically found in the front end:

Buffering: The front end obtains and releases buffers as it receives and transmits data.

Error handling: Line errors, device errors are noted and recorded by the front end.

Control characters: The processor inserts and deletes transmission control characters.

Code translation: Front end provides code translation between dissimilar devices.

Message flow control: Governs the pacing and flow of messages across the attached lines.

Switched services: Provides services to switched, dial-up lines.

Data link control interface: Processor provides data link control (DLC) interfaces to devices using different DLCs (see Chapter 6).

Polling and selection: Front end schedules all polling and selection messages to the attached devices (see Chapter 6).

Chapter 4 provides further information on the communications front end.

4

SOFTWARE AND DATA BASES

COMMUNICATIONS PROGRAMMING

Communications programming (software) performs the logic tasks that are not performed by a host operating system, the applications programs, and the data base management systems. The communications programs reside in the host, the front end, in terminal controllers, terminals, and in other components such as switches. The major portion of the communications software logic is located in the host and the front end. The term CSW will be used in this chapter as an abbreviation for communications software.

The applications programs perform the specific tasks required by the end user. Their requests for communications services are established by a program statement such as a READ, WRITE, GET, PUT, or a call for the execution of a software routine that provides the request to the communications control programs. This results in the applications programmer being able to structure programs much like any other program that accesses disks, printers, or tape files. The applications programmer is relatively free from concern about the communications environment. Consequently, network changes should not require changing the applications programs—only the communications software.

The communications programs interface with the user applications primarily through tables. As an applications system is implemented, parameters are placed into these tables describing the communications requirements of the application program or a group of applications. For example, the following requirements for line handling could be established by the parameters in the tables:

- Line is not opened for outgoing messages until there is a break in the incoming traffic.
- Conversely, input traffic is held up periodically to allow output traffic to be transmitted.
- Send and receive data from/to the user application one batch at a time. Stated another way, manage traffic as if it is a tape file and not an individual message.
- Accept one message and do not permit the use of the line until the user application program sends back a response (held-line discipline).

Communications Programming in the Front End

The IBM 370X (3704/3705) is the most extensively used front end in the industry today. It is used to illustrate functions of communications programming.[1] Figure 4-1 shows a general view of the 3705 components. The Central Control Unit is actually a specialized computer; it addresses storage, executes instructions, and operates under the control of the Network Control Program (NCP). The communications lines are attached to Line Interface Bases (LIBs), which act as the front end's interface to the outside world. The number and speed of the lines depend upon communication scanners; each line attached to the front end is serviced by one of four possible scanners.

Each scanner may be equipped with one to four internal clocks. (See pages 27 and 28 for a review of internal clocks.) This arrangement allows for each scanner to handle up to four different line speeds. The maximum data rate in bits per second of a

[1] For additional information, see (a) *Binary Synchronous and Asynchronous Concepts for Programmers*. IBM Training Document #SR20-7187-0, IBM Corp., 1133 Westchester Avenue, White Plains NY 10604; (b) "The NCP Atlas: Roadmap to IBM's Net Control," by Albert J. Hedeen. *Data Communications*, December 1978, pp. 51–70; (c) *IBM 3704 and 3705 Guide and Reference Manual*. Document #GC30-3008-5 IBM Corp., Systems Communications Division, Box 12195. Research Triangle Park NC 27709. It will be helpful if the reader reviews the material on SNA in Chapter 8; it is closely related to the material in this section.

communications line is determined by the frequency at which the line is scanned (or sampled) by the communications scanner. The data rate of a line can be increased by instructing the NCP to scan a line more often than others; obviously, this also decreases the speed of those lines less frequently scanned. In the absence of scan limits, each line is scanned once per scanning cycle. The architecture allows up to 352 1,800 bps half-duplex lines to be attached and scanned. Higher speeds result in fewer lines scanned and thus restrict the number of lines that can be attached to the machine.

The Network Control Program (NCP) is the major controlling element in the 3705. It is the "operating system" of the front end. It is responsible for receiving orders from the host and translating these orders into the following communications support functions. Notice the similarity of these functions to those of a generic front end discussed in the previous chapter.

- Managing the dialing and answering of switched lines.

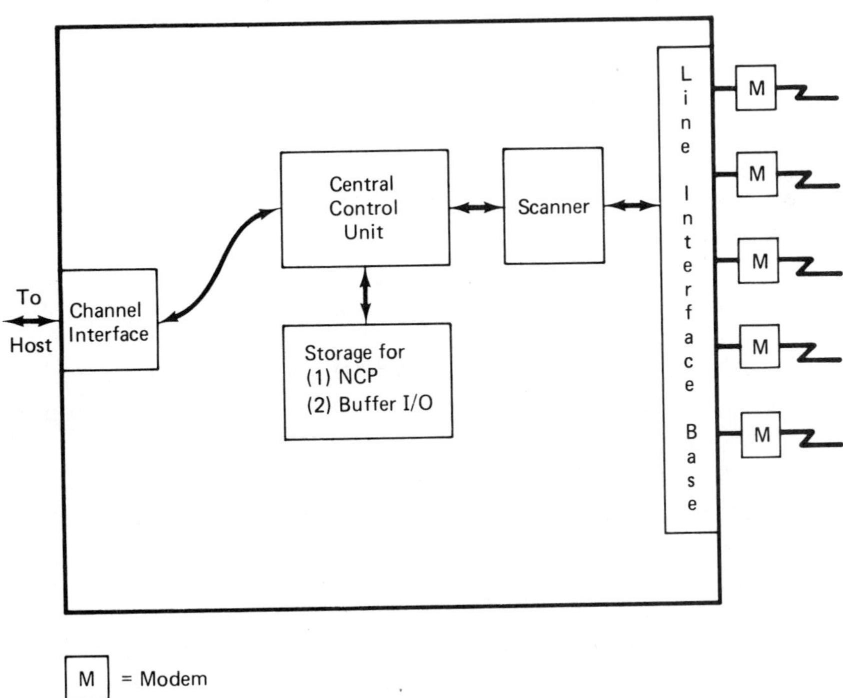

FIGURE 4-1. IBM 3705 Front End.

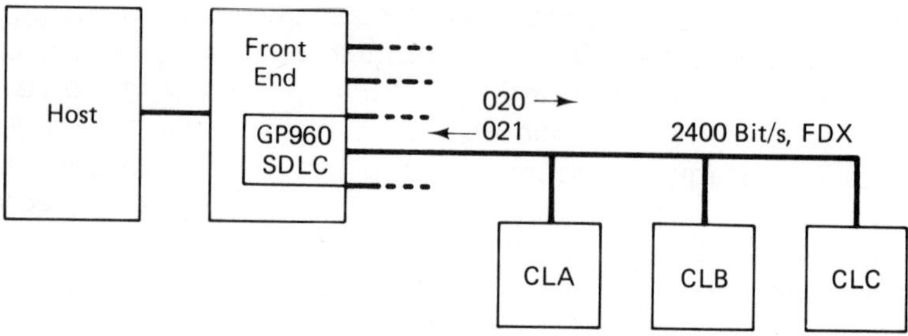

FIGURE 4-2. Nonswitched Line Configuration.

- Responding to a host-resident application program's READ/ WRITE or GET/PUT to send/receive traffic on the attached communications lines.
- Assembling the user bits into an outgoing message and adding communications control characters to the user message.
- Disassembling the incoming message into bits and deleting the control characters.
- Controlling the scanners and LIBs in accordance with the line speeds in bits per second (bit/s) of an attached terminal and other components.
- Translating the codes of dissimilar devices. For example, translating EDCDIC code to ASCII code and vice versa.
- Managing storage buffers; assigning and releasing the storage as messages move into and out of the front end. (Buffer storage is quite important due to a finite amount of memory storage available. Additionally, the manner in which buffer storage is managed can influence performance.)
- Analyzing problems on the communications lines by recording errors, storing statistics, performing diagnostics, and testing lines and attached components.
- Providing alternate paths for failed private point-to-point lines by providing a dial-up backup facility.

Macro Language Programming. Today, many vendors provide tools to aid the organization's technical staff in defining and configuring the communications facilities. The 3705 uses a CSW macro language for this purpose. A simple example is provided here to illustrate this powerful function for the front end processor. Additional macro facilities are usually available for defining the host-specific tasks as well as the entire topology of the network.

Let us assume we wish to configure a nonswitched line with three multidropped stations. The line is to operate at 2400 bps with full-duplex data flow (see Figure 4-2). The NCP macro language parameters are coded to instruct the front end and the NCP how to manage this portion of the network:

Reference Note	Label	Macro Instruction	Instruction Parameters
1	Example	BUILD	MEMSIZE=208, TYPGEN=NCP, SUBAREA=4, MAXSUBA=3, CA=TYPE2, CHANTYP=TYPE2, BFRS=88, TYPSYS=OS, LOADLIB=NCPLIB, OBJLIB=NCPTEMP, SLODOWN=25
2		SYSCNTL	OPTIONS=(MODE, RCNTRL, RCOND, RECMD, RIMM, BHSASSC, ENDCALL)
3		HOST	INBFRS=3, MAXBFRU=10, UNITSZ=88, BFRPAD=28, DELAY=2
4		CSB	TYPE=TYPE2, MOD=0, SPEED=600
5	GP960	GROUP	LNCTL=SDLC
6	L960	LINE	ADDRESS=(020, 021), DUPLEX=FULL, SPEED=2400, POLLED=YES, RETRIES=5, MAXDATA=265, PACING=1
7		SERVICE	Order=(CLA, CLB, CLC)
8	CLA	PU	ADDR=C1, PUTYPE=2, MAXOUT=3, RETRIES=5
9	FALUA1XX	LU	LOCADDR=1, PACING=3
	LUA2XXXX	LU	LOCADDR=2
	LUA3XXXX	LU	LOCADDR=3
	LUA4XXXX	LU	LOCADDR=4
	(Same type of macros for CLB, and CLB: reference notes 8 and 9 apply for all three stations.)		
10		GENEND	
		END	

Reference Note

1 The BUILD macro describes the model of the 3705, its storage size, buffer size, names of various tables, address of the NCP, the channel adapter, and other options.

MEMSIZE=208	Storage capacity of 208K bytes.
TYPGEN=NCP	Program will execute locally and not remotely.
SUBAREA=4	Address of the NCP. Each NCP in a network must have a unique address.
MAXSUBA=3	Specifies maximum number of subareas controlled by the host communications software.
CA=TYPE2	Type of channel adapter.
CHANTYP=TYPE2	Defines the use of the channel.
BFRS=88	Specifies the size in bytes of the buffers for the NCP to use.
TYPESYS=OS	Specifies the host operating system
LOADLIB=NCPLIB	Identifies the disk file that contains the executable NCP program and associated control tables.
OBJLIB=NCPTEMP	Identifies intermediate files for the generation of the NCP code.
SLODOWN=25	Identifies a threshold at which the NCP reduces the amount of traffic it accepts from host and communications lines. Provides a means to avoid saturation. In this example, if 25% or less of buffers are free (i.e., 75% are busy), the NCP enters its slowdown mode.

2 The SYSCNTL specifies certain facilities to be used. The names are extracted from the communications software tables that reside in the host.

3 The HOST macro describes how the front end and the host computer interact. It defines the buffers for receiving data from the host and the timing of data flow across the host front end channel.

INBFRS=3	Specifies number of buffers allocated for each transfer from the host.
MAXBFRU=10	Specifies number of buffers the host allocates for data from NCP.
UNITSZ=88	Specifies size of buffers for data from NCP.
BFRPAD=28	Specifies pad or filler characters for NCP to transmit to

host; the host communications program uses space to insert data or headers.

DELAY=2 — Specifies the interval (in tenths of a second) that NCP delays between the time it has data for host and the time it presents an attention signal to the host.

4 The CSB macro describes the speed, type, and location of the communications scanner.

TYPE=TYPE2 — Specifies the specific model of scanner. The models determine speeds and line groupings.

MOD=0 — Provides specific line interface address of the scanner.

SPEED=600 — Specifies oscillator speed in bit/s—in this case 600 bps.

5 The GROUP macro defines a group of communications lines having certain common characteristics. Using GROUP obviates additional, redundant coding of parameters for each line.

LNCTL=SDLC — Specifies the data link control protocol. (See Chapter 6 for a detailed discussion of data link controls.)

6 The LINE macro describes the characteristics of one communications line attached to the front end. The instruction specifies whether the line is half duplex or full duplex, identifies its address and line speed, defines the manner of clocking (from the front end scanner or the modem), and other features.

ADDRESS=(020,021) — Specifies the line addresses for the full duplex line. The two addresses designate the separate sending and receiving paths.

DUPLEX=FULL — Specifies the physical circuit. This parameter does *not* specify the message flow. (See Chapter 2 for a review of this concept.)

SPEED=2400 — Speed in bit/s of the line.

POLLED=YES — Indicates that this line is controlled by this NCP. An entry of N0 indicates a remote NCP/front end is the master station.

RETRIES=5 — Specifies maximum number of times a garbled message is to be retransmitted.

MAXDATA=265 — Specifies the message size (or block/frame size) transmitted in one data transfer. If this parameter is not coded, the program uses the BFRS parameter to determine message size.

PACING=1 — Specifies number of transmissions the NCP sends to the attached node before requesting a response.

7 The SERVICE macro specifies the order in which the attached stations are to be serviced by the NCP. The stations identified must be physical units (PU).[2]

ORDER=(CLA,CLB,CLC)	Controller with an address of CLA is to be serviced first; controller CLB second and controller CLC third. Notice that the labels of the macros use these parameters.

8 The PU macro defines the physical unit with which the NCP communicates. A physical unit contains logical units.[3]

ADDR=C1	Specifies the address of the unit.
PUTYPE=2	Specifies the type of equipment (controller, terminal, etc.)
MAXOUT=3	Specifies the maximum number of messages the NCP will send to the unit before asking for a response.
RETRIES=5	Establishes the maximum number of retries (retransmissions) that the NCP will attempt in the event of errors.

9 The LU macro specifies exactly what components are associated with the Physical Unit.[3]

LOCADDR=1	Specifies the local address of the logical unit.
PACING=3	Specifies that the NCP is to send 3 messages to this LU before stopping to await a response.

10 The GENEND and END signify the end of the program.

---

[2]IBM communications architecture identifies a *physical unit* (PU) as a hardware control unit such as a terminal cluster controller (i.e., a minicomputer that controls multiple "less-intelligent" terminals). Also, each node in the network must have a physical unit; therefore, a front end can be a physical unit. See Chapter 8 for a discussion on PU.

[3]A *logical unit* (LU) provides the bridge between the end user application program and the communications system. The LU typically supports multiple end users. The idea of the LU is to off load tasks from the user program and to provide a standard interface into the system. Chapter 8 discusses LU in more detail.

Communications Between Host and Front End

It is evident from a review of the front end program that many of the parameters are used to control the flow of messages across the communications lines. The program also defines the topology of devices and lines attached to the front end. As previously mentioned, communications software also resides in the host to coordinate the activities of the applications programs with the front end processor.

The host CSW functions are:

- Start up of network operations.
- Establishing sessions in the host-controlled network between application/terminals/users.
- Routing data between applications/terminals/users and providing alternate routes if a route fails during a session.
- Providing levels of transmission priority.
- Providing session recovery.
- Pacing message flow between the sender and receiver.

It may appear that the communications software in the host and front end have overlapping functions. Such is not the case. Typically, the host software interfaces with the applications programs to determine what actions are requested to satisfy the user program. The host software will perform certain control functions and pass the request to the front end programs which, in turn, will provide additional and complementary functions. The relationship is shown in Figure 4-3:

Event 1: The user application program issues an output instruction. The operating system receives and analyzes the request and passes the request to the host communications software.

Event 2: The host resident communications software examines the user headers and validates and logs the request. Predefined tables provide information to the host CSW such as routing, editing, and error checking. The software also establishes buffers for the application message, inserts any required control characters, and appends communication control headers. Control is passed back to the operating system.

Event 3: A channel program is invoked by the operating system. The channel accepts the message and transmits it across the channel line to the front end.

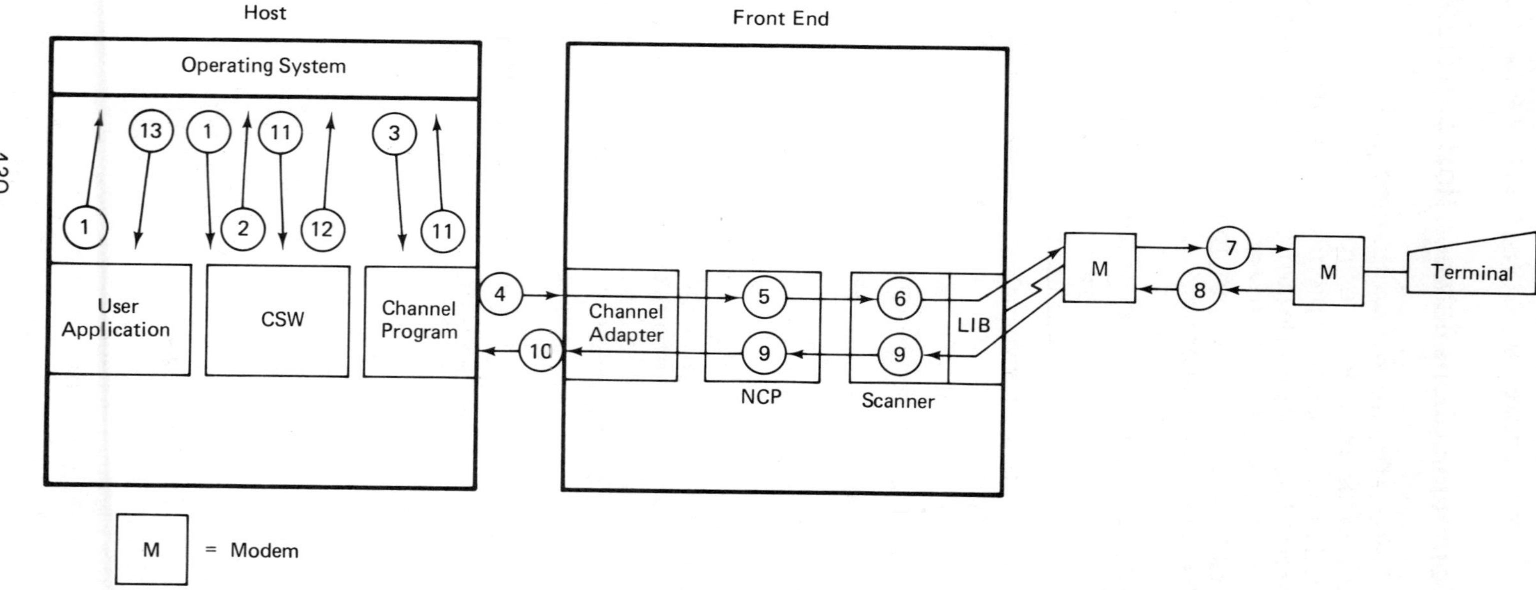

FIGURE 4-3. Data Flow Between Host and Front End.

Event 4: The channel adapter receives and notifies the front-end operating system of the message.

Event 5: The front end operating system examines the communications headers, performs table lookups on the application message, and activates the communications scanner and the appropriate line interface device.

Event 6: The scanner provides the timing to effect the message transfer. The line interface device buffers the data and provides the necessary signals to activate the attached modem's circuits.

Event 7: After modem synchronization, the data is transmitted to the remote device. The remote device accepts the data and relays it to the user.

Event 8: The remote device sends back a reply; the two modems synchronize themselves and data is moved back into the line interfacing through the timing of the scanner.

Event 9: The front end operating system is notified of the arrival of the data; it processes the data as appropriate. (For example, it determines that the message is destined for the host and not another communications line.)

Event 10: The channel interface is activated and the message is transferred to the host channel program.

Event 11: The host is notified of the message and it directs the host CSW to assume responsibility for processing the incoming data.

Event 12: The host CSW examines the headers to determine necessary action and passes the data to the user application's work area buffers. It also logs the traffic, stores statistics, and, if appropriate, eventually terminates the session.

Event 13: The user program is activated by the host operating system. It accesses the data from its work area and continues to execute.

Communications Programming in the Host

Figure 4-2 and the accompanying coding example explain how to establish a nonswitched (leased) line. The example shows the macro code in the front end. To further explain communications software (CSW), we will examine some macro code within the host CSW. In this example, a host mainframe user application requires a dial-out capability from its location in Washington, D. C., to a remote site in New York City (Figure 4-4). Notice the attachment of an

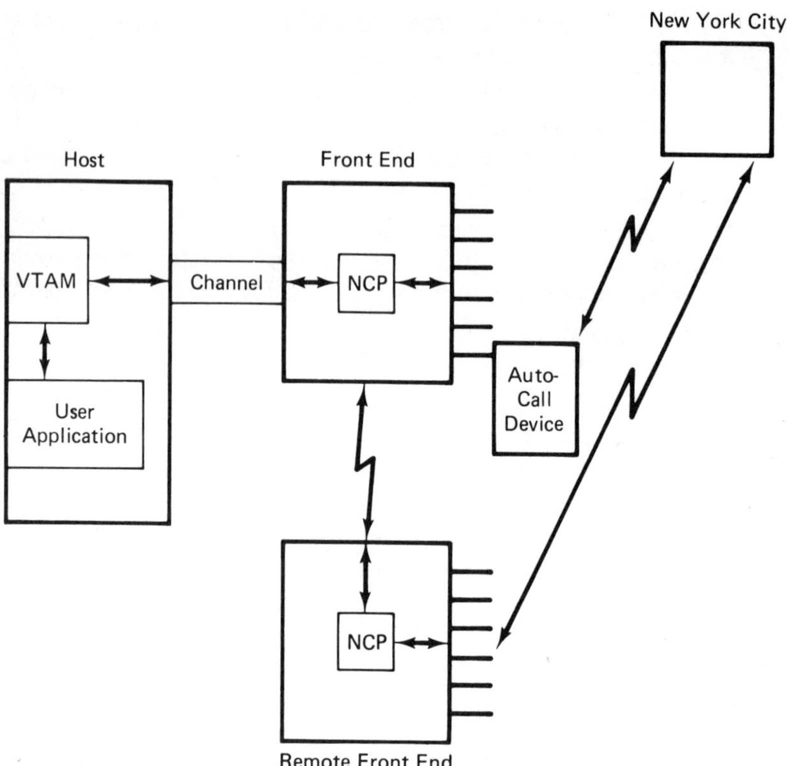

FIGURE 4-4. Switched Line Configuration for VTAM Coding Example.

automatic call unit; this device allows the dialing of a telephone number programatically without human intervention. The user wants more than one dial-out port to preclude blockage during busy line conditions.

IBM's Virtual Telecommunications Access Method (VTAM) is a host resident CSW system. It is used to further illustrate communications programming.[4] The VTAM macro language statements to satisfy the user's alternate dial-up facility are coded as follows:[5]

---

[4]The coding examples of NCP and VTAM are not related. Two separate programs are coded to show examples of nonswitched and switched lines. Other programs and additional statements are required to complete the definitions of these lines. The reader should consult the references for more details.

[5]For more information see *Advanced Communications Function for VTAM*, IBM Manual Number 5665-280, September 1981, IBM Corporation, Department 52Q, Neighborhood Road, Kingston NY 12401.

Reference Note	Label	Macro Instruction	Instruction Parameter
1	Example	VBUILD	TYPE=SWNET, MAXNO=50, MAXGRP=2
2	PU1	PU	MAXPATH=3, ID=X3791XID, DISCNT=Yes
3		PATH	DIALNO=56, PID=1, GID=1, GRPNM=G1, REDIAL=10
4		PATH	DIALNO=9/26756 . . . . . .
5		PATH	DIALNO=17/121226756 . . . . .
		GENEND	
		END	

Reference Note

1 The VBUILD macro defines a network node. The node subcomponents are defined by subsequent PU and LU statements.

TYPE=SWNET Specifies that this is a switched node. All units on this node are connected by a switched link.

MAXNO=50 Maximum number of unique telephone numbers associated with this switched configuration.

MAXGRP=2 Maximum number of group paths defined in GRPNM operand of PATH macro.

2 The PU defines the physical unit for New York City, in this case a communications controller. The label PU1 provides the name of the PU.

MAXPATH=3 Specifies the maximum number of dial-out paths.

ID=X3791XID The CSW uses this number to identify the station during a call procedure. This parameter defines a completely unique ID for the station. The X's are constructed by VTAM from tables and other parameters. The number 3791 specifies the type of communications controller.

DISCNT=YES Specifies that VTAM will physically disconnect ("hang up the telephone") the PU when *all* LU's have finished their sessions with the applications programs.

3 The PATH macro defines the dial-out paths to the physical unit that is defined in the PU statement. VTAM establishes the connection in the order that the PATH statements appear.

DIALNO=56 Specifies the telephone dial numbers to be used to connect to the remote site. This number is an abbreviated number; it allows a shorter number in place of the full dial number.

PID=1 This parameter specifies the path to be used.

GID=1 Identifies a group of paths for purposes of common identification and management.

GRPNM=G1 Specifies the label name of the GROUP macro of the NCP program. This establishes the linkage of VTAM and NCP.

REDIAL=10 Specifies number of times the dialing is retried before returning to VTAM.

4&5 These statements are similar to the first PATH macro. The DIALNO provides information on using internal lines, WATS lines, dial-up lines, etc.

Notice the PATH macro GRPNM parameter identifies the line group defined in the NCP program (not coded here). The PATH statement tells VTAM which NCP line groups and dialing digits to use.

Establishing a Dial-Up Connection Through the Programs. The configuration depicted in Figure 4-4 will be used to illustrate how VTAM and NCP interact to satisfy a user connection request to the New York site.

1. The user application program executes instructions to request the transmittal of output to another user. The application has been defined previously to VTAM tables.
2. VTAM checks the validity of the request, determines if resources are available, performs logging functions, and sends the request to the channel attached NCP. VTAM's information to NCP provides the logical path to be used, the phone number, the line (port) to be used, and the number of timoc to rodial.

3. In the event of an unsuccessful dial, VTAM instructs NCP to try the alternate paths established in the PATH instructions. The call may be forwarded to the remote NCP, established by the third PATH statement.
4. Upon a successful dial-up, an end-to-end session is established in the following manner:
   a. The NCP sends a request for an ID to the remote PU in New York.
   b. Upon receiving the ID from New York, the NCP sends it to VTAM, where it is checked. VTAM then forwards to NCP several parameters (such as message size) to help manage the session.
   c. VTAM issues commands to activate the PU and then to activate the LU that supports the application.
   d. VTAM then "binds" the resources, thereby establishing the session.
5. The application program is now free to send the data across the dial-up line to the New York site.
6. Upon completion of the report transmittal, the application program session is terminated by VTAM. NCP and VTAM update tables to indicate the session is over.

Trends in Software Technology

The examples in our discussions are high-level macro languages. These techniques are seeing increasing use in the industry. Certainly, the macro definitions must be interpreted and decomposed to executable code and thus require powerful computers and/or interpreters. The important benefit of high-level languages is the ease and speed of their use by system programmers.

The NCP and VTAM macro languages are examples of vendor-specific systems. The industry is also progressing toward more general nonprocedural languages and products known as *software development tools*. These tools are used to assist the CSW programmer in specifying, designing, coding, and verifying the software. For example, some products provide methods to graphically establish the user specifications and develop a prototype very quickly. The end result allows the user and designer to check for errors and omissions early in the system life cycle.

DATA BASES

One might question why the subject of data bases is included in a book about communications and networks. The simplest explanation is that networks exist for the purpose of transporting data between computers, terminals, and data bases. Thus, data bases are an integral part of data communications systems and networks. We must understand at least the rudiments of data base technology in order to grasp the issues related to network architectures, communications line configurations, topology layouts, and distributed processing.

The discussion of network data bases is divided into three parts. The basic terms and concepts are covered in this section, Chapter 10 explains distributed data bases, and Chapter 11 introduces network data base design.

Data Storage

Magnetic tape is a widely used storage device. It is quite similar to the tape used in home recorders and consists of microscopic iron particles that can be arranged into magnetized bit patterns. The tape passes READ/WRITE mechanisms that sense the presence or absence of the magnetized bit. The bits are grouped to form a unique code to represent a number, character, or special character. Codes such as EBCDIC and ASCII are commonly used. Magnetic tape is a sequential storage medium, data must be accessed in the order it was placed on the tape.

Network data bases most often reside on disk storage devices. A disk consists of several circular platters (disks), much like a stack of records on a turntable. The disks are separated by a small space and each side of the disk allows for the recording and reading of magnetic bit patterns. READ/WRITE heads are placed between each disk surface in order to process the data stored on the platter. The data is stored on concentric ring tracks around the disk.

Figure 4-5(a) shows a disk unit. The platters or disks rotate around a central spindle, passing across the READ/WRITE (R/W) heads. The disks are divided into tracks and cylinders. The *cylinder* consists of reading surfaces on the same area of each platter. Each recording surface is called a *track*. The cylinder concept is used to improve access efficiency. If data is stored across tracks, the R/W heads must mechanically move across the disk after each track is accessed. On the other hand, storage at the cylinder level necessitates accessing data only by electrically switching on/off the proper set of

R/W heads. A full cylinder can be accessed before the R/W heads are moved.

Data is stored on the disk tracks as shown in Figure 4-5(b). The user data is stored as physical records, preceeded by control and key address areas. The physical record in this example consists of three user records (logical records). The three user records are stored or blocked together for two reasons:

1. The blocking reduces overhead space since the control and key address areas are only needed for each physical record.
2. Access overhead is reduced. A READ request from an application program causes the transfer of the entire physical record with one input operation. Subsequent READs then access the resident block within a memory buffer.

Substantial storage and timing savings can be achieved by blocking. However, if contiguous logical records are not used by the application program, then the data transfer is wasted. For example, if the user wants logical record 1 only, the other two records are transferred into memory but are not used by this program. Consequently, data base design requires a careful assessment of end-user access needs in order to cluster and block the data correctly.

This simple data base structure illustrates a critical point in network data bases (wherein many users are accessing the same logical and physical records): The physical structure format and storage of the data must match the users' access needs. The reader should keep this principle in mind in the later discussions in Chapters 10 and 11.

Organization of Data Bases

Modern data bases are organized with a complex set of software called a *data base management system* (DBMS). These systems offload many of the data-related tasks from the application programmer. (The offloading concept is similar to the rationale for front ends and communications software.) An application can issue a READ, WRITE, GET, or PUT to the DBMS which in turn will locate, format, and access the requested data. The DBMS makes use of an access method to handle the details of the physical access to the data bases (see Figure 4-6). The DBMS provides the user interface to the access method through logical user views called *schemas* and *subschemas*. The logical views are translated into the physical

Rotation
Recording Platter
Tracks
Cylinder
Read/Write Mechanisms

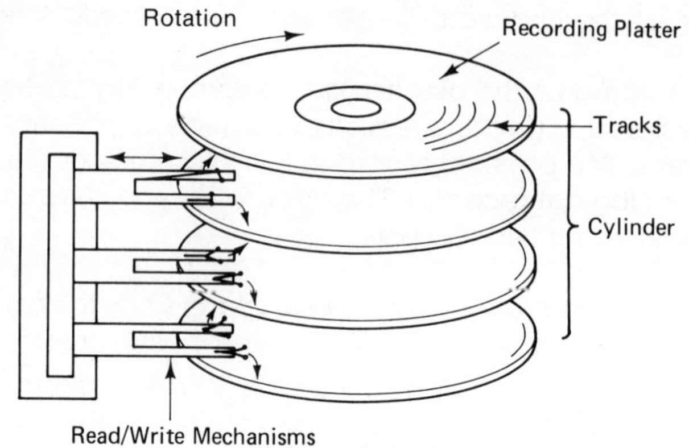

(a) Disk Pack

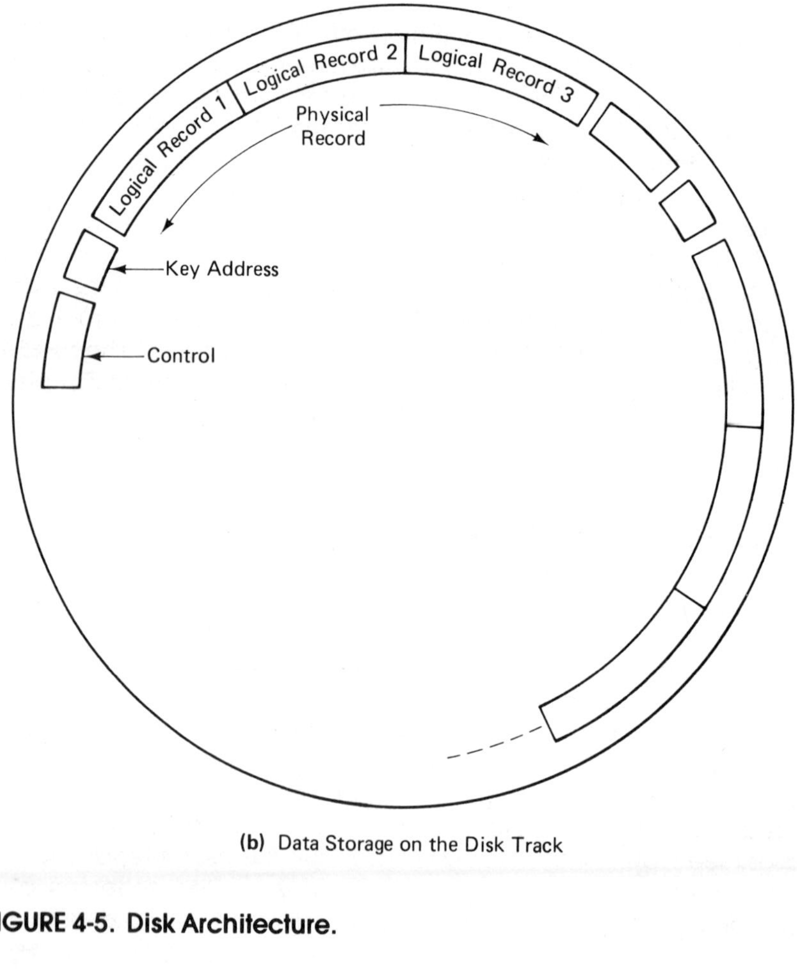

Logical Record 1
Logical Record 2
Logical Record 3
Physical Record
Key Address
Control

(b) Data Storage on the Disk Track

FIGURE 4-5. Disk Architecture.

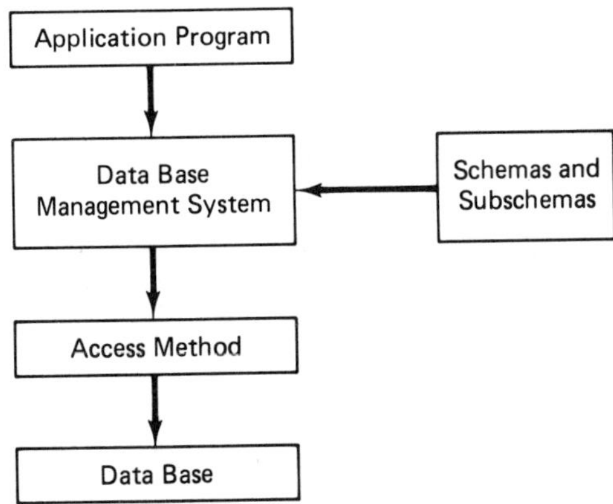

FIGURE 4-6. Data Base Management System Organization.

structure and internal view by the access method. These terms and concepts will be described in the following section.

Physical Access. Data is stored on disk in one of three ways: sequential, indexed, or random (direct). The *sequential storage* simply means that records are stored one after the other on the disk track. A record in the middle of the data track cannot be obtained without first going past the preceding records.

The *indexed method* provides a unique identifier or key for each record. For example, an employee number could be a key for a personnel record. The key is used by the access method to compute a disk address and location through the index. This concept is illustrated in Figure 4-7. The application program executes an instruction that requests an employee record with the key (employee number) of 04420. The request is passed to the DBMS which interprets the request (more on DBMS functions shortly) and passes the task to an indexed access method.

The access method has several indexes available to assist it in locating the record. First, it compares the employee key (EMPNO= 04420) to its high-level index and determines that the record is on one of the cylinders in the 40–79 group. (The key in the index represents the highest possible value of a key within the group.) Next, the access method scans the cylinder index and determines that cylinder 43 contains the record. The track index for cylinder 43 is searched and the record is found to be located on track 10. Finally, track 10 is sequentially accessed through the disk READ/WRITE mechanisms and the control and address areas on the disk. The employee record

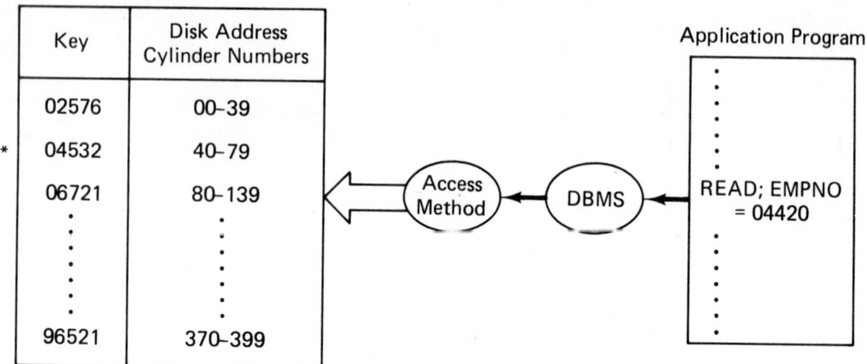

High Level Index

	Key	Disk Address Cylinder Numbers
	02576	00–39
*	04532	40–79
	06721	80–139
	⋮	⋮
	96521	370–399

Cylinder Index

	Key	Cylinder
	02577	40
	03012	41
	03634	42
*	04836	43
	⋮	⋮
	06720	79

Key: Represents Highest Key on Track or Cylinder

Track Index

	Key	Track
	03635	1
	03842	2
	⋮	⋮
*	04436	10
	⋮	⋮

FIGURE 4-7. Indexed Storage.

is found, read into the computer, and placed into the application's workspace.

The efficiency of the indexes in an indexed system is quite important. If possible, they should be memory-resident in order to reduce lengthy input/output (I/O) processing to obtain their values. Moreover, indexed systems often use additional facilities to locate records that are added to a data base. For example, employee record 04420 might contain a pointer (address) to another area of disk that contained a recent addition of the record with EMPNO= 04421. This approach allows a sequential structuring of the files but can create considerable overhead and complexity.

The third major access structure, *random or direct*, relies on the record key to provide the disk address. Typically, a key is "hashed" or subjected to a calculation to obtain the address. For example, an employee record with a key of 02576 could be translated to cylinder and track address by the calculation. The hashing method is attractive because it usually requires only one seek to the disk to obtain a record. Also, it is less complex than the indexed technique.

The users usually have no choice of the access method to support their application. However, the network designers and the network data base administrators must be knowledgeable about the details of vendors' access methods because their performance affects the ability of the network to support the user requirements. Many vendors' DBMS offer all three access methods and even several variations on them. Detailed discussion of access methods is beyond the scope of this book, but the reader is encouraged to pursue the topic further if data access, response time, and retrieval/update costs are of interest or concern.

DBMS Architectures (Models). Above the access method is the DBMS itself. It is structured and presented to the end user in such a manner that the user can view his/her data in a logical sense. This means that the user need not be concerned about track storage, cylinder access, index maintenance, and many other laborious tasks. Rather, the users view the data in the context of their requirements. DBMS architecture is also called a DBMS model, form, or structure. The most commonly used data base models are:

- Hierarchical
- Plex or network
- Relational

Hierarchical model. The hierarchical (sometimes called a tree) model is shown in Figure 4-8. The user sees a tree with hierarchical members of the root, the top of the hierarchy. The members can be thought of as records or segments containing user data.

The lower levels are designated as children to the upper level, which logically enough is a parent level or node. All the second level elements in Figure 4-8 are called twins to each other since they are in the same hierarchical position in the data base. The lower levels usually are considered to contain multiple occurrences of segments.

Hierarchical data bases are useful for some applications and offer a convenient schema and subschema for the partitioning of the data base to distributed sites. For example, a data base of a large organization could be modeled hierarchically with the root segment containing data on the parent company (perhaps a holding company), the second level segments containing data on the subsidiaries, and the third level representing branches and departments of the subsidiaries. Obviously, some user data do not logically appear as a hierarchical organization and other approaches are then used.

Plex or network model. This model is distinguished by a child having more than one parent and presents a more involved structure. A common example is a personnel data base where multiple retirement options exist for multiple employees [see Figure 4-9(a).]

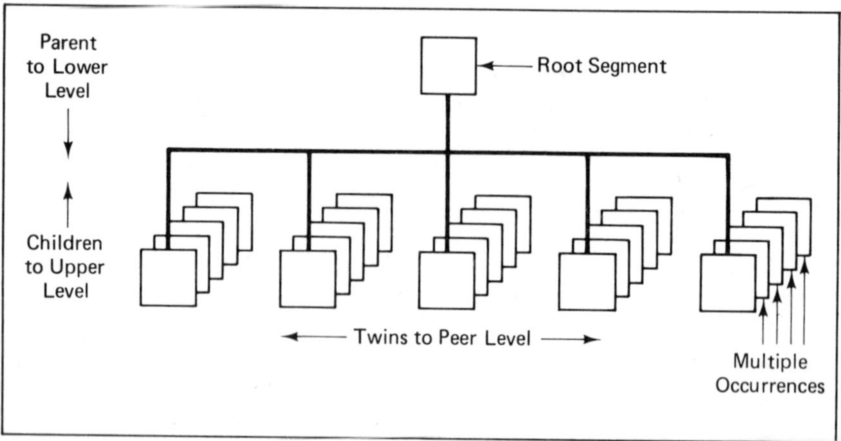

FIGURE 4-8. Hierarchical Model.

Name
Jones
Smith
Green
Baker
Benefits
Options A
Options B
Options C
Options D

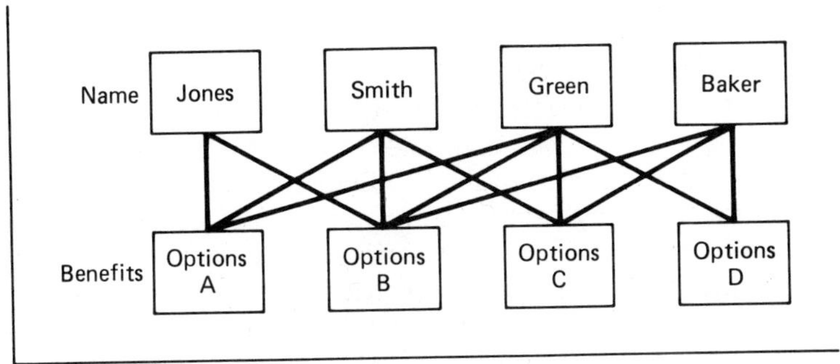

(a) Multiple Relationships

(A)
Name
1
2
Employment History
Salary
Training
Benefits
Promotion
Where
= One to One
= One to Many

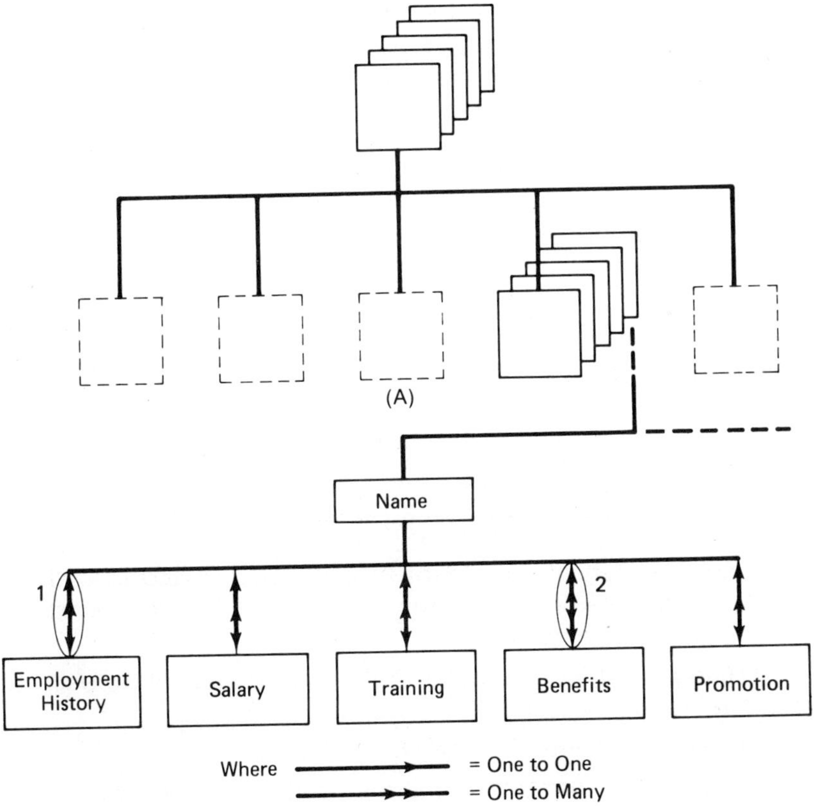

(b) Multiple Relationships Depicted with Arrows

FIGURE 4-9. Plex or Network Model.

The plex model can be depicted with the use of arrows [see Figure 4-9(b)]. The single arrow in circle 1 indicates that an EMPLOYMENT HISTORY segment has one parent, NAME. The double arrow indicates that the NAME segment has many EMPLOYMENT HISTORY segments. The double arrows in circle 2 indicate a many-to-many relationship in which employees choose multiple BENEFIT plan options.

The many-to-many relationships occur frequently in applications and can present very complex user views (not to mention the resulting complex physical storage and access). Recently, automated teller machines (ATMs) have become quite popular. The ATMs allow bank customers to make deposits and withdraw money after banking hours. Many banks have formed consortiums to share the physical ATM facility. Thus, ATMs have several owners and the owners have several ATMs—a typical many-to-many relationship.

Relational model. The relational model has gained considerable popularity and use due to the simplicity of its user views. In its simplest form, the relational data base is presented as an organized number of two-dimensional tables (also called *flat files*). The display form is familiar and understandable to practically everyone. Figure 4-10 shows a typical relational model. The organization consists of an inventory data base.

Relational data base technology uses several unique terms. The following is a list of these terms in the context of Figure 4-10.

Relation: The relation is the table itself (INVENTORY Relation).

Tuple: This describes the row in a relation. The first tuple in the INVENTORY relation contains part #634.

Attribute: The attribute is the name of the type of column. (Also called *role name*.)

Key: A column or a set of columns whose value(s) *uniquely* identify a row. The INVENTORY relation's tuples (rows) can be uniquely identified by COMPONENT PART # and LOCATION. A key consists of one column and a candidate key consists of more than one column.

Prime attributes: Attributes that are members of at least one candidate key.

Element: This describes the field in the tuple.

Degree: The number of columns in the relation. The INVENTORY relation contains six columns.

Cardinality: The number of rows in the relation. The INVENTORY relation contains five rows.

N-ary relations: A table with N columns.

INVENTORY RELATION

	Component Part #	Description	Location	# In Stock	Price	Stock Deletion Date
Tuples →	634	* * *	NY	50	10	7/82
→	788	* * *	LA	200	20	8/82
→	996	* * *	SF	100	15	10/82
→	1500	* * *	KC	350	5	2/83
→	634	* * *	LA	600	10	7/83

FIGURE 4-10. A Relational Model.

Relational data bases adhere to certain rules and conventions:

- The rows (tuples) of a relation are never duplicated.
- Each relation represents a single concept. Any repeating occurrences and resulting many-to-many relationships are placed in smaller relations.
- The columns contain like data elements.
- The ordering of the rows in a relation is not significant, but column ordering is significant.
- Each row has a fixed number of fields, all explicitly named.
- Repeating occurrences may cause redundancy across the relations.
- An attribute in the key cannot be discarded without destroying the key's unique identification of the tuple.

We will see that the relational model can be a very useful concept for network and distributed data bases. Stable data structures are important to data bases in a network and the relational model can provide for more stable structures than the hierarchical and plex models. However, we need some additional information before relating the ideas to networks.

The attractiveness of the relational architecture comes from the theory of normalization; it provides a rigorous discipline for structuring the data in the relations. The primary objective of normalization is to simplify or eliminate update, deletion, and insertion problems in a data base. The major goal of a relational model is to achieve the third normal form:

> An attribute of a relation (attribute y) is functionally dependent upon another attribute in the relation (attribute x), such that each value in x has no more than *one* value in y associated with it. Therefore, x determines y; x is the determinate. A third normal form exists when every determinant is a key. Stated another way, all items in a record are completely dependent upon the key of that record and nothing else.

These terms, concepts, and theories probably sound a bit farfetched so let us provide a pragmatic example to bring things into perspective. Keep in mind that the goal is to provide simplicity to the user. Figure 4-11(a) shows the INVENTORY relation again. However, it does *not* exhibit the third normal form and can present

INVENTORY

Component Part #	Description	Location	# In Stock	Price	Stock Deletion Date
634	* * *	NY	50	10	7/82
788	* * *	LA	200	20	8/82
996	* * *	SF	100	15	10/82
1500	* * *	KC	350	5	2/83
634	* * *	LA	600	10	7/82

(a) Initial Form

PRICE

Component Part #	Description	Price
634	* * *	10
788	* * *	20
996	* * *	15
1500	* * *	5

INVENTORY

Component Part #	Location	# In Stock	Stock Deletion Date
634	NY	50	7/82
788	LA	200	8/82
996	SF	100	10/82
1500	KC	350	2/83
634	LA	600	7/82

(b) Two Relations

PRICE

Component Part #	Description	Price
634	* * *	10
788	* * *	20
996	* * *	15
1500	* * *	5

INVENTORY LOCATION

Component Part #	Location	# In Stock
634	NY	50
788	LA	200
996	SF	100
1500	KC	350
634	LA	600

STOCK DELETION DATE

Component Part #	Stock Deletion Date
634	7/82
788	8/82
996	10/82
1500	2/83

(c) Three Relations

FIGURE 4-11. Normalizing the Network Data Bases.

some serious problems when updates and deletions are applied to the data base.

To illustrate, let us assume the price of COMPONENT PART #634 is to be changed from $10 to $12. In Figure 4-11(a), this update affects more than one row (tuple). The third normal form principle is violated because the x value (COMPONENT PART #) is associated with *more than one* y value (PRICE). This makes updating difficult because all tuples in the relation must be accessed and searched to ensure all values are correctly changed.

The solution to the update problem is to "normalize" the one relation to a PRICE relation and an INVENTORY relation, as in Figure 4-11(b). It can be seen that changing the price of a part will not affect any other value in *either* relation. (We are progressing toward a simpler organization.)

The two-relation data base still presents a violation of the third normal form principle. For example, COMPONENT PART #634 is to be removed from the active data base inventory on July 31, 1982 (7/82). With the present two-relation data base, the deletion of the first tuple of the INVENTORY relation removes *all* information on where the component is located. This may not be a desirable feature of the data base, since the stock still has financial value. To correct the deletion problem, the data model is normalized to create three relations [Figure 4-11(c)]. The INVENTORY LOCATION relation maintains the location of *all* parts, even after they are removed from the active inventory.

The reader may be puzzled about the duplicate COMPONENT PART # in the LOCATION relation of Figure 4-11(c): PART NUMBER 634 is located in New York and Los Angeles. Does this situation create ambiguity and violate the rules of relational models? The answer is no *if* the LOCATION relation uses the two attributes COMPONENT PART # and LOCATION to form the relation's *candidate* key.

Schemas and Subschemas

The structures in Figure 4-11(c) represent the schema of the INVENTORY data base. The schema provides a map or chart of the data, shows the names of the attributes, and establishes the relationships of data elements. It provides the overall view of the data base. It says nothing about the physical structure on disk or the physical access method. The term *subschema* refers to a user or programmer's map of the data base. A subschema is usually a subset of the schema.

For example, inventory control personnel would be interested in the shelf dates of components at various locations in order to answer the question, "What components are to be removed from the active inventory this month; where are they located and how many components will be removed?" This user's view can be satisfied by joining the LOCATION and STOCK DELETION DATE relations; the PRICE relation is not needed.

Typically, organizations have hundreds of subschemas and individual users often have multiple subschemas to satisfy different kinds of retrieval and update requirements. This presents a challenging problem for the data base and network designers: they must provide for a *physical* design that satisfies all user subschemas at all nodes in the network. The problem manifests itself in (a) the physical location of the data on the disk and (b) the physical location of the data within the network. In essence, the problem is translating logical views to efficient physical data structures.

Subschemas and Physical Data Structures in the Network. The first problem of the location of the data on the disk is illustrated in Figure 4-12. User A, the inventory control analyst, wants efficient, fast, and inexpensive access to his/her view. Conversely, User B, in the sales office, wishes the same service to some of the same data found in User A's subschema. Naturally, both users want minimum

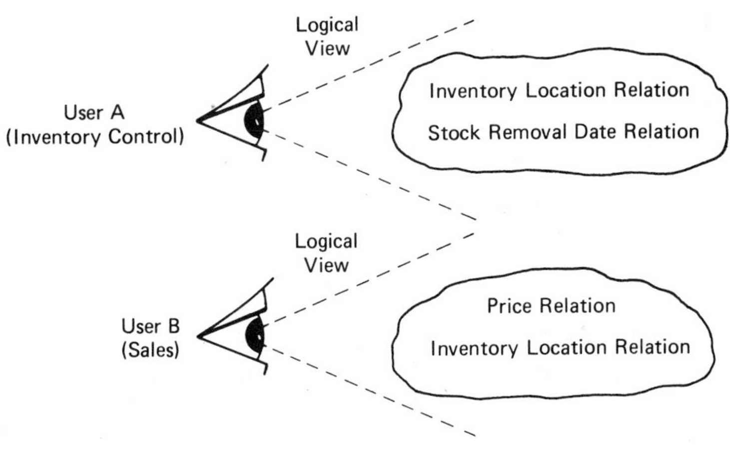

FIGURE 4-12. Subschemas and Physical Storage of Data on Disk.

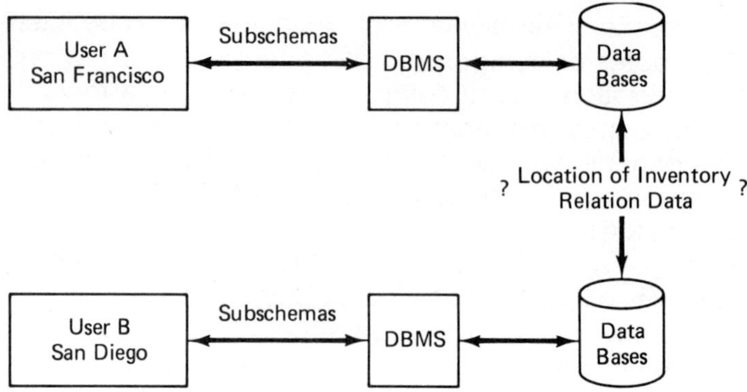

FIGURE 4-13. Subschemas and Network Location of Data.

access overhead to obtain the data. Placing the data in one *physical* block is the most efficient method. However, what is efficient for one user is not necessarily efficient for another. The data base designer must examine both subschemas, the relative priority between the users, their frequency of use, and time of use in order to implement the best physical layout on the disk. In many cases, the common data will be placed in separate physical blocks as a compromise solution.

This kind of situation should be kept transparent to the user community if at all possible. The DBMS systems of today can provide for considerable transparency. However, an organization must be careful not to penalize its users from the standpoint of disk storage costs, DBMS overhead, and extra I/O charges. In fact, the shared data bases should be given discounts (I/O, higher priorities, etc.) to encourage their use. While a DBMS may provide powerful capabilities to the users, if they appear to be expensive to use, the users will access the shared data base only to spin off specialized and redundant files for their exclusive use. Uncontrolled data redundancy presents very serious problems to organizations. Consequently, the translation of the user's subschema to the physical structure is an important task.

The translation of the subschemas to physical locations in the network is of even greater importance. As shown in Figure 4-13, our two users are located in different cities. Where should the commonly used data be stored, in San Francisco or San Diego? Can it be stored in both places? If so, how are the redundant data elements kept in synchronization between the two sites? How are hundreds of subschemas analyzed to determine the proper location of the network data bases? These questions and others are of paramount importance. Their answers determine, to a great extent, the quality and expense of the network and its ability to provide service to the end users.

The value of understanding the concepts of data base technology should now be evident. We will continue this subject in sections of Chapters 10 and 11. Those readers interested in pursuing the subject immediately can skip to these sections. The intervening material is not required to grasp the subject matter.

5

TRANSMISSION IMPAIRMENTS

FACTORS CONTRIBUTING TO ERRORS

The data contained in a message, packet, or frame does not always arrive correctly at the receiving site. For numerous reasons, a bit or several bits can become distorted or garbled during transmission. In this chapter, we explore the major transmission impairments that cause errors in the data, as well as preventive measures to decrease the probability of the errors occurring. In Chapter 6, we will learn how to deal with those error conditions that cannot be completely eliminated.

A typical voice-grade, low-speed line experiences a rate of one errored bit in every 100,000 bits transmitted (or $1{:}10^5$). Some user applications find this error rate acceptable and might choose to ignore an infrequent error. For example, a 120 Hz telegraph line could transmit for several hours and experience only one error during this period. This error rate would most likely be inconsequential for applications that transmit text. However, certain applications cannot tolerate any errors. For example, the loss of one bit in a transmission of financial data could have severe consequences for an accounting system. For these applications, the transmission must be made as error-free as possible.

Treating errors in a data communications system is a more difficult task than one might imagine. Four factors contribute to this situation:

- Distance between components.
- Transmission over hostile environments.
- Number of components involved in transmission.
- Lack of control over the process.

Distance Between Components

Computers and terminals connected by communications links may be located hundreds of miles from each other. The transmission speed of the signals between the sites can be very fast, as in a radio transmission (186 miles per 1 ms), or considerably slower, as in certain wire pairs (10 miles per 1 ms). Whatever the propagation speed may be, the distance introduces a delay.

To illustrate the point, consider a one-way satellite transmission. The up-link and down-link signals require a minimum of 240 ms propagation time (22,300 miles/186,000 miles per second = .120 × 2 = .240) and is most often around 270 ms, depending upon the location of the earth stations. A two-way transmission on a half-duplex link to effect a dialogue between the two sites then requires 540 ms of transmission time. Effective interactive systems should not have a response time of greater than an average of two seconds. Given this satellite path, designers have one-quarter of their window taken away solely from transmission propagation delay.

A shorter transmission delay of 20 to 30 ms on land links between the east and west coasts of the United States can affect performance and error control measures in many systems. For example, updating replicated data bases in a network requires that multiple update transactions be transmitted to all affected sites. The transactions arrive at the sites at different times and, relative to the speed of the computers and channel I/O operations, the arrival delays can be significant. To obtain consistency among the multiple copies, updates must be delayed at those sites receiving the transaction first. If a transaction is garbled during transmission, further delay is introduced while the error is analyzed and corrective action taken. Error analysis is made difficult due to the distance involved and the inherent delays. It is not unusual for an error condition on a transmission path to disappear after a few fractions of a second. Such an error would usually be identified in a centralized mainframe environment. Simply stated, the longer the delay in error analysis, the more difficult it is to identify and resolve the error.

Transmission Over Hostile Environments

When we consider the differences between the data communications and centralized mainframe environments, it becomes clear that data communications systems are more subject to error because of the operations in a hostile environment. A microwave signal is illustrative. During transmission it may encounter varying temperatures, fog, rain, and snow as well as other microwave signals that tend to distort the signal.

On the other hand, the flow of data inside a computer room is subject to strict temperature, humidity, and electromagnetic radiation controls. It is not surprising that the error rate on a channel inside a computer room is several orders of magnitude better than that of a voice-grade communications line.

Number of Components Involved in Transmission

A transmission through a communications system travels through several components and each component introduces the added probability of errors. For example, as a signal moves through the network, it must pass through switches, modems, multiplexers, and other instruments. If the interfaces among these components are not established properly, an error is likely to occur. The components themselves often introduce errors; for instance, some of Bell's older circuit switches can create considerable interference on the line. Moreover, line segments connected in tandem (to form an end-to-end channel) are more prone to error than one stand-alone link. Networks tend to have tandem links.

Lack of Control Over the Process

The classical centralized mainframe operating system (OS) exercises considerable control over its resources. Very little happens without the permission of the OS. In the event an error occurs, the operating system interrupts the work in progress, suspends the problem program, stores its registers and buffers, and executes the requisite analysis to uncover the problem. In a sense, the error and problem are frozen to simplify the analysis. A data communications network may not allow for this type of control. First, it is often impractical to suspend and freeze resources because they may be used by other components. Second, their condition may have changed by the time network control receives the error indication. Third, networks do not always operate under the tight centralized manner found in the centralized mainframe. For example, one computer in a network may not be allowed to control and analyze errors affecting it because they occur in other parts of the system.

UNITS OF MEASUREMENT AND OTHER TERMINOLOGY

The Decibel. The term *decibel* is used in communications to express the ratio of two values. The values can represent power, voltage, current, or sound levels. It should be emphasized that the decibel is (a) a ratio and not an absolute value, (b) expresses a logarithmic relationship and not a linear one, and (c) can be used to indicate either a gain or a loss. A decibel is 10 times the logarithm (in base 10) of the ratio:[1]

$$db = 10 \log_{10} P_1/P_2$$

Where db = Number of decibels.
P_1 = One value of the power.
P_2 = Comparison value of the power.

Decibels are often used to measure the gain or loss of a signal. These measurements are quite valuable for testing the quality of lines and determining noise and signal losses—all of which must be known in order to design the network. For example, suppose a communications line is tested at the sending end and receiving end. The P_1/P_2 ratio yields a reduction of the signal power from the sending to receiving end by a ratio of 200:1. The signal experiences a 23 db loss ($23 = 10_2 \log 200$). The log calculations are readily available from tables published in math books.

[1] A *logarithm* is really an exponent. For example, $2^3 = 8$ and $3 = \log_2 8$ are identical. The log of a number is the power to which some positive base must be raised to equal that number.

Transmission measurements may also need an absolute unit. The dBm is used for this purpose. It is a relative power measurement in which the reference power is one milliwatt (.001 watt):

$$dBm = \log_{10} P/.001$$

Where P = Signal power in milliwatts

This approach allows measurements to be taken in relation to a standard. A signal of a known power level is inserted at one end and measured at the other. A 0 dBm means 1 milliwatt.

Common carriers use a 1,004 Hz tone (referred to as a 1 KHz test tone) to test a line. The 1 KHz tone is used as a reference to other test tones of a different level. The test tone is used to establish a zero transmission level point (TLP). The TLP establishes a point at which the 1 KHz tone is expected. A +6 TLP is a point where the 1 KHz tone would be +7 dBm.

Resistance. The current running through a wire and an electromagnetic transmission in the atmosphere both encounter resistance. Particles in the wire and the atmosphere provide opposition to the signal. The reader may have felt a conducting wire and noticed it was hot—an indication that significant resistance existed. The resistance creates a loss of signal strength, described as *decay* or *attenuation*.

Inductance. A conductor need not be connected directly to another conductor to transfer a signal. For example, an increased current on a wire produces an expanded magnetic field outward from the conductor. This magnetic field can affect other wires and circuits and, in many instances, it can create an induced voltage in another component. Induction coils are used extensively to transfer signals and to step up or step down their voltages.

Capacitance. Materials tend to hold an electric charge even after the voltage source is disconnected. For example, wire cable tends to store charges between the wires in the cable.

Transmission impairments can occur due to "stray" capacitance and inductance. The components in data communications systems all have capacitive and inductive effects on each other. While the effects can be small, they can also cause data errors.

MAJOR IMPAIRMENTS

Transmission impairments can be broadly defined as random or nonrandom events. The random events cannot be predicted. Nonrandom impairments are predictable and, therefore, are subject to preventive maintenance efforts. The following list contains the major kinds of transmission impairments found on voice-grade circuits:

Random

White noise
Impulse noise
Crosstalk
Echoes
Phase changes
Intermodulation noise
Phase jitter
Radio signal fading

Nonrandom

Attenuation
Delay
Harmonic Distortion
Bias and Characteristic Distortion

Random Distortions

White Noise. Chapter 2 discusses how the *nonrandom* movement of electrons creates an electric current that is used for the transmission signal. Along with these signals, all electrical components also experience the vibrations of the *random* movement of electrons. These vibrations cause the emission of electromagnetic waves of all frequencies. The phenomenon is called *white noise* because it contains an average of all the spectral frequencies equally—just as white light does.

White noise is identified by other names (Gaussian, background, thermal, hiss). It forms a constant signal on the communications line against which the data signal must be sent. The power of white noise is proportional to temperature. This has important consequences for satellite transmissions because the transmitters can operate at reduced power if shielded from the sun.

As discussed in Chapter 2, Shannon's Law demonstrates that noise is one determinant of the information capacity of a communications line. The signal-to-noise ratio must remain at a level to keep the data signal separated from the noise. Figure 5-1 illustrates the process. The initial data signal is amplified and transmitted across the line. As will be explained shortly, its strength decreases as it traverses the communications path. Notice, however, that the white noise never drops below a certain level (at point 1). The data signal is amplified at point 2 and brought back to its original strength. The white noise is *also* amplified. In this illustration, the signal did not become mixed with the noise. (Once this happens, the two cannot be separated.) Obviously, the spacing of the amplifiers is an important consideration since improper spacing will allow the mixing of the noise and the data signal.

Impulse Noise. This impairment is a major cause of errors in a data transmission. The sources of impulse noise are many. All unwanted electronic effects, such as voltage changes, dialing noise, dirty electrical contracts, and movement of poorly connected electrical joints are contributing factors to impulse noise. The telephone systems older step-by-step switches are also a major source of impulse noise.

The effect of impulse noise can be seen in Figure 5-2. The original binary bit stream represents the base 10 number 281. In Figure 5-2(a), the data is received as it is sent. If impulse noise is introduced, as in Figure 5-2(b), the bits are altered and the number 281 is changed to represent the number 347, as in Figure 5-2(c).

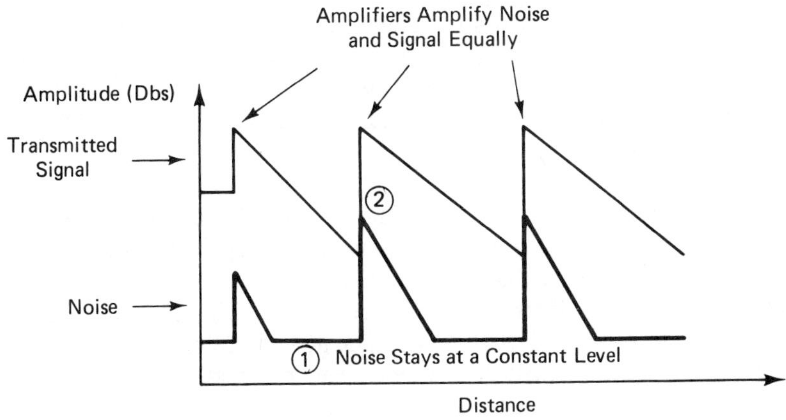

FIGURE 5-1. Signal and Noise Relationship.

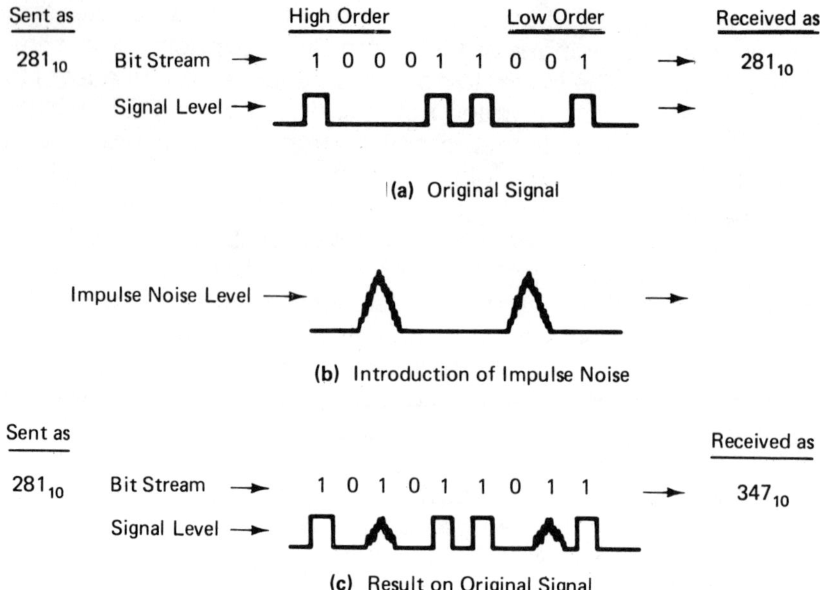

FIGURE 5-2. Effect of Impulse Noise.

The problem is actually more complex than this illustration shows. A one-bit error can usually be detected and corrected at the receiving site. A typical impulse noise lasts approximately 10 ms. Consequently, a 4800 bit/s transmission would have about 50 bits affected by the 10 ms impairment; a 9600 bit/s line would lose about 100 bits. In either case, parity checking would not work and error correction would not be possible. The message would have to be retransmitted.

Cross Talk. Most of us who use the public telephone network have experienced the interference of another party's faint voice on our line. This is cross talk—the interference of signals from another channel. Cross talk is usually a minor irritant and does not present major problems in a data communications network. One source of cross talk is in physical circuits that run parallel to each other in building ducts and telephone facilities. The electromagnetic radiation of the signals on the circuits creates an inductance effect on the nearby circuits.

Echoes. Almost everyone using a telephone has also experienced echoes during a conversation. The effect sounds like one is in an echo chamber—the talker's voice is actually echoed back to the telephone handset. Echoes are caused by the changes in impedances in a communications circuit. (*Impedance* is the combined effect of inductance, resistance, and capacitance on a signal at a particular frequency.) For example, connecting two wires of different gauges could create an impedance mismatch. Echoes are also caused by circuit junctions that erroneously allow a portion of the signal to find its way into the return side of a four-wire circuit.

Figure 5-3 shows one way in which echoes occur on a voice-grade line. At the top of the figure (point 1), the signal comes into a circuit for transfer to a two-wire path (point 2). The signal enters a junction called a *hybrid coil*. This device prevents excessive feedback across the junction of the returning lines (point 3). In addition, the junction contains a balancing network that balances the impedance matches of the lines and the hybrid coils. However,

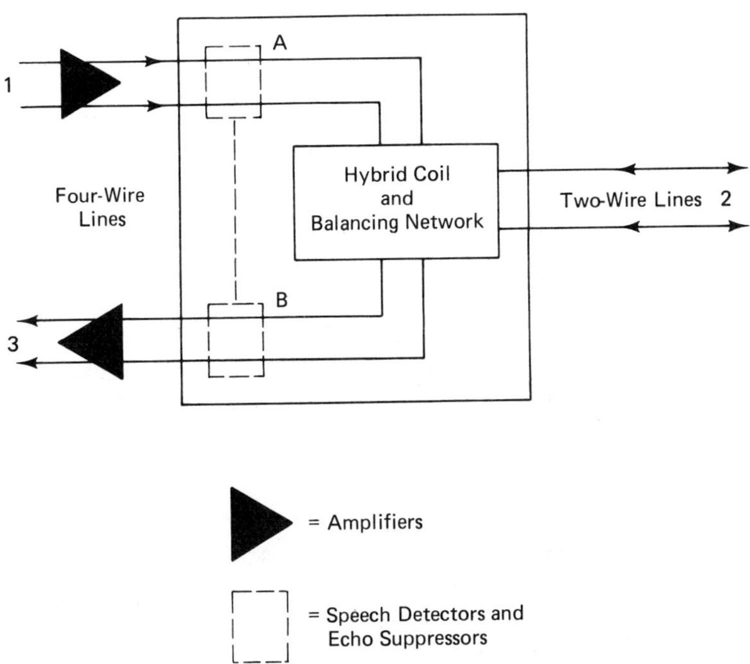

FIGURE 5-3. Echoes.

some of the signal, through inductance, may find its way into the return circuit. It is then strengthened by the amplifier at B, thus creating an echo on the line.

An echo is often not noticed. The feedback on a short-distance circuit happens so quickly that it is imperceptible. Generally, a echo with a delay of greater than 45 ms (.045 seconds) presents problems. For this reason, long-distance lines and satellite links employ devices to reduce the strength of the return signal. These devices, called *echo suppressors*, are activated by speech. For example, in Figure 5-3, the speech detector at A activates the echo suppressor at B.

Echo suppressors cannot be used for data transmission over a voice line. The speech detector has been designed to detect speech signals. Moreover, the delay in reversing the activation of the suppressors at each end often causes the clipping of the first part of a signal. The clipping effect is probably familiar to the reader. It does not usually present serious problems in a voice transmisson, but would cause bit distortions in a data transmission.

Since the telephone network is a primary facility in a data communications network, the echo suppressors must be disabled for data transmissions. This is accomplished by the transmission of a 2000 to 2250 Hz band signal for approximately 400 ms. The suppressor is deactivated until no signal is on the line for about 50 ms. In the event the transmission of messages is not continuous, a signal must be placed on the line to keep the suppressors disabled. Carrier signals can accomplish this purpose. The reader may have had an occasion to dial up a computer. Upon completing the dial, a high-pitch tone can be heard. This tone indicates that the caller has a connection to the computer and it also disables any echo suppressors that may be on the line.

Phase Changes. The phase of a signal is sometimes changed by an impulse noise. The change is usually of a short duration—on the order of a few hundred microseconds (μs). The phase may slip and then return back to the original phase. If it does not, it can create data errors and is especially perceptible in PM modems or modems employing QAM methods.

Intermodulation Noise. Chapter 3 describes the common carriers' frequency multiplexed systems wherein many voice-grade circuits are modulated onto a high capacity link, such as coaxial cable and microwave. These channels can interfere with each other if the equipment is slightly unlinear. Typically, two signals from two separate circuits combine (intermodulate) to form a frequency band reserved for another circuit. The reader has likely heard intermodulation noise during a telephone call; it sounds like a jumble of low-speaking voices, none clearly perceptible.

Intermodulation noise can occur in the transmission of data when a modem uses a *single* frequency to keep the line synchronized when data is not being sent. The single frequency may actually modulate a signal on another channel. This problem can be avoided by either transmitting a variable-frequency signal or transmitting a signal of low amplitude. Intermodulation noise can also stem from the data within the message transmission. A repetitive code in the transmission could create the problem. In Chapter 7 (Digital Transmission), we will examine coding schemes to eliminate a repetitive code transmission.

Phase Jitter. Occasionally, a signal's phase will jitter, causing an ill-defined crossing of the signal through the receiver. Noise-laden signals resemble jitter but are caused by different impairments. Jitter is created by a multiplexed carrier system that creates a forward and backward movement of the individual frequency. Phase jitter is not a very serious problem for most types of transmissions. It is more serious at higher transmission speeds.

Radio Signal Fading. Microwave transmissions are particularly subject to fading. This impairment occurs in two ways. The first, selective fading, occurs when the atmospheric conditions bend a transmission to an extent where signals reach the receiver in slightly different paths (Figure 5-4). The merging paths can cause interference and create data errors. Other channels in the microwave transmission are not affected by a selective fade. Therefore, backup channels are usually provided to allow for protection against this problem.

Flat fading is a more serious problem because it can last several hours and alternate channels will not provide relief. Flat fading occurs during fog and when the surrounding ground is very moist. These conditions change the electrical characteristics of the atmosphere. A portion of the transmitted signal is refracted and does not reach the receiving antenna.

Nonrandom Distortions

Attenuation. The strength of a signal attenuates (decays) as it travels through a transmission path. The amount of attenuation depends upon the frequency of the signal, the transmission medium,

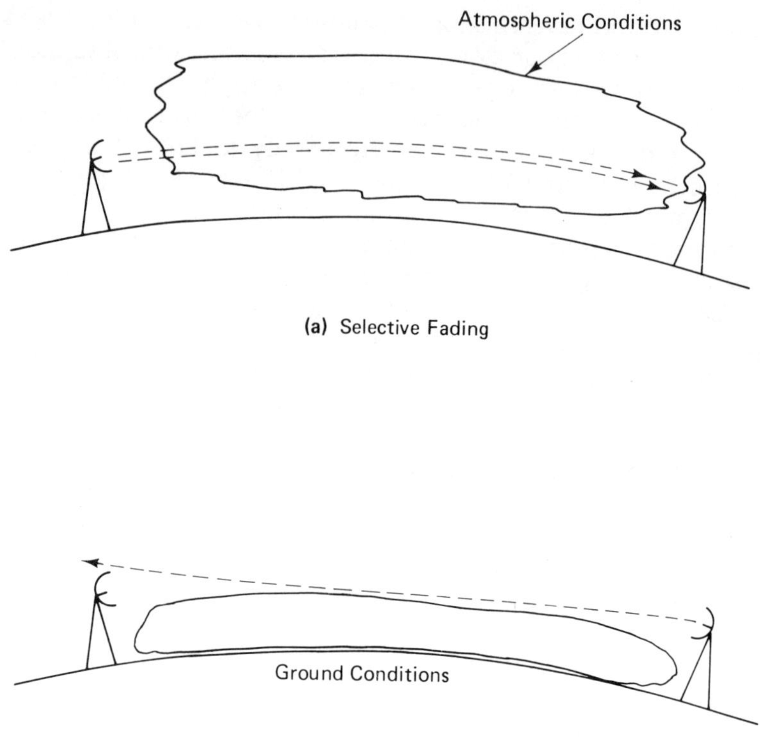

FIGURE 5-4. Microwave Fading.

and the length of the circuit. Unfortunately, signal attenuation is not the same for all frequencies. A private voice-grade line experiences attenuation across the bandwidth of about 10 decibels. Figure 5-5 shows this effect in relation to 1000 Hz. It can be seen that signal loss increases at the higher frequencies. Consequently, those media that use high-frequency bands (coaxial cable, microwave) require the signal to be strengthened (with analog amplifiers or digital repeaters) more often than a low-frequency open-wire pair. Generally speaking, it is desirable to amplify a signal after it has been attenuated by 20 decibels. Given this requirement, an 8 KHz coaxial cable with an attenuation of 10 decibels per mile would require signal amplification every two miles.

Delay. A signal is comprised of many frequencies. These frequencies do not travel at the same speed and, therefore, arrive at the receiver at different times. Excessive delays create errors

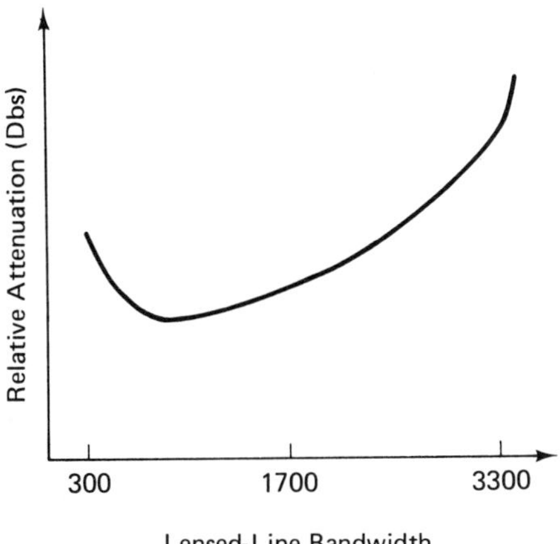

FIGURE 5-5. Attenuation.

known as *delay distortion*. Figure 5-6 shows delay in relation to 1800 Hz. The problem is not serious for voice transmissions because a human readily adjusts to the signal. However, delay distortion creates problems for data transmissions.

Bias and Characteristic Distortion. This problem occurs when the sampling threshold of a digital transmission is not in proper synchronization with the actual digited bit stream. (Chapter 7

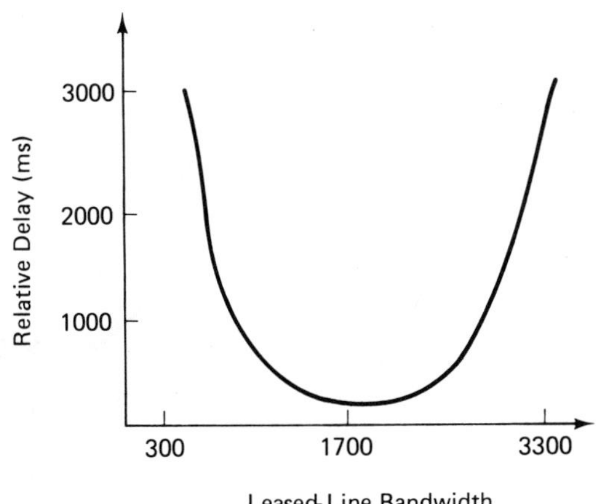

FIGURE 5-6. Delay.

discusses digital transmission.) Digital pulses such as those depicted in Figures 2-4 and 2-7 are actually created with less definition than these pictures suggest. For example, the effect of the building up of the signal strength creates a signal on a telegraph line like that shown in Figure 5-7. A period of time is required for the signal to manifest itself and also for it to diminish once the voltage is removed. The periods of signal buildup and decay are called *transient times*. The slight delay in these changes can create errors if the sampling threshold and sampling periods are not established properly.

ERROR CONTROL METHODS

Given that transmission impairments exist, what can be done to mitigate their effect? In this section we examine several options and choices that can reduce substantially the data errors that result from the transmission distortions.

An individual component in a communications system may be quite reliable and relatively error-free. For example, AT&T sets a goal of channel outage on a long-distance microwave link at 0.0002 annually—less than two hours. Yet this impressive performance means little if all components in the system are not equally error-free. Like the links in a chain, the communications network is only as strong as its weakest component. Figure 5-8 shows a typical, rather simple data communications configuration. Twenty-two components complete the full setup. Let us assume that each component is reliable enough to give 99.9% availability. (Let

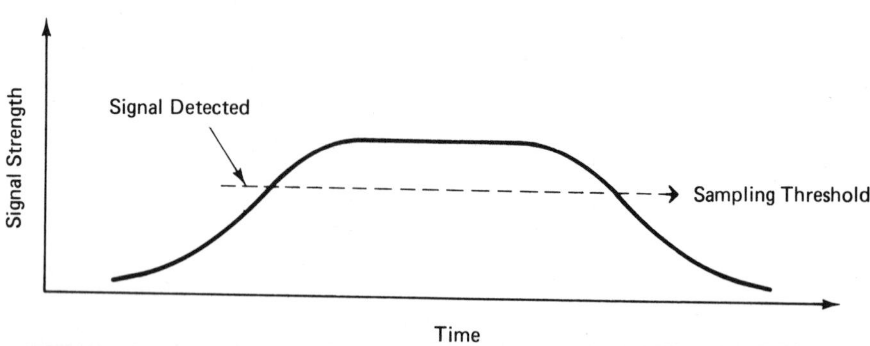

FIGURE 5-7. Signal Buildup.

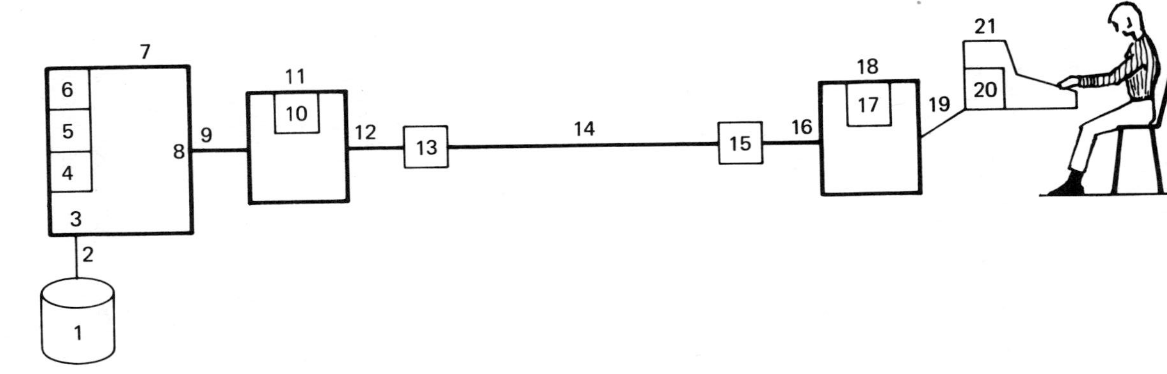

1. Disk Data Set
2. Channel
3. Channel Software
4. Data Base Management System
5. Applications Program
6. Operating System
7. Computer Hardware
8. Channel Software
9. Channel
10. Front End Software
11. Front End Hardware
12. Port Interface
13. Modem
14. Communications Line
15. Modem
16. Port Interface
17. Terminal Controller Software or Microcode
18. Terminal Controller Hardware
19. Cable
20. Terminal Software (possibly)
21. Terminal Hardware
22. Human Operator

FIGURE 5-8. Possible Error Points.

us also exclude point 22, the human operator, since we can never expect that kind of availability.) A 99.9% availability might be considered adequate for certain systems, but this figure is for each individual component. Since the components are in tandem, the overall availability of the system must be computed by multiplying each availability number together, yielding .979. Obviously, 97.9% availability is substantially different from 99.9%. If one component is down, it makes little difference, from the standpoint of availability, if the others are up and running.

Murphy's Law states that if anything can go wrong, it will. In developing error control methods for a data communications system, some of Murphy's corollaries should all be kept in mind:

- Nothing is as easy as it looks.
- Everything takes longer than you think.
- If there is a possibility of several things going wrong, the one that will cause the most damage will be the one to go wrong.
- The probability of anything happening is in inverse ratio to its desirability.
- Murphy was an optimist.

Perceived Availability

Murphy's Law notwithstanding, the primary objective in devising error control methods is to provide perceived availability to the user. This means that an error or malfunction in the network remains invisible—the user perceives that the system is fully operational. From the user's point of view, perceived availability is achieved by the network providing optimum performance in:[2]

Mean time between failure (MTBF).

Mean time to recover (MTTREC).

Mean time to repair (MTTREP).

MTBF should be increased to the greatest extent possible. This is accomplished in two ways. First, failures are reduced to the

---

[2] Richard A. Lang and Richard E. Pigman, "Making DDP Network Failures Invisible," *Data Communications*, April 1980.

maximum extent possible and, second, the scope of effect of the failures is kept as isolated as possible.

MTTREC is kept as low as possible. In the event of a failure, the recovery should be fast. For example, redundant components should assume network functions without perceptible delay. In the event a component fails, MTTREP requires rapid diagnosis of the problem and facilities that provide rapid corrective action.

The MTBF, MTTREC, and MTTREP performance factors should be continuously monitored. The installation should establish performance thresholds against which the system is measured. The statistics should provide for trend analysis as well in order to identify potential trouble areas for preventive maintenance.

Conditioning

The common carrier provides measures to improve the quality of leased lines in the form of line conditioning. The following impairments are addressed with conditioning:

C—Conditioning
  Attenuation distortion
  Delay distortion

D—Conditioning
  Noise
  Harmonic distortion

Conditioning is purchased at several levels. C-conditioning is available in five varieties and D-conditioning has two levels. C1, C2, and C4 apply to private leased lines; C3 and C5 are used on lines in a larger configuration, such as a dedicated network and Bell's newer switching systems. D1 applies to point-to-point lines; D2 is used on two- or three-point lines.

C-conditioning provides measures to diminish the problems of attenuation and delay but it does *not* remove the impairments. Rather, it provides for more consistency across the bandwidth. For attenuation, the common carrier introduces equipment that attenuates those signals in the bandwidth that tend to remain at a higher level than others. Thus, attenuation still occurs but is more evenly distributed across the channel. The same idea is used in delay. The faster frequencies are slowed so that the signal is more consistent across the band.

Each C-conditioning level establishes performance thresholds (see Figure 5-9). For example, in type C1, delay distortion shall not exceed a maximum difference of 1,000 μsecs between 1,000 and 2,400 Hz, and attenuation of a 1,000 Hz reference signal between 300 and 2,200 Hz shall not exceed −2db to +6db.

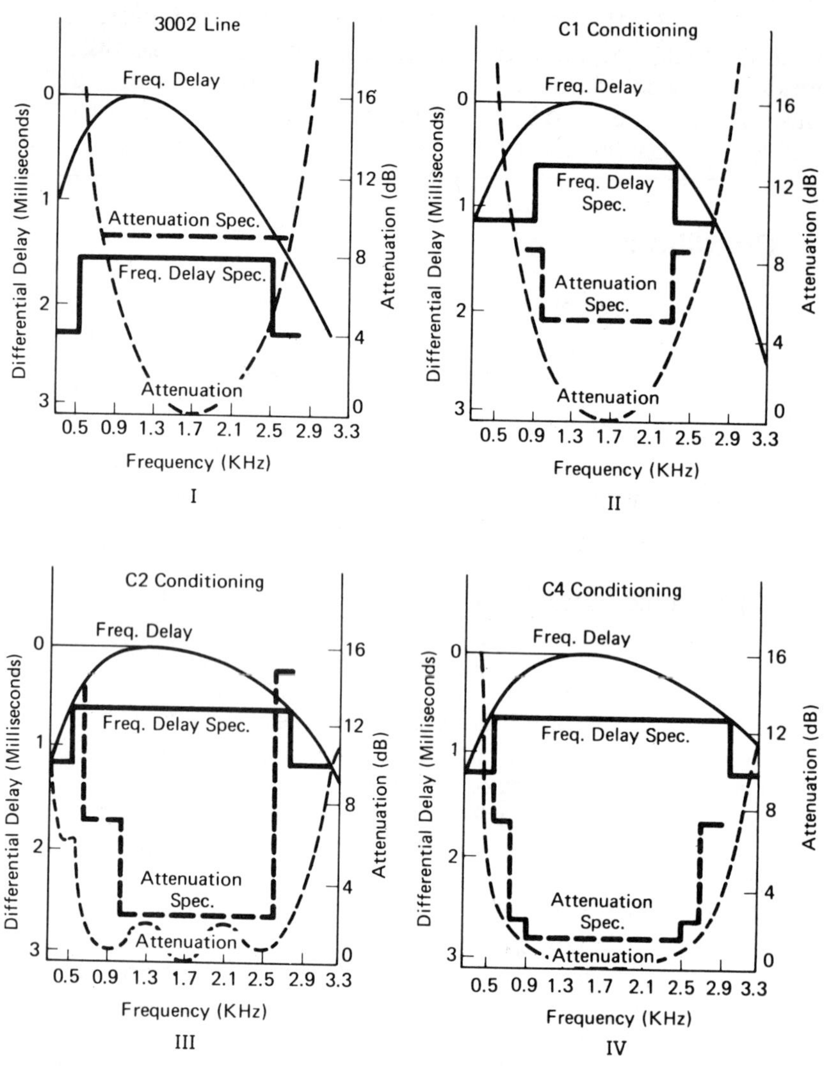

FIGURE 5-9. Conditioning. (From *Data Communications: A User's Guide* by Kenneth Sherman. Reston VA: Reston Publishing Company, 1981, p. 210. Reprinted with permission.)

D-conditioning specifies limits for lines operating with 9.6 kbit/s speeds with respect to the signal-to-noise ratio and harmonic distortion. Unlike the impairments addressed in C-conditioning, these problems cannot be reversed but their effects can be diminished by the carrier selecting high quality circuits and switches for the leased line connection. Figure 5-10 summarizes both conditioning offerings.

The reader should also be aware that many modems on the market today obviate purchasing the common carriers' C-conditioning services. These modems provide for manual or automatic equalization of the signal across the bandwidth.

<table>
<tr><th></th><th>Basic AT&T Internal Control</th><th>Areas of C-Condition Control</th><th>Areas of D-Condition Control</th></tr>
<tr><td>Attenuation Distortion</td><td>X</td><td>X</td><td rowspan="2">Same as Basic</td></tr>
<tr><td>Envelope Delay Distortion</td><td>X</td><td>X</td></tr>
<tr><td>Signal to Noise Ratio</td><td>X</td><td rowspan="2">Same as Basic</td><td>X</td></tr>
<tr><td>Harmonic Distortion</td><td>X</td><td>X</td></tr>
<tr><td>Impulse Noise</td><td>X</td><td colspan="2" rowspan="4">Same as Basic 3002</td></tr>
<tr><td>Frequency Shift</td><td>X</td></tr>
<tr><td>Phase Jitter</td><td>X</td></tr>
<tr><td>Echo</td><td>X</td></tr>
<tr><td>Phase Hits
Gain Hits
Dropouts</td><td colspan="3">Not Controlled</td></tr>
</table>

(a) Conditioning Control

Levels	
C1	Point to Point
C1	Multipoint
C2	Point to Point
C2	Multipoint
C4	Point to Point
C4	Multipoint
D1	Point to Point
D2	2 or 3 Point

(b) Conditioning Levels.

FIGURE 5-10. Conditioning.

Testing and Monitoring

The majority of communications networks have a network management center (NMC). The NMC is responsible for the reliable and efficient operation of the network. It monitors and tests the communications systems and repairs or replaces failed components. The NMC is also responsible for day-to-day preventive maintenance operations in the network. It uses a wide variety of tools and techniques to keep the network running smoothly and to provide perceived availability to the user. The NMC may use the following equipment; all described here have overlapping functions.

Typical Diagnostic Equipment

Transmission Test Monitor. This instrument monitors the data stream on a communications line. The vendors' products vary but a test monitor usually provides the following functions:

- Simulates a data stream from a modem, terminal, or front end.
- Operates and detects parity errors; tests operation and timing of modem connections (e.g., the RS232-C RTS, CTS, CD, DTR, and RI pins).
- Verifies that polls are functioning and the timing is correct.
- Counts carrier signal dropouts.
- Sends test data patterns and verifies their accuracy.

Network Control System. This machine provides the same functions as the test monitor. It also provides facilities for backing up failed components. It allows alternate components to be switched into the network and it patches a variety of test equipment to check the fault. The control system may also provide qualitative analysis of line impairments such as harmonic distortion, white noise, line drops, and impulse noise. It may also have the capability to switch and monitor both digital and analog facilities.

Pattern Generator. A simpler device than the test monitor and control system, the pattern generator usually consists of two devices; one generates signals and the other analyzes them. The generator produces bit patterns (usually 63,54,2047 bits) of various codes and parities. It can also introduce bias and characteristic distortions in variable amounts. The analyzer receives the bit patterns and observes any problems to determine if the modem or terminal is functioning properly.

The error rate is described by a bit error rate. (BER = number of bits received in error/total bits received) and is used to determine the probable error rate of a physical block of bits (BLER = number of blocks received with at least one bit in error/total blocks received). We have used the terms *message*, *packet*, or *frame* to describe a block of bits.[3]

The diagnostic equipment industry is quite diverse. The offerings provide some or all of the functions described above. Again, no exact distinction exists in classifying and describing the equipment. The reader is faced with many choices in this area of data communications networks.

Loopbacks

A failed communications link is tested by placing the modems in a loopback mode. Practically all modems can be put into loopback tests with a switch on the modem (see Figure 5-11). The loopback signals are analyzed to determine their quality and the bit error rate resulting from the tests. The loopbacks can be sent through the local modem to test its analog and digital circuits; this test is called a *local loopback*. If the bit error rate is not beyond a specified level, the next step is likely to be a *remote analog loopback* that tests the carrier signal and the analog circuitry of the remote modem. The remote modem must be placed in the loopback mode in order for this test to be completed. Care should be taken in drawing conclusions from the remote analog loopback tests since the looped signal is being tested at twice its specification. Due to this problem, it is often advisable to send the signal through the remote modem digital circuitry in order to boost the signal power. This test is called a *remote digital loopback*.

---

[3]The block size of a message is an important design consideration since the larger block sizes are more likely to contain an errored bit. Smaller-size message blocks are less likely to encounter an error but decrease overall throughput of user data bits due to the additional overhead of control headers and trailers in each message block. Thus, the designer must determine the optimum block size from the standpoint of BLER rate and overhead/throughput, as well as delay/response time. (Refer to page 91 for a review of block size and delay.)

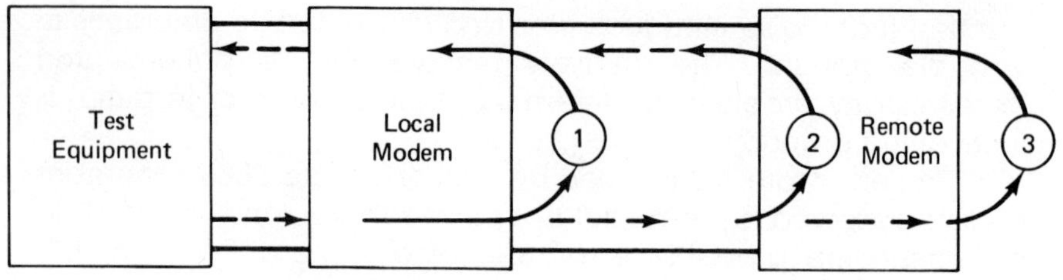

(1) Local Loopback

(2) Remote Analog Loopback

(3) Remote Analog/Digital Loopback

FIGURE 5-11. Loopback Tests.

Tests are also performed with equipment to verify the proper functioning of the components' protocol logic, primarily in the terminal (DTE) and not the modem (DCE). The equipment generates the protocol's control messages to the attached terminals and tests the content and timing of the responses. The testers also verify that the attached terminals are properly checking for transmission errors and, in many models, perform tests on the terminal's RS232-C Request to Send channel (RTS). Other protocol tests also generate bit patterns for BER and BLER tests.

6

DATA LINK CONTROLS

DATA ACCOUNTABILITY AND LINE CONTROL

Since certain errors are inevitable in the system, a method must be provided to deal with the periodic data distortions that occur within a message. The data communications system must provide each site with the capability to send data to another site. The sending site must be assured that the data arrives error-free at the receiving site. The sending and receiving sites must maintain complete accountability for all messages. In the event the data is distorted, the receiving site must have the capability of notifying the originator to resend the erroneous message or otherwise correct the errors.

The movement of these messages to and from the many points within the network must flow in a controlled and orderly manner. This means that the sending and receiving sites must know the identification and sequencing of the messages being transmitted among all users. The connection path between sites is usually shared by more than one user (as in a multipoint configuration); consequently, procedures must provide for the allocation and sharing of the path among the many users.

Data link controls (DLCs) provide for these needs. They manage the flow of data messages across the communications path or link. DLCs consist of a combination of software and

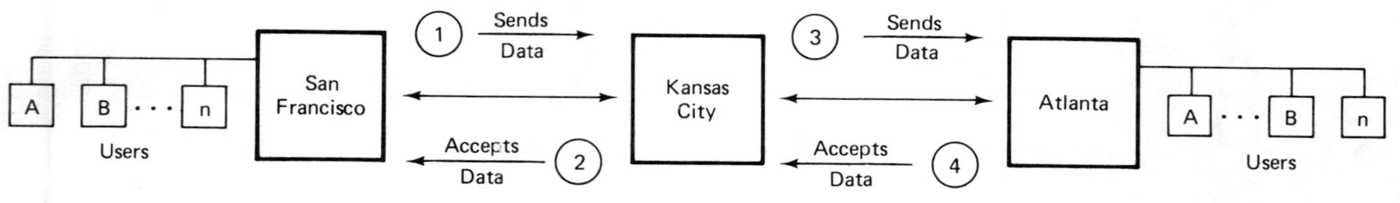

FIGURE 6-1. Message Flow in a Network.

hardware and are located at each site in the network. The DLC is concerned with providing the following functions to the network:

- Synchronizing the sender and receiver.
- Controlling the sending and receiving of data.
- Detecting and recovering transmission errors between two points.
- Maintaining awareness of link conditions.

A multinode network that has intermediate points between a session of two users would operate as depicted in Figure 6-1. User terminal A in San Francisco sends data across the network to user terminal B in Atlanta as part of user A/B session. The data could, for example, be passed to an intermediate node at Kansas City (Event 1). The DLC in Kansas City receives the data, checks for errors, and sends a receipt acknowledgment (ACK) of the data to San Francisco (Event 2). Kansas City then assumes responsibility for this data. In Event 3, Kansas City sends the data to Atlanta which checks the data and sends an acceptance response to Kansas City (Event 4). Thus, the DLC relays the message through the network, much like the passing of a baton in a relay race.

The DLC does not provide the user with end-to-end accountability. User A did not receive any indication of receipt of data in Atlanta, the final end point. Since data link controls do not provide for end-to-end access and flow control, a higher level of control is required to provide for session-to-session accountability and control. This topic is discussed in Chapter 8.

CLASSIFICATION OF DATA LINK CONTROLS

Data link controls can be described and classified by (a) message format, (b) line control method, (c) error handling method, and (d) flow control procedure.

Message Format

Asynchronous. The message format of a DLC is either asynchronous or synchronous. (Refer to Chapter 2 for an explanation of asynchronous and synchronous transmission.) Asynchronous formats originated with older equipment that had limited capabilities, but are still widely used due to their simplicity. Their main disadvantage is the overhead of the control bits; for example, a ratio of three control bits to eight user character bits is not unusual. Notwithstanding, many variations and improvements have been made to asynchronous protocols and, in spite of the overhead, DLCs with asynchronous formats still remain one of the dominant data link controls. Asynchronous techniques are found in practically all systems that use teleprinter and teletype terminals.

Synchronous Byte-Oriented. The DLCs with synchronous formats are further distinguished as byte- or bit-oriented formats. The synchronous byte-oriented DLC uses the same string of bits and bytes to represent data characters and control characters. For example, the control field EOT (End of Transmission) might occur in a user data stream, in which case the DLC logic could mistakenly interpret the data as control information. Recent versions of synchronous byte techniques have logic provisions to handle this problem.

Synchronous Bit-Oriented. The more advanced synchronous DLCs use the bit-oriented approach. In this method, line control characters are always unique and cannot occur in the user data stream. The logic of the DLC examines the data stream before it is transmitted and alters the user data if it contains a bit configuration that could be interpreted as a control field. Of course, with this approach, the receiving DLC has the capability to change the data stream to its original contents. The bit-oriented DLC achieves code transparency. This means the logic is not dependent upon a particular code such as ASCII or EBCDIC.

Line Control Method

Polling/Selection. The most common method of line control is through the use of polling/selection techniques. This process is illustrated in Figure 6-2. A site in the network is designated as the master or primary station. This site is responsible for the sending and receiving of messages between all secondary or slave sites on the line. In fact, the secondary sites cannot send any messages until the master station gives approval.

Let us suppose site B wishes to send data to site A on the San Francisco master station line. The master station begins this process by sending a polling message to B (Event 1). In effect, the poll message says, "Site B, have you a message to send?" Site B

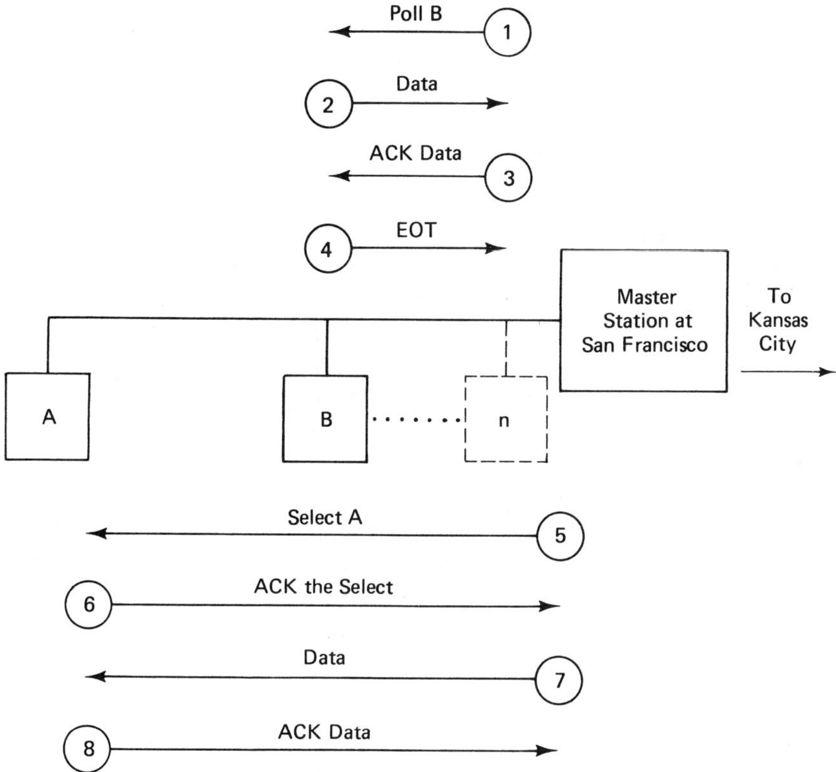

FIGURE 6-2. Polling/Selection Techniques.

responds by sending a message to the master site (Event 2). The master site checks the message for errors and responds back to B with a positive acknowledgment (ACK) as Event 3.

There are many variations of the polling/selection technique. For example, some vendors' implementations allow B to continue sending messages, eventually terminating the process with an End of Transmission (EOT) message (Event 4). Others provide an EOT indicator within the final message itself, which eliminates Event 4 and an overhead message.

The selection process begins in Event 5. Now the master site informs A that it has data destined for A by sending a selection message. This message means, "Site A, I have data for you; can you receive?" Site A must respond with a positive acknowledgment (ACK) or negative acknowledgment (NAK) as in Event 6. The receiving station may not be able to accept a message. For example, it may be busy or its storage buffers may be full. If A can receive the message, the master site transmits it (Event 7). The message is checked for errors at A and an acknowledgment is relayed to the master station (Event 8). User A could then respond to user B by going through the master station and executing the process again.

It's useful to compare Figure 6-1 with Figure 6-2. In Figure 6-2, the master site in San Francisco controls the user terminals within its local environment. In Figure 6-1, the intermediate node in Kansas City controls the San Francisco and Atlanta points. This means that the polling/selection approach allows a site to be a master to one part of the network and a slave to another part.

The polling/selection protocol is very widely used for several reasons:

- The centralized approach allows for hierarchical control. Traffic flow is directed from one point, which provides for simpler control than a noncentralized approach.
- Priorities can be established among the users. Certain terminals can be polled or selected more frequently than others, thus giving precedence to certain users and their applications.
- Sites (terminals, software applications, or computers) can be readily added by changing polling/selection tables within the DLC logic.

There is a price to pay for all these features. The polling/selection DLC incurs a substantial amount of overhead due to the requirement for polling, selection, ACK, NAK, and EOT control messages. On some networks, the negative responses to polls (in which a terminal or user application is solicited for data but has nothing to send) can consume a significant portion of the network capacity. We shall see that the more recent implementations of the polling/selection DLC use some very clever methods to reduce the number of overhead messages.

Timeouts. Timeouts allow the link control station to check for errors or questionable conditions on the line. A timeout occurs when a polled station does not respond within a certain time. The nonresponse condition evokes recovery action on the part of the controlling station.

The timeout threshold is dependent upon three factors: (a) signal propagation delay to and from the polled station, (b) processing time at the polled station, and (c) turnaround delay at the polled station (raising the clear-to-send circuit) on a half-duplex line.

These factors are highly variable and depend upon the line type, line length, modem performance, and processing speed at the polled terminals' site. A timeout threshold for a network using leased full-duplex lines operating within a distance of 150 miles between the polling and polled site might range between 30 to 60 ms. The threshold for the same arrangement, using satellite links, could be as great as 900 ms due to the longer propagation time and additional delays at intermediate points (earth stations). Local networks operating within small confines might use a timeout threshold of a very few milliseconds or microseconds. It is evident that the personnel tasked with the software or chip design of the link control method must work closely with the communications engineers, or have an understanding of the characteristics of electronic signaling, in order to design the appropriate timeout logic.

Hub polling. A variation of polling/selection is known as hub polling. This approach is used on a multipoint line to avoid the delay inherent in the polled terminals turning around a half-duplex line to return a negative response to the poll. In this operation, the master station sends a poll to a terminal on the line. This terminal turns the line around with a message to the master station if it has data to send. If it has no data, it sends the polling message to the next terminal on the line. If this terminal is busy, idle, or has nothing to send, it relays the polling message to the next appropriate station. The transmission of this poll continues in one direction without additional turnarounds. Eventually, a terminal will be found that has data for the master station. Thus, hub polling eliminates the line turnaround time that occurs if each terminal receives a poll from the host.

Contention. Contention is another widely used link control method. It differs significantly from polling/selection since there is no master station. With contention, each site has equal status on the line and the use of this path is determined by the station that first gains access during an idle line period. Contention DLCs must provide for a station to relinquish the use of the path at an appropriate interval of time in order to prevent line domination from one site.

An example of contention controls is provided in Figure 6-3. The path is multidropped with sites A, B, C, and D on the line. Each site, using a line signal sensing device, determines if a message is traveling on the path (through sensing the electrical signal). If there is a message, the sites defer to it. Since the messages are being transmitted at very high speeds, the waiting periods are usually quite short.

Figure 6-3 shows that site D has transmitted a message destined for A. Sites B and C monitor the line, determine if it is occupied, and wait a brief period before sensing again. If both B and C have a message to transmit, each will attempt to gain access to the path after the message from D is received at A. Assuming site C gains access first, site B must wait until C has completed the transmission of its data. Like the polling/selection method, user A will check for errors and send an ACK or NAK message to D when the path becomes available. Remember that site A recognizes its message by address detection.

The contention control method experiences occasional collisions of messages. This occurs when more than one site senses an idle line and transmits messages at approximately the same time. The messages' signals intermix and become distorted. In these situations, the sensing devices must be capable of detecting the collision and must so indicate to the DLC logic. The logic must then direct the stations to resend the messages that have been distorted.

Contention control is widely used today, primarily because of its relative simplicity and the absence of a master station. The polling/selection approach suffers from the vulnerability of the

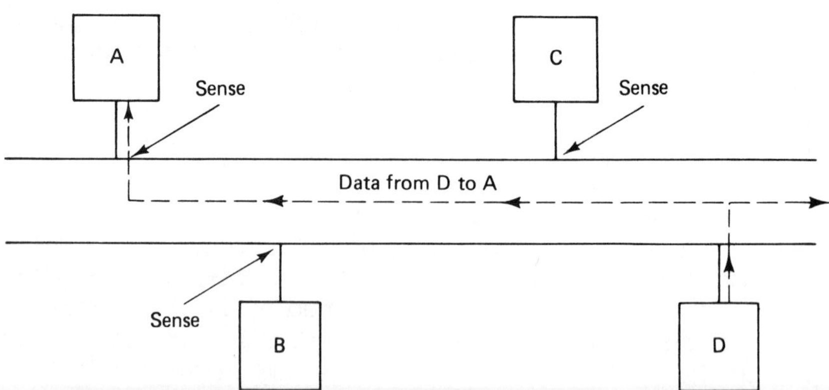

FIGURE 6-3. Contention Data Link Control.

primary site to failure, which could bring down all sites on the link. Since the contention method does not rely on a controlling site, a failed site does not prevent the other sites from communicating with each other. Moreover, the many overhead messages (polls, selections, etc.) found in polling/selection methods do not exist in the contention DLCs.

One major disadvantage of a contention DLC is the inability to provide priorities for the use of the transmission path. Since all stations are equal, none have priority over others—even though some user stations and applications may require greater use of the facilities. Since many of the contention DLCs are used on local networks with very high-speed paths, the equal allocation may not be discernible to the station with more frequent access needs.

Another potential disadvantage of the contention network is the distance limitation placed on it. For example, if two sites are located at a remote distance from each other, it is possible for both stations to transmit, turn themselves "off," (i.e., go on to other activities), and never detect the signal or the collision due to the propagation delay of the signals traveling on the line. The contention logic must be designed to handle this situation.

Time Slots. Time slot control avoids the collision problem found in the contention DLC by reserving times of access to the communications path. For example, in Figure 6-4 each site or station is given a slot of time on the link. During the period that the station has access to the path, it typically can send one or a predetermined multiple number of messages. The next station then gains access and transmits its messages across the communications link.

Time slot link controls are simple and are found in many applications and networks today. Their principal disadvantage is the wasted line capacity that occurs when a station's time slot is not used because it has nothing to transmit. Many time slot DLCs now avoid this problem with additional capabilities. The discussions on multiplexers, local networks, and satellite transmission techniques provide more information on time slot control methods.

Error Handling Method

The method of detecting and correcting errors in a message is a key selection criterion for a data link control. Some vendors have more efficient techniques than others. The majority of offerings provide for one of the methods discussed here.

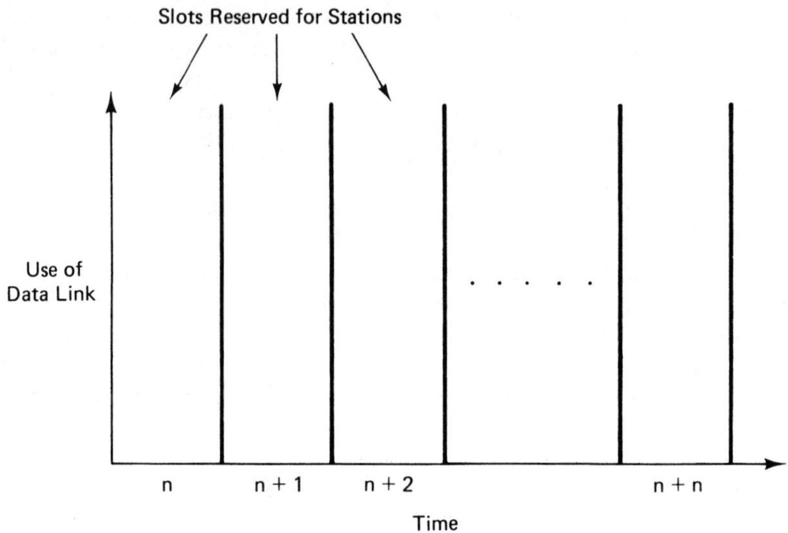

FIGURE 6-4. Time Slot Controls.

Certainly, one option is to simply ignore errors. For instance, the transmission of textual data may not require that every bit of the text arrive error-free. Assuming a bit error rate of $1{:}10^5$ (i.e., one errored bit in every 100,000 bits), a 400-page book could be transmitted with but 35 Baudot code characters in error. A $1{:}10^5$ error rate is not pushing the state of the art by any means, so ignoring errors is a viable option.

However, many applications cannot afford errors. Financial data systems, such as an electronic transfer of funds between customers' bank accounts, must have completely accurate data messages arrive at the end point. In these applications, the data link controls must detect and handle any errors that occur on the path.

It is sometimes difficult for individuals to understand that the reliability and speed of a communications path through a network is appreciably different from the flow of data in a conventional mainframe environment where all components are located in one room or one building.[1] A computer-to-disk channel can operate at

[1] Andrew S. Tanenbaum, *Computer Networks*. Englewood Cliffs NJ: Prentice-Hall, 1981, p. 56..

data transfer speeds of 10^7 to 10^9 bps with an error rate of 1 in 10^{12} or 10^{13} bits slot—very fast and very reliable. In contrast, a dial-up line transmits data at a rate of 10^3 to 10^4 bps, at an error rate of 1 in 10^5 bits. The combined bit transfer times error rates for the inhouse environment is 11 orders of magnitude better than a dial-up line in a network. Consequently, error handling techniques in a data communications system necessarily require considerable attention.

Bit-Checking Techniques. Most methods used to provide for data error detection entail the insertion of redundant bits in the message. The actual bit configuration of the redundant bits is derived from the data bit stream.

Vertical redundancy check (VRC). VRC is a simple technique. It consists of adding a single bit (a parity bit) to each string of bits that comprise a character. The bit is set to 1 or 0 to give the character bits an odd or even number of bits that are 1s. This parity bit is inserted at the transmitting station, sent with each character in the message, and checked at the receiver to determine if each character is the correct parity. If a transmission impairment caused a "bit flip" of 1 to 0 or 0 to 1, the parity check would so indicate. However, a two-bit flip would not be detected by the VRC technique, which creates a high incidence of errors in some transmissions. For example, multilevel modulation (where two or three bits are represented in a signal cycle) requires a more sophisticated technique. The single-bit VRC is also unsuited to most analog voice-grade lines because of the groupings of errors that usually occur on this type of link.

Longitudinal redundancy check (LRC). LRC is a refinement of the VRC approach. Instead of a parity bit on each character, LRC places a parity (odd or even) on a block of characters. The block check provides a better method to detect for errors across characters. It is usually implemented with VRC and is then called a two-dimensional parity check code (see Figure 6-5). The VRC-LRC combination provides a substantial improvement over a single method. A typical telephone line with an error rate of $1{:}10^5$ can be improved to a range of $1{:}10^7$ and $1{:}10^9$ with the two-dimensional check.[2]

---

[2]Dixon Doll, *Data Communications, Facilities, Networks, and System Design*. New York: Wiley Interscience, 1978, p. 263.

Bits in Characters	Characters 1	2	3	. . .	n	LRC ↓
1	0	1	0			0
2	1	0	0			0
3	1	0	1			1
4	0	0	1			0
5	0	1	0			0
6	1	0	1			1
7	0	0	1			0
VRC →	0	1	1			1

FIGURE 6-5. VRC-LRC and Two-Dimensional Parity Check.

Hamming code. The Hamming code is a more sophisticated variation of the VRC. It uses more than one parity bit per byte or character. The parity bit values are based on various combinations of the user character, and the parities are inserted in between the bits of the character. For example, in one Hamming approach, a byte of bits b_1 b_2 b_3 b_4 b_5 b_6 b_7 carries a 10-bit code as p_1 b_1 p_2 b_2 b_3 p_3 b_4 b_5 b_6 b_7. The parity bit p_1, is set odd- or even-based on the values of b_1, b_3, b_5, and b_7; p_2 is based on b_2, b_3, b_6, b_7; and p_3 is based on b_4, b_5, b_6, and b_7. The Hamming code achieves better results than a VRC or an LRC; however, it does carry more overhead.

Cyclic redundancy checking (CRC). Several other techniques are in use today. One that is quite widely used is Cyclic Redundancy Checking (CRC). Figure 6-6 shows an example of CRC use. The CRC approach entails the division of the user data stream by a predetermined binary number. The remainder of the number is appended to the message as a CRC field. The data stream at the receiving site has the identical calculation performed and compared to the CRC field. If the two values are identical, the message is accepted as correct.

The divisor polynomial $X^{16} + X^{15} + X^2 + 1$ is often used (11000000000000101) in data communications. It can detect, ". . . all possible single-error bursts not exceeding 16 bits, 99.9969% of all possible single bursts 17 bits long, and 99.9984% of all possible longer bursts".[3] This provides for 1 bit error per every 10^{14} bits transmitted and is much better than the other methods.

---

[3]Anthony Ralston (Ed.), *Encyclopedia of Computer Science*. New York: Van Nostrand Reinhold, 1976, p. 385.

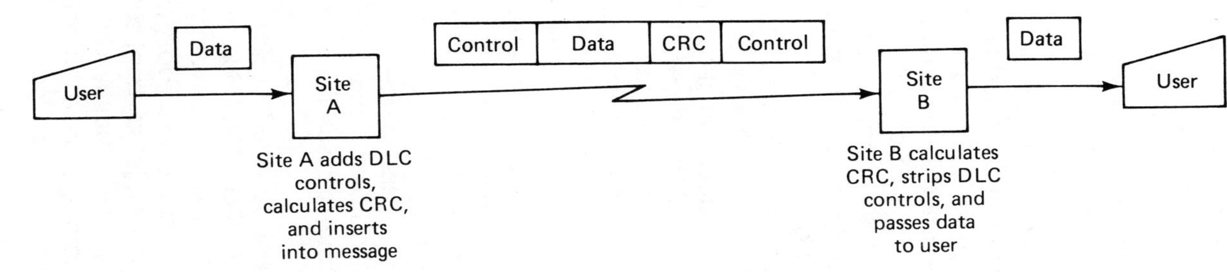

FIGURE 6-6. Cyclic Redundancy Checking.

Flow Control

It should be recognized that a DLC must manage the transmission and receipt of perhaps thousands of messages in a short period of time. The DLC must move the data traffic efficiently. Communications lines should be evenly used and no station should be unnecessarily idle or saturated with excessive traffic. Thus, flow control is a critical part of the network.

Stop-and-Wait. Figure 6-7 depicts the stop-and-wait DLC. This DLC allows one message to be transmitted (Event 1), checked for errors with techniques such as VRC or LRC (Event 2), and an appropriate ACK or NAK returned to the sending station (Event 3). No other data messages can be transmitted until the receiving station sends back a reply. Thus, the name *stop-and-wait* is derived from the originating station sending a message, stopping further transmission, and waiting for a reply.

The stop-and-wait approach is well-suited to half-duplex transmission arrangements since it provides for data transmission in both directions, but only in one direction at a time. Moreover, it is a simple approach requiring no elaborate sequencing of messages or extensive message buffers in the terminals.

Its major drawback is the idle line time that results when the stations are in the wait period. Most stop-and-wait data link controls now provide for more than one terminal on the line. The terminals are still operating under the simple arrangement. They are fairly inexpensive with limited intelligence. The host or primary station is responsible for interleaving the messages among the terminals (usually through a more intelligent device that is in front of the terminals) and controlling access to the communications link. Nonetheless, a point-to-point line is underutilized with this approach; a high speed stop-and-wait satellite channel may use only 3 to 5% of its total capacity.

The simple arrangement depicted in Figure 6-7 also creates serious problems when the ACK or NAK is lost in the network or on the line. If the ACK in Event 3 is lost, the master station times out and retransmits the same message to the secondary site. The redundant transmission could possibly create a duplicate record in the secondary site user data files. Consequently, data link controls must provide a means to identify and sequence the transmitted messages with the appropriate ACKs or NAKs. The logic must have a method to check for duplicate messages.

A typical approach to solve this problem is the provision for a sequence number in the header of the message. The receiver can

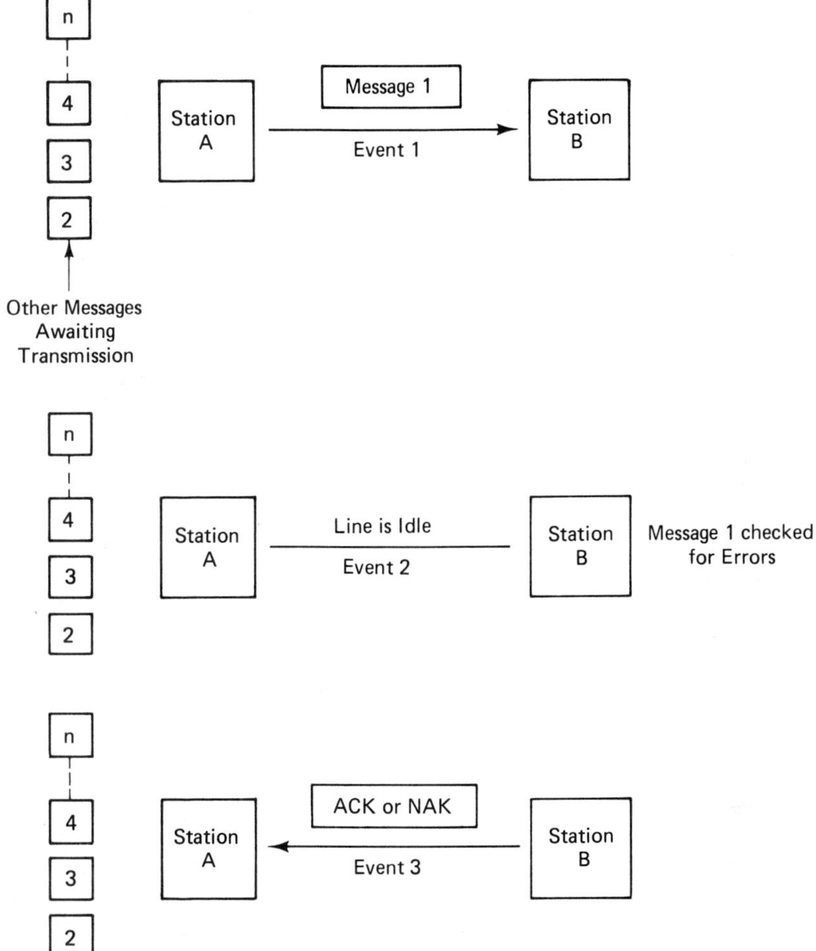

FIGURE 6-7. Stop-and-Wait Data Link Control.

then check for the sequence number to determine if the message is a duplicate. The stop-and-wait DLC requires a very small sequence number since only one message is outstanding at any time. The sending and receiving station need only use a one-bit alternating sequence of 0 or 1 to maintain the relationship of the transmitted message and its ACK/NAK.

Figure 6-8 shows how this arrangement works. In Event 1, the sending station transmits a message with the sequence number 0 in the header. The receiving station responds with an ACK and a

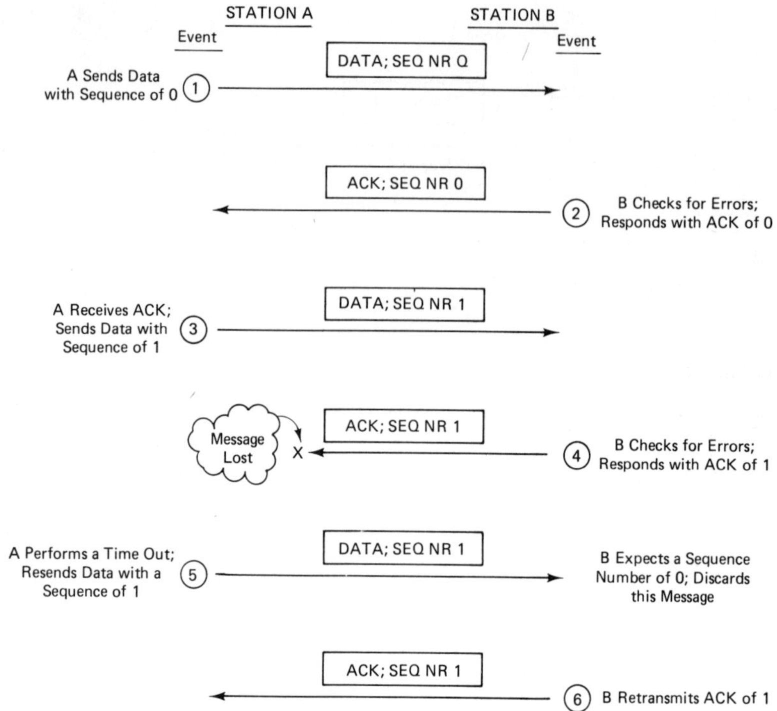

FIGURE 6-8. Stop-and-Wait Alternating Sequence.

sequence number of 0 (Event 2). The sender receives the ACK, examines the 0 in the header, flips the sequence number to a 1, and transmits the next message (Event 3). The receiving station receives and acknowledges the message with an ACK 1 in Event 4. However, this message is received garbled or is lost on the line. The sending station recognizes that the message in Event 3 has not been acknowledged. It performs a timeout and retransmits this message (Event 5). The receiving station is looking for a message with a sequence number of 0. It discards the message since it is a duplicate of the message transmitted in Event 3. To complete the accountability, the receiving station retransmits the ACK of 1 (Event 6).

Sliding Window Control. The inherent inefficiency of the stop-and-wait DLC resulted in the development of techniques to provide for the overlapping of data messages and their corresponding control messages. The newer data link controls employ this method.

The data and control signals flow from sender to receiver in a more continuous manner, and several data and control messages can be outstanding (on the line or in the receiver's buffers) at any one time.

These DLCs are often called *sliding windows* because of the method used to synchronize the sending sequence numbers in the headers with the appropriate acknowledgments. The transmitting station maintains a sending window that delineates the number of messages (and their sequence numbers) it is permitted to send. The receiving station maintains a receiving window that performs complementary functions. The two sites use the windows to coordinate the flow of messages between each other. In essence, the window states how many messages can be outstanding on the line or at the receiver before the sender stops sending and awaits a reply. For example, in Figure 6-9(a) the receiving of the ACK of message 1 allows the San Francisco site to slide its window by one sequence number. If a total of 10 messages could be within the window, San Francisco could still transmit messages 5, 6, 7, 8, 9, 0, 1. (Keep in mind that messages 2, 3, 4 are in transit.) It could not transmit a message using sequence 2 until it had received an ACK for 2. The window wraps around to reuse the same set of numbers.

Go-Back-N. The Go-Back-N method is a sliding window technique. It allows data and control messages to be transmitted continuously. In the event an error is detected at the receiving site, the erroneous message is retransmitted as well as all other messages that were transmitted after the errored message.

Figure 6-9 shows the message flow of the Go-Back-N method. In Figure 6-9(a), messages 2, 3, and 4 are transmitted on the line to Kansas City and an ACK of previously received message 1 is sent back to San Francisco. Notice the full-duplex transmission scheme. In Figure 6-9(b), messages 4 and 5 are on the path; Kansas City has now received messages 2 and 3. It determines that message 3 is in error and transmits a NAK to the originating station. San Francisco responds to the NAK in Figure 6-9(c) by retransmitting message 3 as well as messages 4 and 5.

One might question why messages 4 and 5 are retransmitted since it could mean a duplication of effort and result in wasted resources. These concerns are valid but the approach also provides a simple means to keep the messages in the proper sequence between the two points, which in turn simplifies the software or chip logic and decreases the length of certain control fields in the message. An error condition on the line (such as a rain storm on a microwave path) might affect not just one message, but the

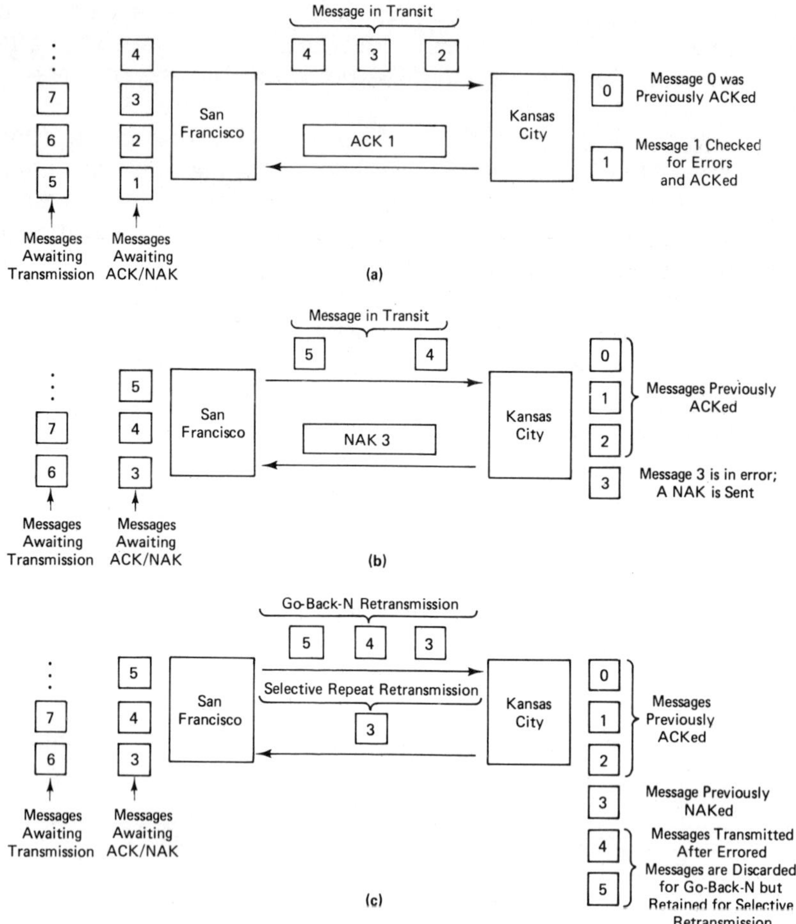

FIGURE 6-9. Sliding Window Data Link Control.

subsequent messages that are traveling down the path as well. For example, a 20 ms distortion on a 50 kbs line will distort 1000 bits, possibly in more than one message. Consequently, these messages may be retransmitted anyway, because they are in error.

Selective Repeat. The Selective Repeat method provides for a more refined approach. In contrast to the Go-Back-N, the only messages retransmitted are those that are NAKed. In Figure 6-9(c), message 3 only is resent from the originator in San Francisco.

Studies reveal that the Selective Repeat DLC obtains greater throughput than the Go-Back-N. However, the differences are not great if the comparison is made on a reliable transmission path. The Selective Repeat DLC requires additional logic to maintain the sequence of the resent message and merge it into the proper place on the queue at the receiving site. Consequently, the Go-Back-N is found in more data link controls than is the Selective Repeat DLC.

The window size of these data link controls is an important element in the determination of message accountability and efficient line utilization. Due to delay in the propagation of a signal, a message requires a certain amount of time before all bits of the message arrive at the receiver and the acknowledgment is returned to the sender. The window size should allow for a continuous flow of data. The returned ACKs should arrive before the sender has transmitted all messages within its window. This timing allows the sliding of the sending window and prevents the sender from waiting for the ACK. In other words, the window size should keep the line busy. It can be seen that the window size is an important design decision.

Many varieties of data link controls are found in the industry today. Generally, each vendor has chosen to develop its own version of a DLC. The result is incompatibility between vendor products. However, due to the influence of the larger vendors and international standards, many companies have now developed compatible DLCs in order to market products that will interface into the other companies' components. This has helped foster standard data link controls but, as we will see in later discussions, multiple standards, methods, and procedures continue to present significant problems to the users of these products.

EXAMPLES OF DATA LINK CONTROLS

The previous discussion focused on the types of DLCs in existence today. We will now examine three products that use various combinations of the formats, line control, error handling, and flow control methods. IBM products are cited in the exmaples. This is not meant to imply endorsement of IBM's offerings; it is done in recognition that IBM's products dominate the market. Many other vendors provide "IBM—compatible" controls and, thus, resemble these three illustrative products.

The three data link controls and their classifications are as follows. We will first cover the simpler, less powerful DLC and move to the more sophisticated approaches:

- 2740 Asynchronous Control
  - Asynchronous format
  - Byte-oriented format
  - Parity checking error detection
  - Polling/selection line control
  - Stop-and-wait flow control
- Binary Synchronous Communications Control (BSC)
  - Synchronous format
  - Byte-oriented format
  - Parity checking or CRC error detection
  - Polling/selection line control
  - Stop-and-wait flow control
- Synchronous Data Link Control (SDLC)
  - Synchronous format
  - Bit-oriented format
  - CRC error detection
  - Polling/selection line control
  - Go-Back-N flow control

2740 Asynchronous control. The IBM 2740 terminal uses the asynchronous half-duplex transmission technique. It has options for buffers but the 2740 logic still provides start and stop bits around each character in the transmission. The model 1 operates at 134.5 bit/s; the model 2 has options for speeds of 75 to 600 bit/s. The 2740 is a rather old machine and is used here to illustrate an early DLC. It uses an older IBM code called Extended Binary Coded Decimal (EBCD). It is a six-bit code with a seventh bit inserted for parity checking. In addition, the code provides for functions to control the 2740 printer—for example, new line, backspace, and upper case.

The code set is the same for data and control characters. This problem is approached by using special characters to inform the terminal logic that the transmission is either user data or control characters. The EOT (End of Transmission) character places the link and all terminals in the control mode and the EOA (End of Addressing) character places the system in the text (or user data) mode.

The 2740 uses vertical redundancy checking (VRC) for error detection. The parity bit is turned on to create odd parity of the character. The receiving logic checks each character for an odd number of 1 bits and signals the transmitter if it detects an error condition. The system also uses the Longitudinal Redundancy Check (LRC).

Figure 6-10 shows an exchange of transmissions between a host and a 2740 terminal. The picture shows data moving directly

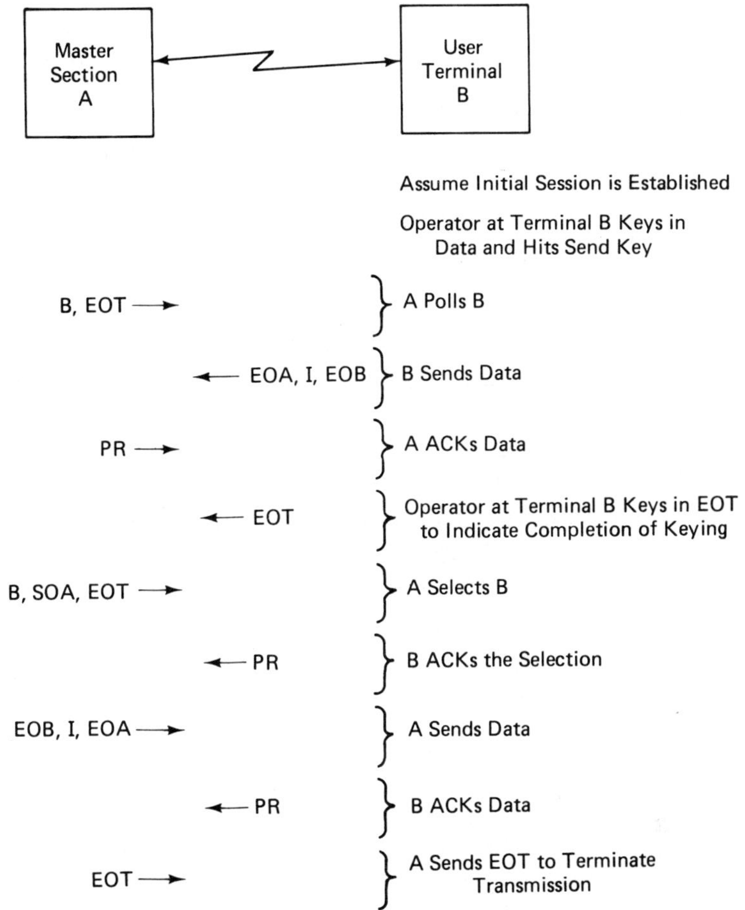

FIGURE 6-10. 2740 Asynchronous Transmissions.

into and out of the terminal. A 2740 control unit is actually involved in this process. The following legend is used to trace the transmission flow and format:

C, A, I, C

Where: C = Control character(s)
A = Address of terminal
I = User information field

The legend does not imply any order of bit transmission. In actual transmission, a control character precedes the address and user data. The legend also does not show the start/stop bits; remember that each character has these bits around them. Figure 6-10 shows an arrow to indicate the direction of the transmission.

For purposes of understanding Figure 6-10, an explanation of the control character is provided.

EOT (End of Transmission): Puts system into control mode. EOT also indicates a poll when transmitted from the master station.

EOA (End of Addressing): Puts system into text mode. EOA is also a positive acknowledgment (ACK) to a poll.

EOB (End of Block): Signals the end of a block of user data (text). Always followed by the LRC check.

PR (Positive Response): Positive acknowledgment (ACK) of transmission. PR is also a positive acknowledgment to a master station's selection transmission.

SOA (Start of Addressing): Signals the beginning of an addressing (i.e., selection) operation.

Figure 6-10 illustrates the overhead of the asynchronous method. Each character (not shown) may contain the start-stop bits. Moreover, nine separate transmissions are required to transmit two user data messages and no other transmissions take place during this process. On the other hand, it is a simple arrangement and well-suited to low-speed, teleprinter applications.

Binary Synchronous Communications Control (BSC)

BSC is one of the most widely used data link control methods in the U.S. Until the advent of SDLC (discussed next), BSC was IBM's major DLC offering. Many vendors offer a BSC-like project; many others offer BSC emulation packages. The product is also called Bisync, a short term for its full title. The method is intended for half-duplex, point-to-point, or multipoint lines. A complete description of BSC can be found in IBM's "Binary Synchronous Communications—General Information" (GA 27-3004).

BSC operates with EBCDIC, ASCII, or Transcode. All stations on a line must use the same code. If ASCII code is used, error checking is accomplished by a Vertical Redundancy Check (VRC) and a

Longitudinal Redundancy Check (LRC). If EBCDIC or Transcode are used, error checking is accomplished by Cyclic Redundancy Checking (CRC).

BCS uses the alternating sequence acknowledgment discussed earlier. The receiving station replies with an ACK0 to the successful reception of even-numbered messages and an ACK1 to odd-numbered messages. A reception of two consecutive identical ACK characters alerts the transmitting station to an exceptional situation.

The Bisync message format is shown in Figure 6-11. The SYN characters are used as described in Chapter 2. Other control characters' functions are as follows:

EOT (End of Transmission): Signifies end of transmission, and also can place system in control mode (similar to 2740 method).

SOH (Start of Header): This field is not defined in BSC logic but is user or application dependent. The header is optional.

STX (Start of Text): Places system in user data (or text) mode. (Note the similarity to the 2740 method.) Also used for polling and selection indicators.

ETX (End of Text): Signifies end of user data (text).

ETB (End of Block): If user data is divided into multiple blocks, the ETB is sent at the end of each block except the last one.

ACK0 (Positive Acknowledgment): This field positively acknowledges (ACKs) even-numbered text blocks.

ACK1 (Positive Acknowledgment): This field positively acknowledges (ACKs) odd-numbered text blocks.

NAK (Negative Acknowledgment): This is used to negatively acknowledge polls/selects and other control messages.

ENQ (Enquiry): On a point-to-point line, this asks if the station can accept a transmission. On a multipoint line, ENQ is used to initiate polling and selection.

DLE (Data Link Escape): Provides a method to use different codes.

SYN	SYN	SOH	Header	STX	User Data	ETX or ETB	Error Check Field

FIGURE 6-11. Bisync (BSC) Message Format.

Figure 6-12 shows an exchange of transmissions in a BSC environment. The following legend is used to trace the transmission flow and format:

C, A, I, C

Where: C = Control characters
A = Address of station
I = User information field

The legend does not imply any order of bit transmission. In actual transmission, SYN characters precede the message and the control characters, address, and I field (if appropriate) follow. In most instances, control fields follow the message as well. Figure 6-12 shows arrows to indicate the direction of the transmission. It also shows upper- and lower-case letters in the A field: upper-case depicts a poll; lower-case depicts a select.

BSC provides other features and is a substantial improvement over the asynchronous approach. The following additional control fields show these facilities:

WACK (Wait Before Transmitting): This field provides a positive ACK of the message and also requests holding up any transmissions until the sending station sends an ENQ and receives an ACK.

RVI (Reverse Interrupt): This provides the capability to stop the current transmission sequence to service higher priority work.

TTD (Temporary Text Delay): This allows the transmitting station to delay its sending of a message and still retain control of the line.

BSC was designed in the 1960s and has served the industry well. The early applications usually had a human operator at the remote station and the half-duplex approach was sufficient. As applications and data communications became more sophisticated, more powerful data link control techniques were needed.

Synchronous Data Link Control (SDLC)

Synchronous Data Link Control (SDLC) is IBM's major DLC offering today. The product was introduced in 1973 to support communications lines among teller terminals in banks. The offering is discussed in this book for two reasons: (a) it is widely used in the U.S. and (b) it is quite similar to the international standard HDLC. The product is in many of the IBM's communications components and will eventually replace the BSC offering.

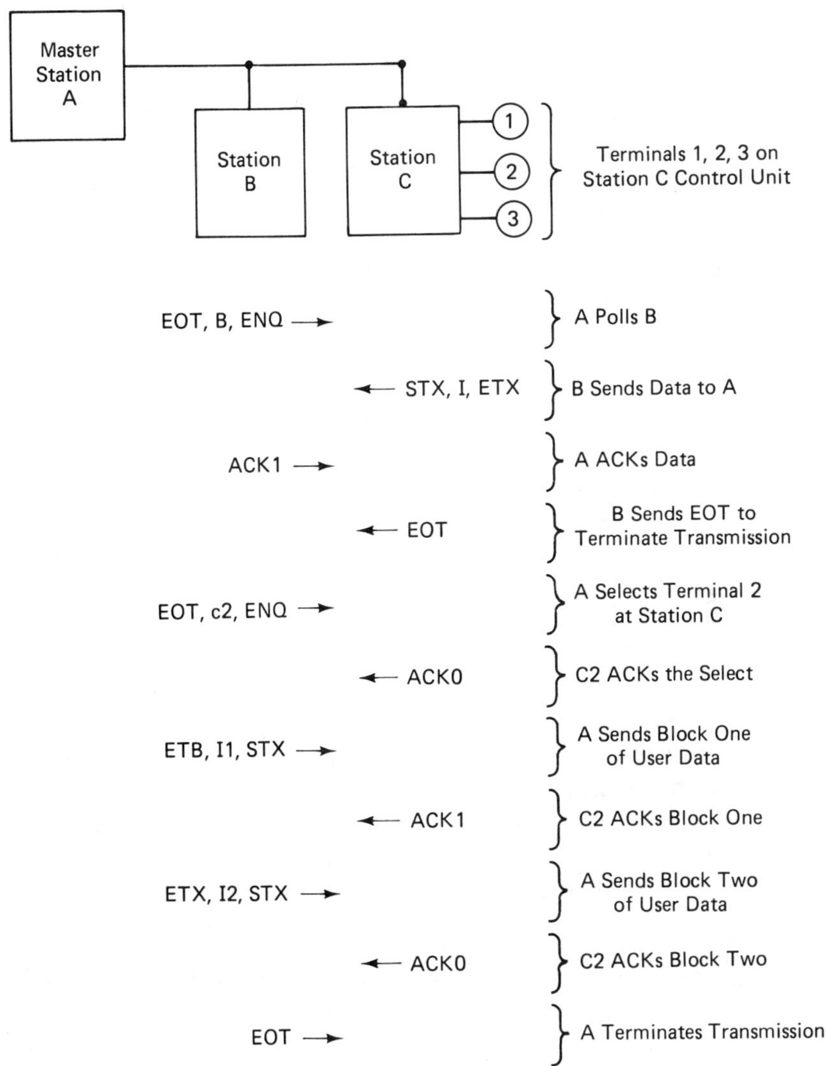

FIGURE 6-12. Multipoint Bisync Transmissions.

SDLC uses the synchronous, bit-oriented, Go-Back-N methods. It controls a single line configured as point-to-point, multipoint, or loop. It operates on half-duplex, duplex, switched, or private lines. It also provides for duplex, multipoint operation in which a station can transmit to one station while receiving from another.

SDLC uses variations of the polling/selection technique. The primary station is responsible for the control of the line. It initiates all transmissions (such as a poll) from the secondary stations with a command. The stations reply with a response. The primary station can also be a secondary station to another primary station.

SDLC Frame. SDLC messages are transmitted across the line in a specific format called a *frame* (see Figure 6-13). The beginning and ending flags each consist of an eight-bit byte pattern of 01111110. These fields serve as references for the beginning and ending of the message, like a SYN field. The ending flag may serve as a beginning flag for the next frame. Multiple flags may be repeated between frames to keep the line in an active state.

Flag	Address	Control	User Data	FCS	Flag

FIGURE 6-13. SDLC Frame.

SDLC is code transparent and the only unique bit stream is the flag field. The logic will not allow the 01111110 pattern to be transmitted in other parts of the frame. At the transmitting end, SDLC examines the frame contents (flag fields excluded) and inserts a 0 after any succession of five consecutive 1s within the frame. The receiving site receives the frame, recognizes the two flags, and then removes any 0 that follows five consecutive 1s. Consequently, SDLC is not dependent on any specific code such as ASCII or EBCIDIC. It does require that all fields in the frame be in multiples of eight bits after the stuffed 0s have been removed.

The address field follows the beginning flag. The address identifies the secondary station. SDLC also allows addressing a number of stations on the line (group address) as well as all stations (broadcast address). A poll message identifies the polled secondary station. A response also contains the address of the secondary station.

The control field defines the function of the frame and, therefore, invokes the SDLC logic at the receiving and sending stations. The field is an eight-bit byte and can be in one of three formats.

- *Unnumbered Format* (U) frames are used for control purposes such as initializing secondary stations, disconnecting stations, testing stations, and controlling modes of responses from stations.
- *Supervisory Format* (S) frames are used to positively acknowledge (ACK) and negatively acknowledge (NAK) user data (information frames). Supervisory frames do not carry user data. They are used to confirm received data, report busy or ready conditions, and to report frame numbering errors.
- *Information Transfer* (I) frames contain the user data.

The next field of the frame is the frame check sequence field (FCS). This field contains a 16-bit sequence that is computed from the contents of the address, control, and information fields at the transmitting stations. The receiver performs a similar computation to determine if errors have been introduced during the transmission process. The receiver will not accept a frame that is in error.

SDLC Control Field. The control field is further described in Table 6-1. The low-order bits (right-most) identify the frame format (11 for unnumbered; 10 for supervisory; 0 for information). The remainder of the bits define the specific type and function of the frame:

UI (Unnumbered Information): This command allows for transmission of user data in an unnumbered (i.e., unsequenced) frame.

RIM (Request Initialization Mode): The RIM frame is a request from a secondary station to a primary station for an SIM command.

SIM (Set Initialization Mode): This command is used to initialize the primary-secondary session. UA is the expected response.

SNRM (Set Normal Response Mode): This places the secondary station in a NRM (Normal Response Mode). The NRM precludes the secondary station from sending any unsolicited frames. This means the primary station controls all message flow on the link.

TABLE 6-1. SDLC Control Field.

Command Response	Format	Control Field			Command	Response	I-Field Prohibited	Resets Nr and Ns
UI	U	000	P/F	0011	X	X		
RIM	U	000	F	0111		X	X	
SIM	U	000	P	0111	X		X	X
SNRM	U	100	P	0011	X		X	X
DM	U	000	F	1111		X	X	
DISC	U	010	P	0011	X		X	
UA	U	011	F	0011		X	X	
FRMR	U	100	F	0111		X		
BCN	U	111	F	1111		X	X	
CFGR	U	110	P/F	0111	X	X		
RD	U	010	F	0011		X	X	
XID	U	101	P/F	1111	X	X		
UP	U	001	P	0011	X		X	
TEST	U	111	P/F	0011	X	X		
RR	S	Nr	P/F	0001	X	X	X	
RNR	S	Nr	P/F	0101	X	X	X	
REJ	S	Nr	P/F	1001	X	X	X	
I	I	Nr	P/F	Ns 0	X	X		

DM (Disconnect Mode): This frame is transmitted from a secondary station to indicate it is in the disconnect mode.

DISC (Disconnect): This command from the primary station places the secondary station in the normal disconnected mode. This command is valuable for switched lines; the command provides a function similar to hanging up a telephone.

Note: The Initialization, Normal Response, and Normal Disconnected Modes are the three allowable modes in SDLC.

UA (Unnumbered Acknowledgment): This is an ACK to the SNRM, DISC, or SIM command.

FRMR (Frame Reject): The secondary station sends this frame when it encounters an invalid frame. This is not used for a bit error indicated in the Frame Check Sequence field but for more unusual conditions such as a frame that is too long for a secondary station's buffers or an erroneous control field.

BCN (Beacon): Explained later during discussions on loop configurations.

CFGR (Configure): Explained later during discussions on loop configurations.

RD (Request Disconnect): This is a request from a secondary station to be disconnected.

XID (Exchange Station Identification): This command asks for the identification of a secondary station. It is used on switched facilities to identify the calling station.

UP (Unnumbered Polls): Explained later during discussions on loop configurations.

Test (Test): This frame is used to solicit testing responses from the secondary station.

RR (Receive Ready): Indicates secondary or primary station is ready to receive. It is also used to ACK or NAK information frames.

RNR (Receive Not Ready): Indicates secondary or primary station is temporarily busy. It is also used to ACK or NAK information frames.

REJ (Reject): This frame can be used to explicitly request the transmission or retransmission of information frames. This command or response is very useful when the frame sequencing becomes out of order.

I (Information): This frame contains the user data. It can also ACK and NAK received frames.

Frame Flow Control. SDLC uses the sliding window technique to manage the flow of frames between the sender and receiver. Each station maintains a send (Ns) and receive (Nr) count. A sending station counts each outgoing frame and transmits the count in the Ns portion of the control field. The receiving station receives and checks the frame. If the frame is error-free and properly sequenced it advances its Nr count by one and sends an ACK with the appropriate Nr count to the transmitter.

The stations' Nr and Ns fields are set to 0 at session initialization. The flow of frames and the incremental counting at each end of the Nr and Ns fields provide the capability for the receiving station's Nr count to be the same as the transmitting station's Ns field of the *next* frame to be transmitted.

The Nr and Ns counts in the control field are each three bits, thus allowing a counting capacity of 8 using the binary numbers of 0 through 7. As in most sliding window techniques, the window wraps around to 0 after the count reaches 7. Up to seven unconfirmed frames can be outstanding at one time between the two stations.

SDLC provides for inclusive acknowledgment. For example, an ACK of a frame with an Ns count of 5 can also be an ACK of frames 4, 3, 2, 1, 0, 7, and 6. This ACK essentially says, "I have received and accepted all frames up to and including the frame with Ns = 4. The next frame I receive from you should have an Ns = 5."

The P/F (poll/final) bit in the control field provides the following function: (a) the primary station turns the bit on to indicate a poll to the secondary station and (b) the secondary station turns the bit on to indicate the last frame of a transmission.

Figures 6-14 and 6-15 provide examples of SDLC exchanges of data and control frames. The following legend is used to trace the transmission flow and format:

F, A, C, Ns, P/F, Nr, FCS, F

Where: F = Flags
A = Address of secondary station
C = Command/Response Acronym
Ns = Count of sending sequence field
P/F = P: poll bit on
$\bar{P}$: poll bit off
F: final bit on
$\bar{F}$: final bit off
Nr = Count of receiving sequence field
FCS = Frame check sequence field

The legend does not imply any order of bit transmission. In actual transmission, the frame is sent in a serial bit fashion with the bits transmitted left to right as shown in Figure 6-13. The figures show an arrow to indicate the direction of the transmission. A dash (—) in any of the fields indicates that the field is not used for that particular type of frame.

Loop Transmission. SDLC also has the hub polling capability. The arrangement, called *loop transmission*, provides for the primary station (loop controller) to send command frames to any or all the stations of the loop. Each secondary station decodes the

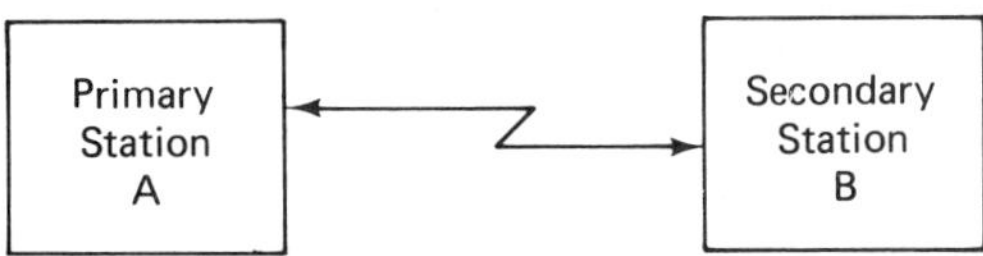

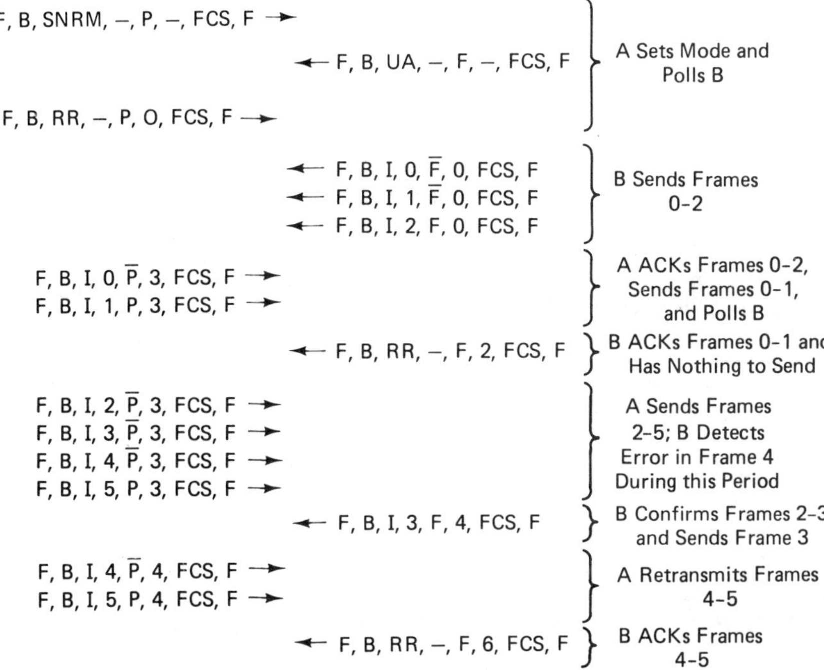

FIGURE 6-14. SDLC Transmission with one Station. (Adapted from *IBM Synchronous Data Link Control*, Manual GA27-3093-2.)

address field of each frame and accepts the frame, if appropriate. The frame is also passed to the next station (down-loop station.)

After the loop controller has completed the transmission of the command frames, it sends eight consecutive 0s to signal the secondary stations of the completion of the frames. It then transmits continuous 1s to indicate it is in a receive mode and awaits the receiving of the 1s to ascertain that the loop is complete.

The following additional formats are used for loop transmission:

UP (Unnumbered Polls): This frame is useful for loop operations because it gives the secondary station the option of responding

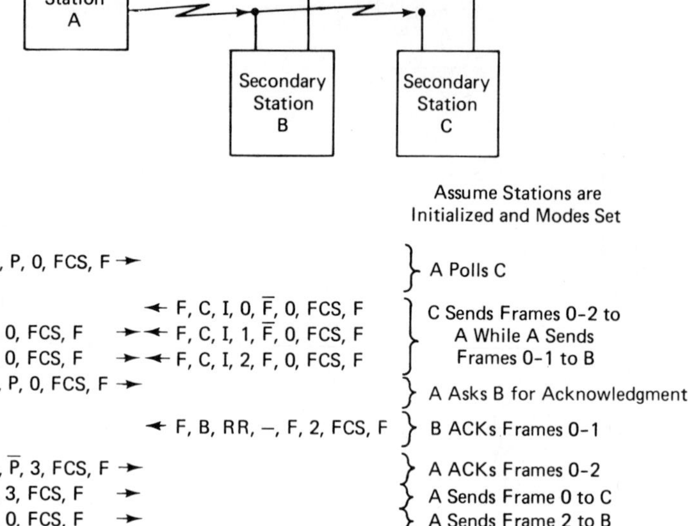

FIGURE 6-15. SDLC Multipoint Transmission. (Adapted from *IBM Synchronous Data Link Control*, Manual GA27-3093-2.)

to the poll. The poll can then be transmitted around the loop and used by all stations without regard to any previous sequence (Nr or Ns) binding.

CFGR (Configure): This command provides for several testing and configuration features (clearing functions, placing a station in receive only mode, placing stations off line).

BCN (Beacon): This frame is used to trace down problems with the carrier signal—to determine what part of the path is causing the problem. It causes the secondary station to suppress transmission of the carrier or to begin transmitting the carrier again after suppressing it.

The unnumbered polls serve to evoke frames from the secondary stations on the line and can be sent from the primary station after the continuous 1s have completed the loop. Figure 6-16

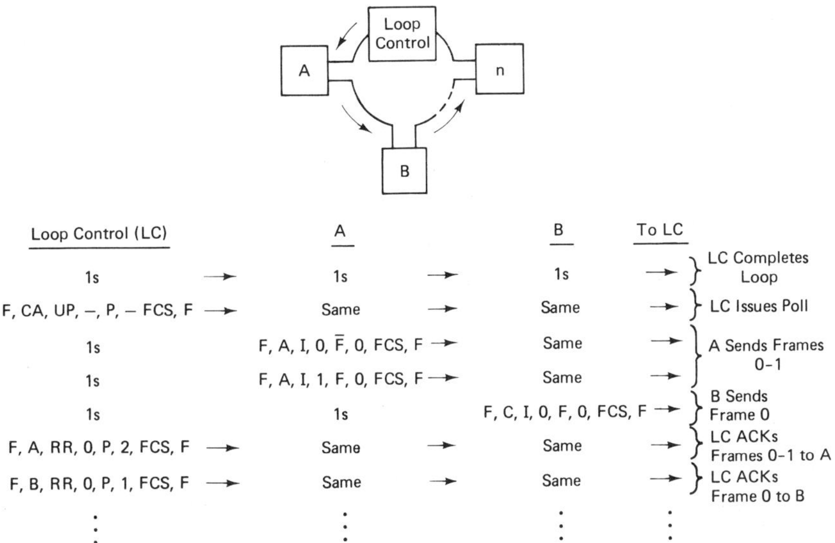

FIGURE 6-16. SDLC Loop Transmission.

shows an SDLC loop operation. The legend shown earlier still pertains. In addition, the CA means a common address intended for all stations.

COMMUNICATIONS SATELLITES AND DATA LINK CONTROLS

From Chapter 2, the reader may recall that geosynchronous satellites incur an end-to-end signal propagation of several hundred ms and, under certain conditions, as long as 900 ms. Certain DLCs do not execute efficiently on satellite links. In some instances, the propagation delay can cause the DLC software to wrap around itself, continuously executing its time-out code waiting for the incoming message. Additionally, a Stop-and-Wait technique (such as Bisync) would spend a good deal of valuable channel time awaiting the next block of data. On the other hand, the sliding window techniques such as SDLC are well-suited to satellite transmission due to the continuous sending and receiving of messages between the two sites.

Satellite vendors now offer components that allow a user site to maintain the older data link controls. The components, called *delay compensation units* (DCU), terminate the user's DLC locally and build another data block into a delay-insensitive DLC. The DCU responds locally to the terminal or computer as if it were the remote site. Most vendors provide DCUs that operate on the EIA RS232-C interface with data rates up to 9.6 Kbs or 56 Kbs.

SUMMARY

The progress in developing and using improved link control techniques has been slow due to the investments in the machines and software that support the older approaches. Nonetheless, the newer products such as SDLC are in wide use today; they offer significant improvements over a 2740 or BSC product. In spite of the increased complexity of the new DLC techniques and the requirement for larger machines to support them, the new DLCs provide opportunities to reduce overall costs due to more efficient line utilization, increased reliability, and better error control.

7

DIGITAL TRANSMISSION

The data communications industry is evolving toward the use of all digital networks. These networks carry digital bit pulses instead of the conventional voice-oriented analog signal. In the near future, networks will be integrated to transmit digital images of data, voice, facsimile, graphics, television, and any other image. It might surprise the reader to know that many of AT&T's long-distance facilities use digital technology today.

ADVANTAGES OF DIGITAL TRANSMISSION

If Ma Bell had it to do over again, it is probable that analog technologies would never have been used. Digital transmission schemes are more attractive for several reasons. First, long-haul digital signals are more error-free than analog signals. As depicted in Figure 5-1, line noise and other distortions are periodically amplified with the analog signal. Digital transmission entails the use of regenerative repeaters wherein only the absence of a pulse (binary 0) or the presence of a pulse (binary 1) need be detected. Thereafter, the signal is completely reconstructed. The repeaters create a signal as good as the original transmission (see Figure 7-1). Consequently, digital signals can experience more distortion and cross talk and a higher noise-to-signal ratio than analog signals.

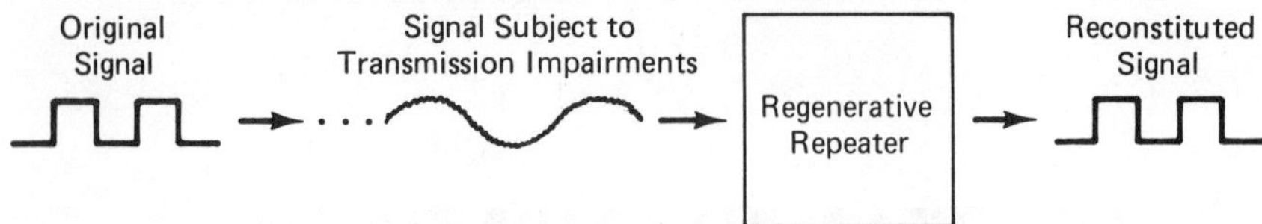

FIGURE 7-1. Digital Signals and Signal Regeneration.

Second, the rapidly declining costs of circuitry and processors make digital schemes increasingly attractive. The ideas and theories of digital transmission have been around for many years; the introduction of inexpensive large scale integration (LSI) circuitry now makes the concepts cost-effective.

Third, many transmission types can be accommodated with a digital facility. Television, data, voice, telegraphy, facsimile, and even music can be multiplexed together. The digital network treats all signals as binary values in the form of digital pulses of a positive, zero, or negative voltage. This capability is quite valuable because organizations will be able to implement one network for *all* transmissions. Digital transmissions use a data transmission unit to send the binary bits directly onto the network. Analog transmissions (such as voice and television) use an analog-to-digital converter (A/D) to translate the signal to binary images at the transmitter. At the receiver, a digital-to-analog device (D/A) converts the signal back to an analog form (see Figure 7-2).

Fourth, digital transmission is inherently more secure than analog transmission. Encryption is relatively easy. The older analog scramblers are not very effective against the decoding ability of the computer. Moreover, scrambled speech can often be deciphered by a trained listener. Digital encryption costs less than analog scrambling.

Fifth, the newer satellite transmission schemes use digital transmission to increase reliability and signal throughput. The schemes are based on time division multiplexing using digital bit streams. The approach will see increasing use as satellites use higher bands in the electromagnetic spectrum.

Sixth, newer transmission technologies such as optic fibers benefit from digital transmission. For example, lasers are used to transmit pulses of light to represent binary bits.

Seventh, recent advances in voice digitization point to reduced bandwidth requirements for sending voice-generated signals. Speech can be digitized at 2,400 bit/s, which permits four simultaneous conversations on a voice-grade circuit. Digitizing speech at this bit rate is expensive but the decreasing costs of LSI circuitry point toward more cost-effective digitization arrangements.

Last, switching and control signaling can be accomplished more effectively with digital facilities. The two components can be more fully integrated because equipment can be made common for both. Switching and signaling are more reliable and efficient using digital techniques.

HOW DIGITAL TRANSMISSION WORKS

It was established in 1937 that the periodic sampling of a signal at a rate twice the highest frequency in the sample

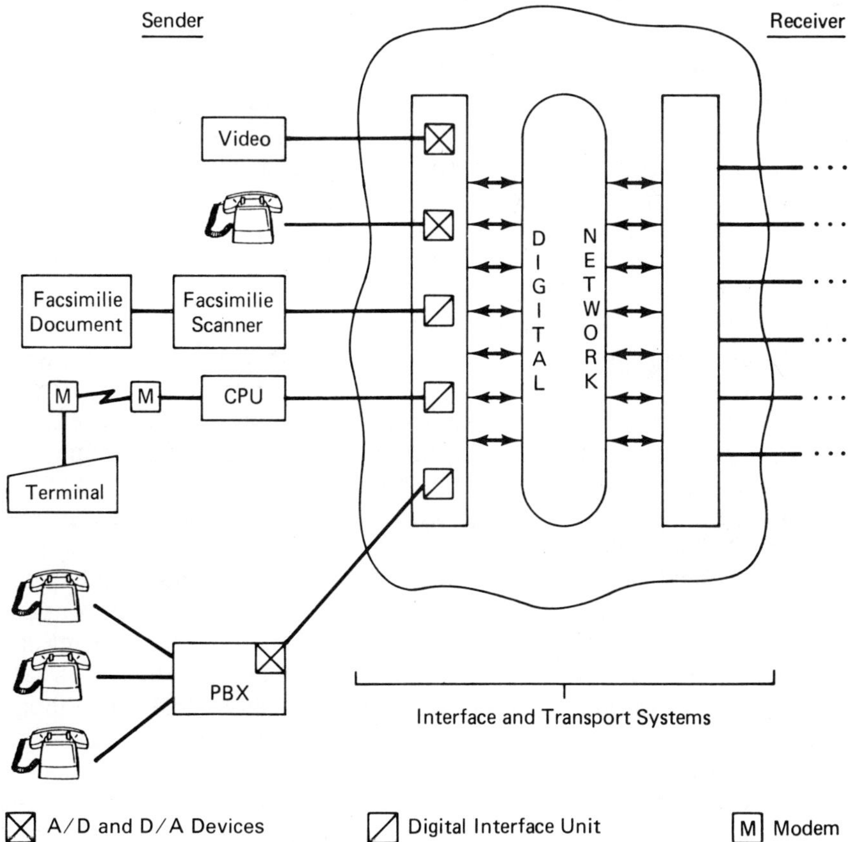

FIGURE 7-2. Integrated Digital Networks.

would provide all the components required to capture the signal and reconstruct it at a later time. A voice-grade line would, therefore, require a minimum sampling rate of 6,600 times a second since the voice-grade band is about 300 to 3300 Hz. Channels actually occupy a 4 KHz bandwidth, so the Bell T1 carrier system uses a rate of 8,000 samples per second.

Digitizing an analog signal consists of three steps: sampling, quantizing, and encoding. The technique is shown in Figure 7-3. Each analog signal is sampled every 125 μsecs (1 second/8,000 samples = .000125). The sample is assigned a level based on the signal amplitude at the time of the sample. The assignment of a level is based on a scale of 128 values and is called a *quantizing scale*. The number 128 permits seven binary bits to be used to represent any one of the possible sample levels. The quantized levels are coded with the seven bits and then converted to a digital pulse stream for transmission. In addition to the seven-bit code, an eighth bit is added for supervisory and signaling purposes. This complete process is termed *pulse code modulation* (PCM) and is the most common digitizing technique in use today.

Common carriers transmit 24 voice channels together with time division multiplexing techniques. Bell's T1 carrier system provides the multiplexing by sampling the 24 channels at a combined rate of 192,000 times per second (8,000 times per second per channel × 24 channels = 192,000). Figure 7-4 shows how the 24 channels are multiplexed into a frame. The frame contains one sample from each channel, plus an additional bit for frame synchronization. Thus, the complete frame is 193 bits (8 bits per channel × 24 channels + 1 sync bit = 193 bits). Since a frame represents only one of the required 8,000 samples per second, a Bell T1 system operates at 1,544,000 bits per second to accommodate all 8,000 frames (193 bits per frame × 8,000 frames = 1,544,000). Each sample in the frame has a TDM slot of 5.2 μsecs (1 sec/192,000 = .0000052).

The 193rd bit performs a function similar to the START bit in an asynchronous transmission and a SYNC byte in a synchronous transmission (see Chapter 2). It is used to establish and maintain synchronization between the sending and receiving sites. The bit alternates as 1 or 0 in each succeeding frame. Since this alternating pattern seldom occurs very long in the data/voice frames, the receiver can "sync" on these framing pulses. The timing mechanisms at the sender and receiver are synchronized to permit the connection of the correct channel of the frame to both ends at the same exact times.

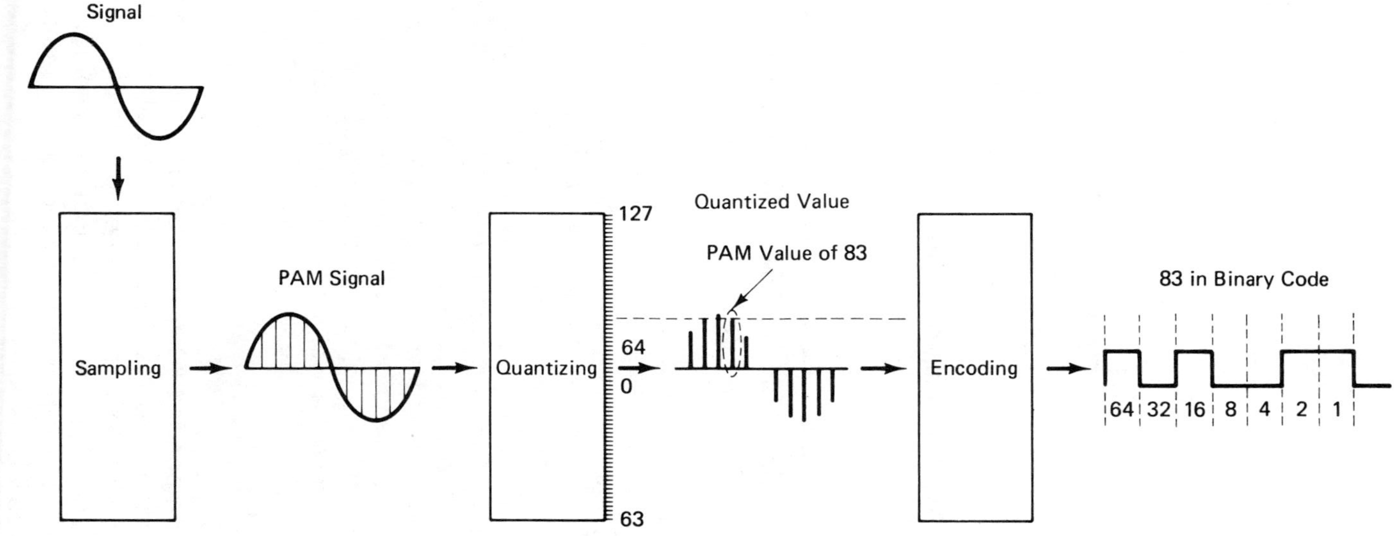

FIGURE 7-3. Sampling, Quantizing, and Encoding.

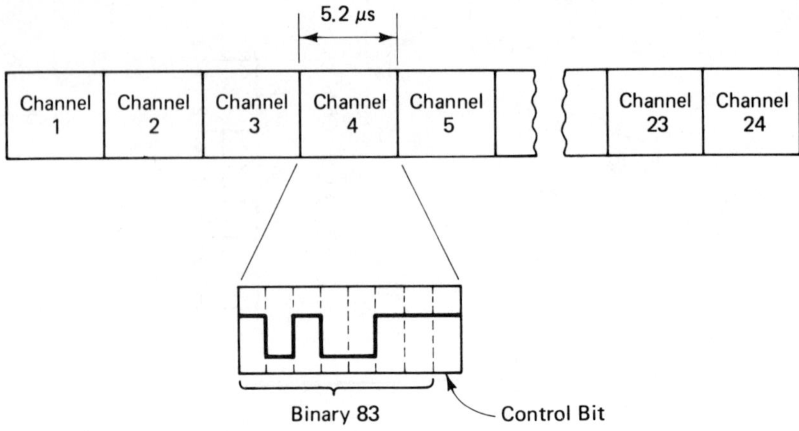

FIGURE 7-4. The Multiplexed Frame.

DIGITIZATION SCHEMES AND VOICE DIGITIZATION RATE

Voice digitization rate (VDR) describes the number of bits required to represent the voice signal. From the preceding description of digital techniques, we know that a voice channel is sampled at 8,000 times per second, with seven bits representing each sample. A 56 Kbit/s VDR is required to carry the coded signal (8,000 × 7 = 56,000). Actually, a higher VDR rate is required to *assure* high voice quality and signal control. The 56 Kbit/s VDR is high and entails a large bandwidth with associated high transmission costs. Yet, a lower VDR requires more expensive translation devices and usually results in a poorer quality of voice reproduction. Current technology encompasses schemes to reduce the VDR but they must be weighed against the cost of signal conversions and the quality of the signal. Most of these schemes use PCM-based techniques.

Companded PCM. A lower VDR can be achieved by taking advantage of two common characteristics of the human ear and speech. First, the ear is more sensitive to low sound levels. Second, speech occurs more frequently at the lower levels. Ear sensitivity is actually logarithmic, experiencing incremental insensitivity at higher levels of sound.

Companded PCM uses a device called a *compander* (compressor-expander) to boost the smaller level signals and attenuate (or hold constant) the larger amplitudes. The process uses nonuniform quantizing levels to give more steps to the smaller amplitude signals (see Figure 7-5); the compressed levels give more gain to these signals. The compander gives more steps to the lower signals to reduce the effect of quantizing noise. A voice signal can never be reproduced exactly because the quantizing steps introduce a discrete, nonanalog function to an analog process. The resulting inaccuracy is called *quantizing noise*. Lower level signals are particularly susceptible to quantizing error so a compander is used to provide more steps to those signals.

Companded PCM, also called log PCM, yields VDRs of 48 to 64 Kbit/s. Bell's D2 system uses this technique to achieve a VDR of 64 Kbit/s using 256 (2^8) quantizing levels; the D2 channel bank provides excellent quality in voice reproduction.

Differential PCM. This scheme is a variation of PCM. An analog voice signal yields consecutive samples that are close to each other on a quantizing scale. Consequently, differential PCM (DPCM)

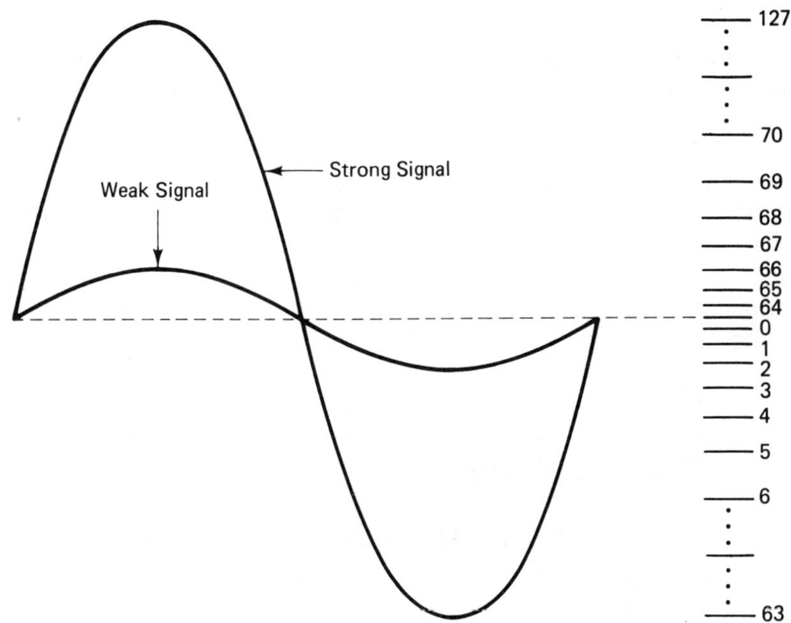

FIGURE 7-5. Companded PCM.

transmits binary pulse streams to represent the difference between consecutive samples and not the sample itself. DPCM yields VDR's ranging between 32 and 48 Kbit/s.

A DPCM codec (coder-decoder) samples an input and uses the signal to apply a predictive weight to the next sample. The value actually transmitted is the difference between the value of the current input and its predicted value. A more sophisticated version, the adaptive DPCM (ADPCM), dynamically alters the quantizing levels based on the input signal's amplitude. The ADPCM yields an effective VDR of 24 Kbit/s.

Delta Modulation. This technique is a variation of DPCM. It encodes the transmitted binary bit stream from the changes of the input samples, but the differences between the samples are transmitted as one of two levels represented by a 1 or 0. The delta modulation sampling device detects changes in successive samples by providing a feedback mechanism of a previous sample to the current sampling gate. Delta modulation is a simple and inexpensive process and achieves a VDR rate of 16 to 32 Kbit/s. Moreover, by adding a compander, the quality of the signal can be improved by providing more quantizing levels to the smaller signals; this technique is called *adaptive delta modulation* (ADM) or *companded delta modulation* (CDM).

Analysis-Synthesis Techniques. These techniques are not widely used in common carrier systems due to their cost. However, they are attractive from the VDR standpoint—generally 2.4 to 9.6 Kbit/s. Analysis-synthesis does not preserve the analog speech waveform as in the other methods. Rather, certain characteristics of the human voice are encoded and transmitted. The codes contain information on resonance frequencies of the vocal tract for particular positions of the tongue, lips, and other speech-related organs. Analysis-synthesis will likely see increasing use as the cost of its speech digitizers (vocoders) decrease.

DIGITAL PULSE CODES

The codes commonly used in data communications are listed below. Other codes are available but these digital schemes are found in most systems today.

- Nonreturn to zero (NRZ).

- Return to zero (RZ).
- Biphase.

Nonreturn to Zero (NRZ)

Figure 7-6(a) shows the NRZ pulse code. The signal level throughout a bit cell (or bit duration) remains stable. In this example, the signal level remains high for a bit 1 and goes low for bit 0. The code could also use a high signal for a 0. The choice depends on the probability of the occurrence of 1s or 0s in the transmission. The alternative chosen should provide the code that yields the greater number of signal transitions in order to improve synchronization.

NRZ is widely used in data communications systems. The asynchronous process described on pages 27 through 29 uses the NRZ coding scheme. The Universal Asynchronous Receiver Transmitter chip (UART) uses NRZ code. The UART has become an informal industry standard.

While NRZ makes efficient use of bandwidth (a bit is represented for every signal change, or baud), it does suffer from the lack of self-clocking capabilities. Self-clocking is best achieved when each successive bit cell undergoes a signal level transition. A continuous stream of 1s or 0s would not create a level transition until the bit stream changed to the opposite binary number. Consequently, the receiver may not know where to begin bit sampling or what group of bits constitute a character (byte). The NRZ scheme requires an independent clocking mechanism, often a separate transmission. This approach solves the non-self-clocking deficiency of NRZ but incurs additional synchronization problems if the data signal and clocking signal "drift" from each other as they traverse down the communications path. The clocking and synchronization problems with NRZ codes can be diminished through the randomized NRZ code. The NRZ signal is passed through a component that randomizes the bit stream to increase the number of signal level transitions.

Return to Zero (RZ)

In return to zero (RZ) code, the cell stays high for a part of the bit duration and returns to a low level for the remaining time. The standard RZ code, RZ-L, represents a 1 with a high level at the beginning of the cell and goes low during the second half. A 0 is

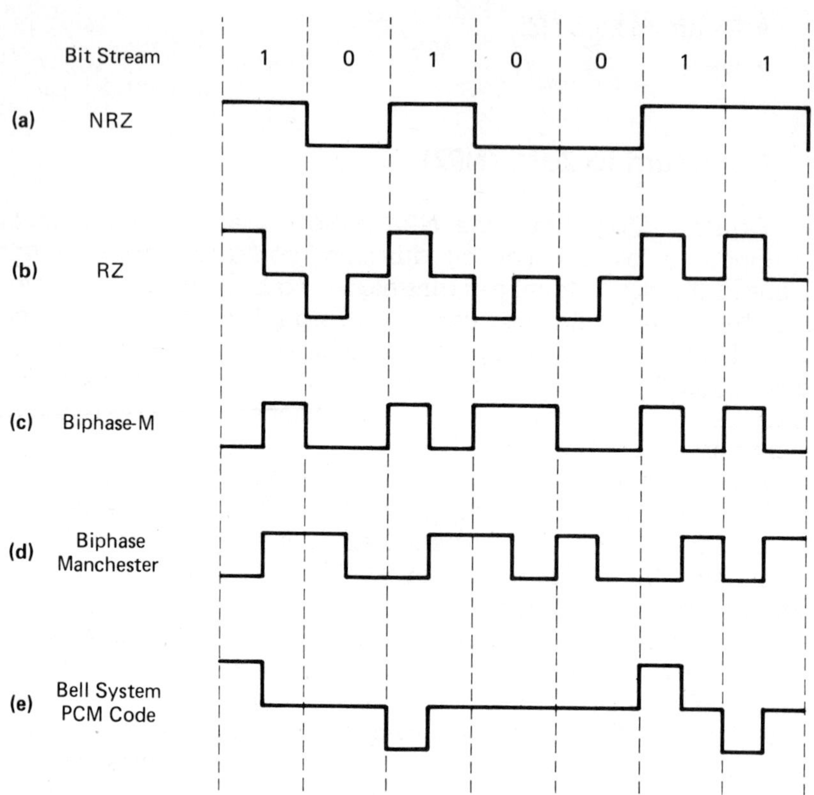

FIGURE 7-6. Digital Pulse Codes.

represented with first a low signal, then a high signal. Figure 7-6(b) illustrates the RZ code. A second form of RZ is ratio encoding which uses two-thirds of the bit cell to represent a high signal level and the remaining one-third to represent a low signal level. The most common forms of RZ use equal periods to represent both levels.

RZ codes provide a transition in every bit cell. Consequently, they have good synchronization characteristics. RZ is a self-clocking code because of its bit cell transition properties. However, since RZ experiences two signal transitions within each cell, its information carrying capacity is not as great as the NRZ code. To achieve the synchronization capabilities, RZ requires a baud rate that is twice the bit/s rate.

Biphase

Biphase codes have several variations and as many names: phase encoding, frequency encoding, and frequency shift encoding. All biphase codes have at least one level transition per bit cell, which is similar to RZ codes. However, most of the biphase codes have the 1s or 0s defined by the direction of the signal level transition.

A biphase code is shown in Figure 7-6(c). This is known as a biphase-M code. Notice that each cell has a level transition and the following characteristics: (a) the 0 cells alternate between low and high and the signal remains for the entire cell; (b) the 1 cells go from low to high *except* when preceded by a low level 0 cell, then go high to low until preceded by a high level 0 cell; and (c) the 1 cells have a transition at midcell. The biphase-M code has high signal transition density that makes it very effective for synchronization purposes. However, it does require more bandwidth.

Biphase codes are found extensively in magnetic recording and in data communications systems utilizing optic fiber links. The codes are used in applications requiring a high degree of accuracy; the code is self-clocking.

One of the most widely used variations of biphase is the Manchester code. It is used in several data communications systems, notably the local area network, Ethernet (discussed in Chapter 9). The Manchester code is illustrated in Figure 7-6(d). It provides a low level to high level in each bit cell to represent a 1, and a high to low level to represent a 0. The Ethernet data rate is 10 mbit/s, so each bit cell is 100 ns long (1 sec/10,000,000 bit/s = .0000001). The double-level transition does require a baud rate that is twice the data rate. However, with the wide bandwidth of coaxial cable (and optic fibers), this does not present a serious problem.

COMMON CARRIER PULSE CODES

AT&T and other common carriers use a variation of the return to zero (RZ) coding scheme [see Figure 7-6(e)]. A high level signal represents 1 bit, *but* the next 1 bit is a low negative signal. In other words, successive 1 bits have opposite polarity. The 0 bits are represented as a low nonnegative level, or an absence of a pulse signal. In addition, the 1 pulses have a level transition at midcell. This scheme is called *bipolar transmission*; the use of alternate polarities to represent the pulse train is known as *alternate mark inversion* (AMI).

The bipolar approach presents problems when a long string of 0s is in the digital frame. The components in the system have no way to synchronize with bit cell level transitions. To avoid this problem, the system substitutes a special signal for a long string of 0s. One such code is high density binary 3 (HDB3), which alters a bit stream with more than three consecutive 0s. The fourth 0 is transmitted with the same polarity as the previous 1; to avoid overriding the AMI scheme, the first 0 may be changed to a 1 to maintain consistency in the AMI concept. Figure 7-7 illustrates a bipolar AMI code without and with HDB3 insertion.

REGENERATIVE REPEATERS

Earlier discussions in this chapter explained that digital regenerative repeaters provided for more error-free transmission than analog signals. The equipment also permits higher data rates. The bandwidth of a communications channel places a restriction on the data rate. If bit pulses are transmitted over a line at too high a rate, the signal becomes distorted and, the faster the data rate, the more the signal is impaired. However, if repeaters are placed close enough together on the line, a high bit rate can be transmitted. All that is needed is to recognize the absence or presence of a binary pulse. Regenerative repeaters permit the

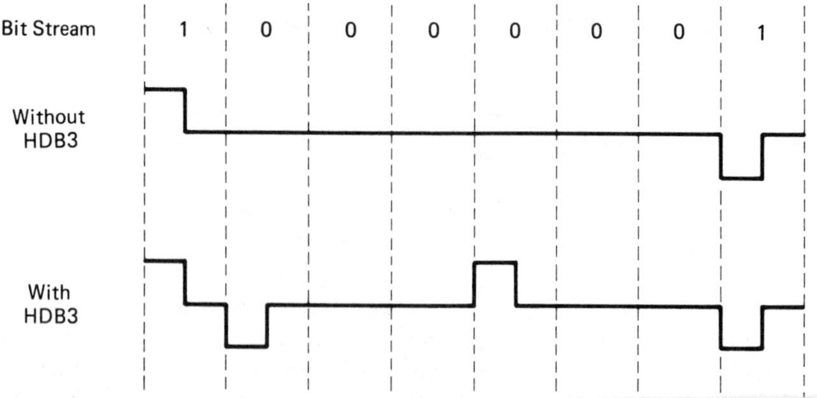

FIGURE 7-7. High Density Binary 3 Insertion.

data rate to be high and the distortion to be severe. If repeaters were placed every few hundred feet, a very high bit rate could be achieved.

The repeaters on the Bell T1 system are spaced 6,000 feet apart. They replace loading coils, which are analog components on the line. The loading coils are used to keep impedance and attenuation constant over the length of the line. They are designed for voice-frequency signals and are not found in present digital carrier systems.

AT&T DIGITAL HIERARCHY

The Bell System digital hierarchy is shown in Figure 7-8. It has five levels of digital data rates that are combined to form the hierarchy by channel banks and multiplexers. The levels are designated T carriers:

T1	1.544 mbit/s	24 PCM voice channels
T1C	3.152 mbit/s	48 PCM voice channels
T2	6.312 mbit/s	96 PCM voice channels
T3	44.736 mbit/s	672 PCM voice channels
T4M	274.176 mbit/s	4032 PCM voice channels

The T1 carrier was developed in the 1960s by Bell Labs to operate over twisted pair cables on interoffice and toll trunks. Regenerative repeaters are placed every 6,000 feet using No. 22 gauge cable pairs. The T1 system carries 24 voice channels. The T1 carrier uses a D1 channel bank (now becoming obsolete). The device uses 128 quantizing steps for the pulse code modulation. The T1C was later developed as an upgrade to T1 lines. It is similar to T1.

The T2 carrier system was implemented in 1972 for use on the intertoll trunks. It carries 96 voice channels or one video channel. The T3 system serves as a bridge between the other carriers. Utilizing the M13 multiplexer, it provides a 44.736 m/bits data rate.

The T4M system is the largest digital carrier. It transmits 274.176 m/bits over a coaxial cable and provides for 4,032 voice channels. A typical coaxial cable conduit contains 18 cables, so the T4M system can handle a considerable amount of traffic.

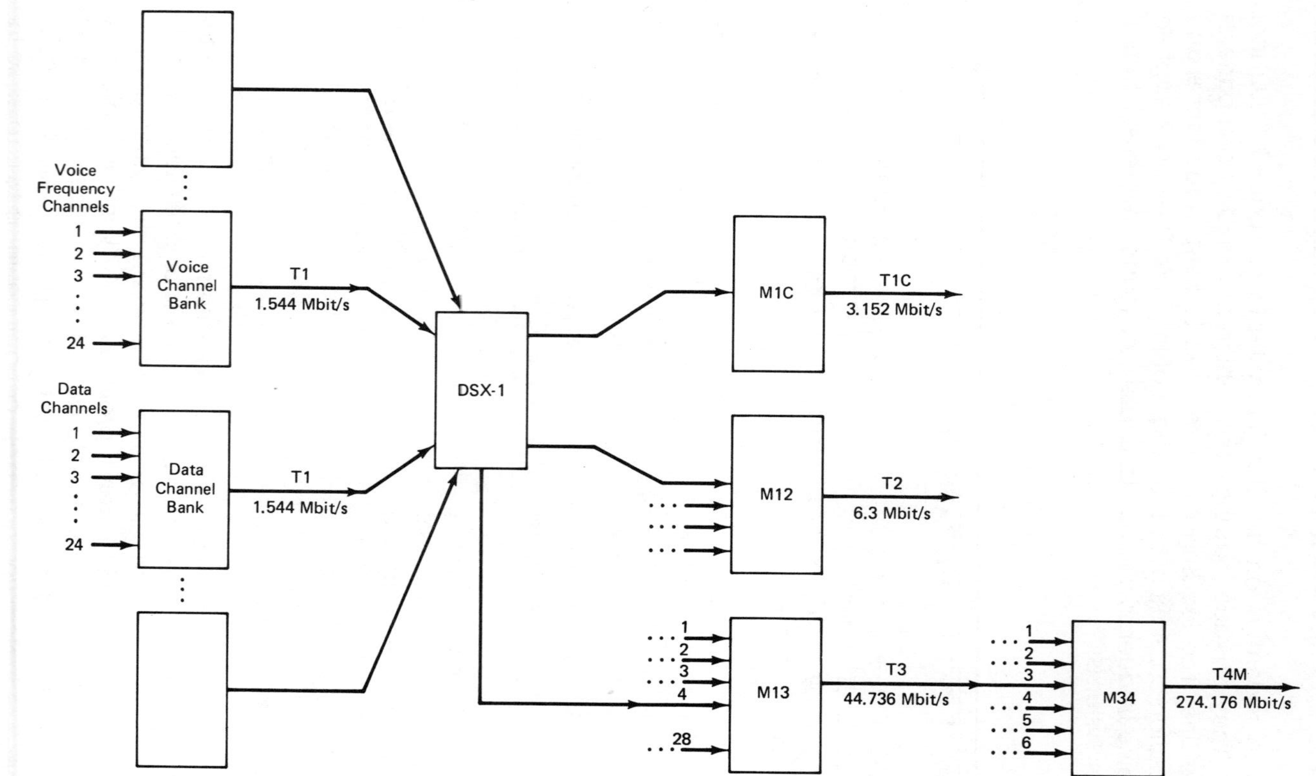

FIGURE 7-8. AT&T Digital Hierarchy.

As with so many other data communications systems, other countries and the CCITT have developed systems that are different from those of the United States. The CCITT recommendation is somewhat similar to the AT&T system. The basic carrier speed is 1.544 mbit/s and both employ a 193-bit PCM frame. However, the synchronization bit alignment in the frames differs and the functions are also different. Moreover, the CCITT standard provides for either 64 Kbit/s or 62.6 Kbit/s data rates. Also, the frame lengths vary significantly. Obviously, integrating international digital networks will require interface and conversion facilities.

DIGITAL PBXs

Digital technology is playing a key role in the booming PBX (private branch exchange) industry. The PBX, in existence for many years, was developed to switch telephone calls in office buildings. Earlier PBXs used crossbar or step-by-step methods to switch analog calls. Today, modern PBXs use computer-based control to perform a multitude of functions—including voice digitization and digital switching.

The digital PBX is replacing its analog counterpart because it is cheaper and easier to maintain and switches native digital devices (terminals, processors) more efficiently. Major suppliers continue to supply analog PBXs (such as AT&T's Dimension) but the orientation is changing.

The digital PBX converts voice signals for digital switching using variations of pulse code modulation (PCM). One popular method is *delta modulation* (discussed earlier), a variation of companded PCM. Delta modulation samples at a higher rate than conventional PCM and transmits one of two levels with a binary 1 or 0. The levels represent a speech waveform in a staircase manner, taking advantage of the high correlation between successive waveform samples.

Some vendors employ the Bell T1 carrier techniques; others are using the CCITT Standard. However, even the PBXs using T1-like techniques usually require additional conversion circuitry.

The newer PBXs offer the following user-initiated telephone features:

- *Call forwarding:* Routing an incoming call to another phone in an office.
- *Automatic callback:* Receiving a call back when a busy extension becomes free.

- *Conference calls:* Connecting more than two parties.
- *Speed calling:* Using an abbreviated dialing number.
- *Automatic recall:* Informing an incoming caller of connection status.
- *Call break-in:* Allowing certain stations to interrupt a call.
- *Satellite operations:* Operating unattended in a remote location.

In addition, some PBXs are integrating power data communications functions:

- Providing support for asynchronous and synchronous transmissions.
- Performing protocol and code conversions.
- Providing RS232-C connections.
- Handling voice and data simultaneously.
- Providing programmable logic.

EMERGING DIGITAL TECHNOLOGY

In the last few years, linear predictive coding (LPC) has been used for voice digitization. Like DPCM, it uses predictive techniques on samples but also periodically updates the predictor parameters. Speech is synthesized into frames corresponding to the samples; the frames contain data on the amplitude, pitch, or frequency of the sample.

LPC is used in a technique called *constructive synthesis* to produce digitized voice. This technique does not preserve the shape of the original voice waveform. Instead, the speech signal is broken down into subcomponents called *phonemes*, which are the basic units of a language. All utterances can be represented by phonemes (for example, two labial phonemes, p and b, are found in *pit* and *bit*). The U.S. style of English language has 42 phonemes.

Constructive synthesis systems accept nonhuman input (such as CRT text), break the words into phonemes, and store them in LSI ROM memory. Later, the phonemes are strung together according to a set of rules to form the speech output. The signal is stored on an LPC speech chip. The phonemes are usually further divided into allophones to enhance the quality of the voice.

A more sophisticated version of constructive analysis is *analysis-synthesis*. This technique analyzes a human voice, encodes the signal, and then uses a prestored vocabulary with LPC analysis to produce the stored digitized voice signal. The primary difference between the two is that analysis-synthesis uses actual speech as input and constructive synthesis uses stringed phonemes. Analysis-synthesis yields better quality speech, but it is more expensive and time consuming. For example, a one-second utterance using analysis-synthesis techniques requires over 100 times as many bit storage cells.

The LPC and synthesis techniques have paved the way for many exciting and useful digital applications. Automobile instrumentation, children's learning tools, automatic answering devices, error messages, and "talking" factory assembly lines are some examples. The technology is nowhere near its maturity. (LPC methods are now being refined to produce even better signals at less cost.) Computer aided instruction (CAI), using these techniques, will substantially alter our concepts of teaching and learning.

8

NETWORK ARCHITECTURES

INTRODUCTION

The term *architecture* is commonly used today to describe networks. Paraphrasing the dictionary definition, an architecture is a formation of a structure. Stated another way, it is a system of structure. A network architecture describes what things exist, how they operate, and what form they take. An architecture encompasses hardware, software, data link controls (DLCs), standards, topologies, and protocols. With the exception of protocols, all of these terms have been described in previous chapters.

Like architecture, the term *protocol* is borrowed from other disciplines and professions. In basic terms, a protocol defines how network components establish communications, exchange data, and terminate communications, just as a diplomatic protocol defines the rules for social parlance.

Data link controls (DLCs, Chapter 7) certainly qualify as one form of protocol—at the line (or link) level. Other protocols are also needed in the network to provide proper communications "parlance" beyond the individual line DLC. As we shall see, these higher level protocols are an integral part of the network architecture.

This chapter highlights two network architectures. The Open Systems Interconnection (OSI) standard is included because of its probable impact in the future. The second example is IBM's

Systems Network Architecture (SNA); it is included because of IBM's position in the industry and the growing use of SNA.

LAYERED PROTOCOLS

Modern networks are implemented using the concept of layered protocols. The early networks providing communications service were relatively simple and did not use layers. Terminals were connected to a computer in which several software programs controlled the terminal transmission and placed the data onto a telephone line. The line was usually attached to an interface unit within or connected to the computer.

As organizations became larger, more complex, and more geographically dispersed, the supporting communications software and hardware assumed more tasks and grew in size and function. Unfortunately, many of these components grew haphazardly. The system often became unwieldy and difficult to maintain. In some instances, telecommunications programs became complex monoliths. When these systems were changed, the resulting output sometimes had predictable results.

The older networks often had several different protocols that had been added in a somewhat evolutionary and unplanned manner. The protocols in the networks had poorly defined interfaces. It was not uncommon for a change in the network architecture at one site to adversely affect a seemingly unrelated component at another site. Often, the components in a network were simply incompatible. The concept of layered protocols developed largely as a result of this situation.

The basic purpose of layered protocols is to reduce complexity, provide for peer-to-peer layer interaction across nodes, and allow changes to be made in one layer without affecting others. For example, a change to a routing algorithm in a network control program should not affect the functions of message sequencing which is located in another layer in the protocol. Layered protocols also permit the partitioning of the design and development of the many network components. Since each layer is relatively self-contained, different teams (perhaps dispersed at various distributed sites) can work on different layers.

Layered functions also owe their origin to several concepts generally called *structured techniques*. These ideas provide an impetus to design hardware or software systems that have clearly defined interfaces. The systems contain modules that perform one function or closely related functions (sometimes called *cohesive-*

ness or *binding*). In addition, these techniques can produce a system in which a change to a module should not affect any component in a system that the changed module does not control. This approach, called either *loose coupling* or *atomic action*, is discussed further in Chapter 11.

Layered network protocols also allow interaction between functionally paired layers in different locations. This concept aids in permitting the distribution of functions to remote sites. In the majority of layered protocols, the data unit passed from one layer to another is usually not altered. The data unit contents may be examined and used to append additional data (trailers/headers) to the existing unit.

For example, layer B might examine a field in a message that was inserted by layer A. The field might contain a logical address of the recipient of the message; layer B would translate this logical address to a node address. Then, perhaps layer C would interpret the node address into an actual physical communications line address. If the line became inoperable, layer C would make the necessary changes to a line address without affecting layers A or B. This is one advantage of layered network protocols: The network can be reconfigured without affecting those components that work with logical or virtual addresses. At the receiving end, layer C would pass the message to layer B (from the communications line); layer B would examine the logical address and institute the actions for the message to go to the proper logical recipient.

The relationship of the layers is shown in Figure 8-1. Each layer contains entities that exchange data and provide functions in a *logical* sense with peer entities at other sites in the network. Entities in adjacent layers interact through the common upper and lower boundaries in a *physical* sense by passing parameters such as headers, trailers, and data parameters. An entity in a higher layer is referred to as N + 1 and an entity in a lower layer is N − 1. The services provided by higher layers are the result of the services provided by all lower levels. The primitives are standard names used to communicate among the layers.

Typically, each layer (except the lowest) adds header information to data. The headers are used to establish peer-to-peer sessions across nodes and some layer implementations use headers to invoke functions and services at the N + 1 or N − 1 adjacent layers (see Figure 8-2). At the transmitting site, an end user invokes the system by passing data, primitive names, and control messages to the highest layer of the protocol. The system passes the data physically through the layers, adding headers and invoking func-

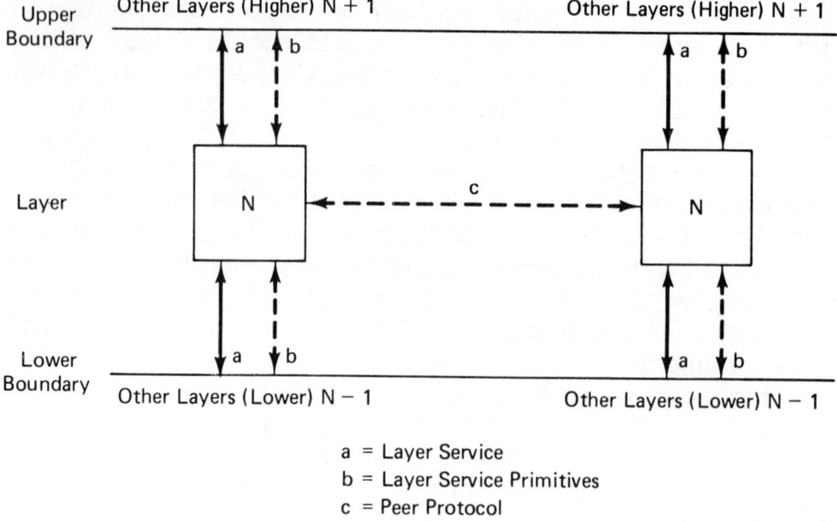

FIGURE 8-1. OSI Layers. (From *Reference Model of Open Systems Interconnection for CCITT Applications*. Proposed Draft Recommendation. CCITT SG VII. Special Rapporteur on Layered Models. Melbourne, Australia, March 26, 1982.)

tions in accordance with the rules of the protocol. At the receiving site, a reverse process occurs. The header and control message invoke services and a *peer-to-peer* logical interaction of entities across the nodes. Generally speaking, layers in the same node communicate with parameters passed thorough primitives and peer layers across nodes communicate with the use of the headers. It is important to emphasize once again that layer (N) behavior should be atomic, exhibiting strong functional binding within itself and loose coupling to layers N + 1 and N − 1.

ISO OPEN SYSTEMS INTERCONNECTION

The ideas just discussed are found in the ISO's Open System Interconnection Model.[1]

[1] *Reference Model of Open Systems Interconnection for CCITT Applications.* CCITT SG VIII. Special Rapporteur on Layered Models. March 26, 1982. Melbourne, Australia.

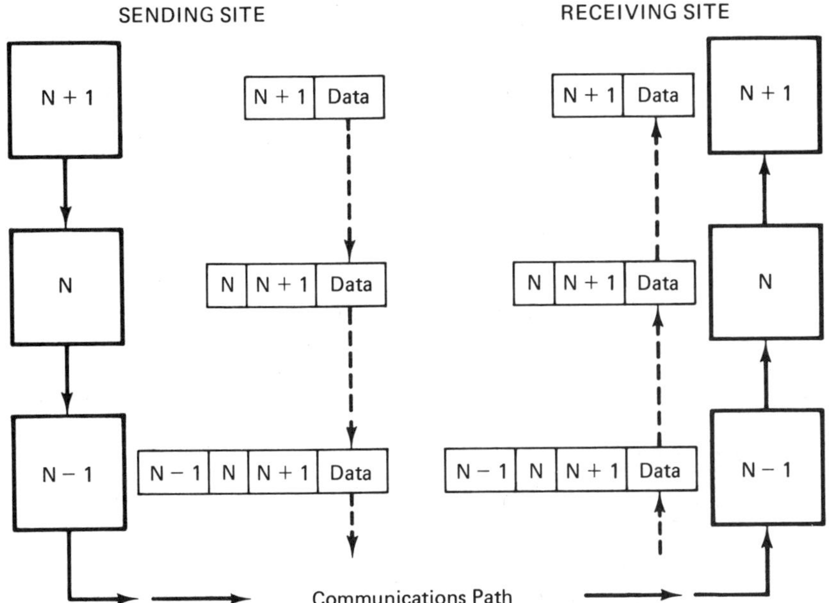

FIGURE 8-2. Layer Interaction.

ISO's Technical Committee 97 formed a subcommittee, SC 16, in 1977 to develop standards for a model of Open Systems Interconnection (OSI) and for the exchange of information between distributed systems. The idea of OSI is to provide protocols for different vendor's/manufacturer's products to connect with each other—thus allowing an open systems interconnection of user applications. The work has proceeded and several layers of the model have now been defined. Figure 8-3 shows the seven layers of the OSI model. The following discussion of the functions in each layer reflects the level of standards and agreements reached thus far. Several layers are fairly well-defined. Others are in process of definition within the ISO working groups.

The ISO layered concept uses the principles explained in the previous section. As illustrated in Figure 8-4, peer entities and functions communicate logically across nodes. The physical flow of the data, headers, and parameters moves through each layer at each node. Each layer's services are defined concisely with specifications of its functions to the peer layer and to the N + 1 and N − 1 layers.

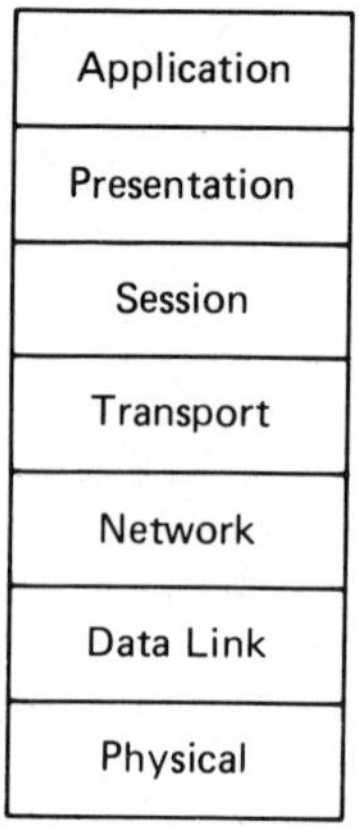

Application
Presentation
Session
Transport
Network
Data Link
Physical

FIGURE 8-3. OSI Layers.

Physical Layer (X.21)

The physical layer provides for the transparent bit transmission between the data link entities. Its purpose is to activate, maintain, and deactivate physical connections between the data terminal equipment (DTE) and the data circuit-terminating equipment (DCE). Physical level standards have been widely used for years. CCITT has established X.21, which specifies the functions at this level

Application
Presentation
Session
Transport
Network
Data Link
Physical
Application
Presentation
Session
Transport
Network
Data Link
Physical
Communications Path
Physical
Logical

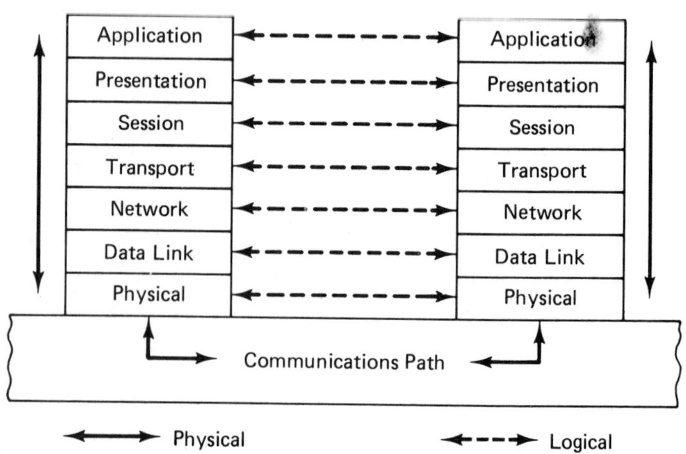

FIGURE 8-4. OSI Layer Interaction.

for leased circuits. (X.21 also specifies circuit-switching functions at the network layer.) RS232-C and RS449 (Chapter 3) are other examples of physical level standards. In the future, X.21 will be used for further enhancements. Since the dominant American Standard, RS232-C, has been used to explain how this level works, we will address, in a general manner, the ISO Standards in this section.

X.24 specifies the functional characteristics of the standard. The DTE/DCE interface provides for two data circuits (T, R), two control circuits (C, I), two timing circuits (S, B), and three common return circuits (G, Ga, Gb). Table 8-1 summarizes the interface.

Two sets of electrical specifications are available. X.26 (like EIA RS-423) and X.27 (like RS-422) describe unbalanced and balanced circuits respectively. X.27 provides for up to 10 Mbit/s across the interface.

X.21 uses the functional and electrical specifications of X.24, X.26, and X.27 (see Figure 8-5). In a leased line operation, the physical level X.21 standard specifies changing the C and I circuits from OFF to ON to initiate data flow between DTEs. The DTEs must be using the same data link (N+1) protocol. The T and R circuits are used to transmit data between the DTE and DCE. The C circuit

TABLE 8-1. X.24 Functional Characteristics.

Circuit Designation	Circuit Name	Data From DCE	Data To DCE	Control From DCE	Control To DCE	Timing From DCE	Timing To DCE
G	Signal ground or common return						
Ga	DTE common return				X		
Gb	DCE common return			X			
T	Transmit		X				
R	Receive	X					
C	Control				X		
I	Indication			X			
S	Signal timing					X	
B	Byte timing (optional)					X	

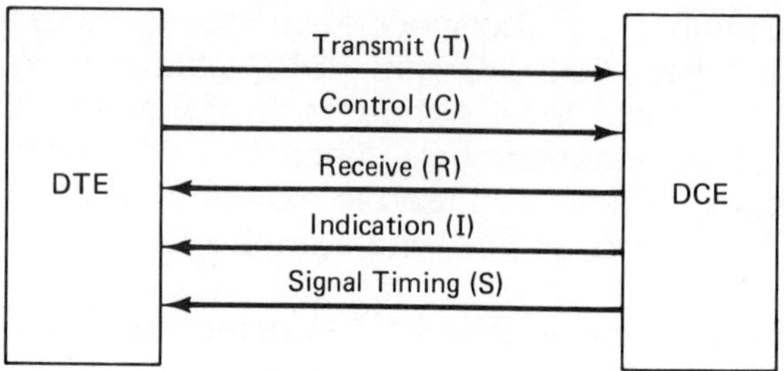

FIGURE 8-5. X.21 Interface.

is used to control data flow and Call Requests (for switched service). The C remains ON for leased circuits. The S circuit provides timing to the DTE and the I circuit serves as an alternate indicator to the DTE. For a more detailed examination of the functions of the physical level, refer to pages 69 through 77.

Data Link Layer (HDLC)

Chapter 6 describes the functions of the data link layer. In that chapter, IBM's synchronous data link control was explained. SDLC is very closely related to ISO's standard, HDLC (High Level Data Link Control). IBM contends that SDLC conforms to a defined operational subset of HDLC—the unbalanced normal class of procedure. This simply means that IBM's approach provides for a primary or master station. HDLC's balanced mode allows for combined stations in which all stations can send *and* receive commands and responses.

The differences between SDLC and HDLC are as follows:

- SDLC supports a subset of HDLC modes, namely the unbalanced normal class.
- HDLC permits any number of bits in the I field (see page 199). SDLC permits any number of eight-bit bytes.
- HDLC does not provide a TEST command and response for link testing.
- SDLC provides for loop operations with the CFGR configure command and the CFGR and BCN (Beacon) responses. HDLC does not include loop configurations.
- Certain other commands/responses vary. For example, IBM does not use the Request Disconnect (RD) response.

The reader can refer to pages 175 through 193 for a detailed discussion of the data link functions. Be aware that HDLC and SDLC are quite similar, but specific implementation planning requires further analysis.

Network Layer (X.25)

The network layer is implemented by the X.25 packet-switching standard. This specification describes how packet-type data is transferred across the data terminal equipment (DTE)/data circuit-terminating equipment (DCE) interface. See pages 89 through 93 for a generic description of packet technology. X.25 establishes the packet format, packet control identifiers, call setup, data flow management, packet windows, call termination, and many other features. The X.25 standard is rapidly gaining acceptance in the industry and will see implementation in many networks and vendor products.

Figure 8–6 illustrates how an end user-to-end user session is established under X.25 protocol. The source site transmits a control packet to request a session or connection. The Call Request packet is transmitted to the sink site where it is either accepted or rejected. X.25 defines limits to the number of end user sessions allowed at one time. Assuming the request is accepted, a Call Accepted packet is returned to the requester. Both of these control packets contain identifiers to provide for a session or call binding. After call establishment, the packets containing user data are exchanged, with X.25 defining the packet flow control and window rules. After all data have been transmitted, a Clear Request control packet is sent to the receiving site and the session can be terminated by a Clear Confirmation packet.

The packets and sessions between the users at the DTE are identified by logical channel numbers. Each packet contains fields to identify a channel group (0-15) and an individual (channel) number (0-255) within the group. X.25 defines 0-4095 logical channels at each packet node; each logical channel operates independently of others. A "free" logical channel must exist in order to complete a session establishment, i.e., a virtual call. The actual range of logical channels used for virtual calls can be established by the specific network implementation.

Certain virtual calls may have preassigned channel numbers to preclude the overhead of the call setup. This X.25 feature is the permanent virtual circuit and is similar to the idea of a leased or private circuit. The end-to-end connection is made for an indefinite period.

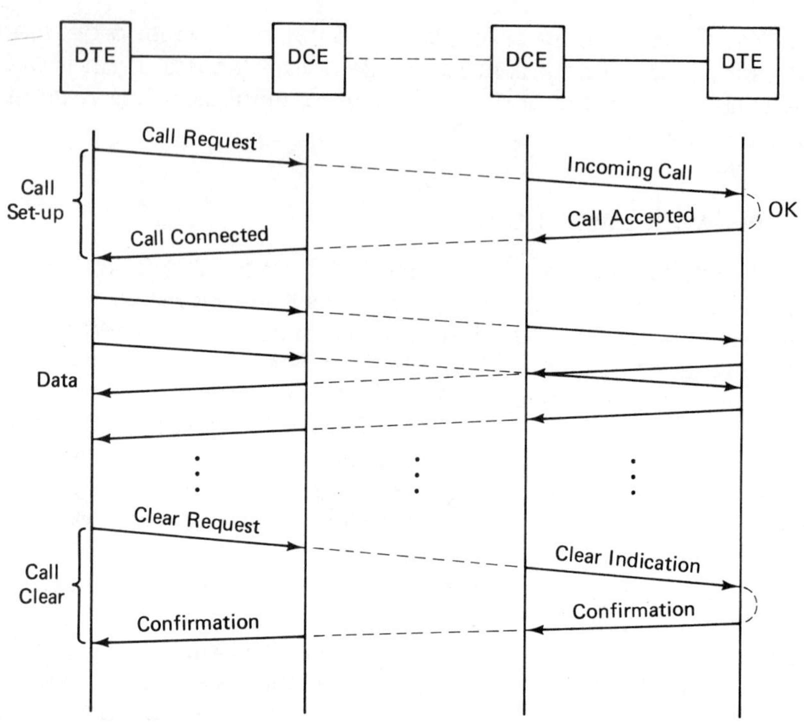

FIGURE 8-6. X.25 Session.

X.25 also provides for applications with short, one-message flows such as point-of-sale, funds transfers, and credit checks. These applications would not benefit from the overhead of a switched virtual call (Call Request, Call Accepted, one packet of data, Clear Request, Clear Confirmation) nor would their occasional use of the network warrant a permanent virtual circuit. The standard meets this need with its datagram and fast select specifications.

The datagram is a completely independent packet containing all control and user data needed to move the packet to the end destination. The datagram does not require any call setup or clearing. The fast select feature provides similar capabilities to the datagram. It may be used in one of two ways: (a) The Call Request/Incoming Call packet contains user data. The subsequent Call Accept/Call Connect may also contain user data. Thereafter, the conventional data transfer and call clearing procedures apply. (b) The Call Request/Incoming Call packet contains user data. The receiving DTE responds with a Clear Indication packet containing user data. The sending DTE must respond with a Clear Confirmation packet containing no user data. One might wonder why X.25 contains both datagram and fast select, since they are quite similar. Both were brought into the standard to accommodate differing views of the member nations sitting on the standards committees.

Figure 8-7 illustrates the logical channel assignments for virtual calls, permanent virtual circuits, and datagrams. Actual assignment is determined by the implementing network. The channels are assigned as follows (refer to Figure 8-7 for explanation of the acronyms):

LC1 to LIC: Range of logical channels assigned to permanent virtual circuits and datagrams.

LIC to HIC: Range of logical channels assigned to one-way incoming virtual calls.

LTC TO HTC: Range of logical channels assigned to two-way virtual calls.

LOC to HOC: Range of logical channels assigned to one-way outgoing virtual calls.

The packet formats are concisely defined by X.25. The formats depend on the type of packet (Call Request, data, etc.). Figure 8-8 shows the format for a Call Request/Incoming Call packet. The general format identifier field indicates the format of the packet. The logical channel group number and the logical channel number fields identify the channel assignment of the call. The third byte or octet contains the identifier of the packet type.

The packet type identifiers for call setups, clearings, and data flows are:[2]

		Bits							
DTE to DCE	DCE to DTE	8	7	6	5	4	3	2	1
Call Request	Incoming Call	0	0	0	0	1	0	1	1
Call Accepted	Call Connected	0	0	0	0	1	1	1	1
Clear Request	Clear Indication	0	0	0	1	0	0	1	1
Clear Confirmation	Clear Confirmation	0	0	0	1	0	1	1	1

[2]The list does not contain all the packet identifiers. Other packets that deal with interrupts, flow control, restart, reset, and diagnostics are beyond the scope of this overview.

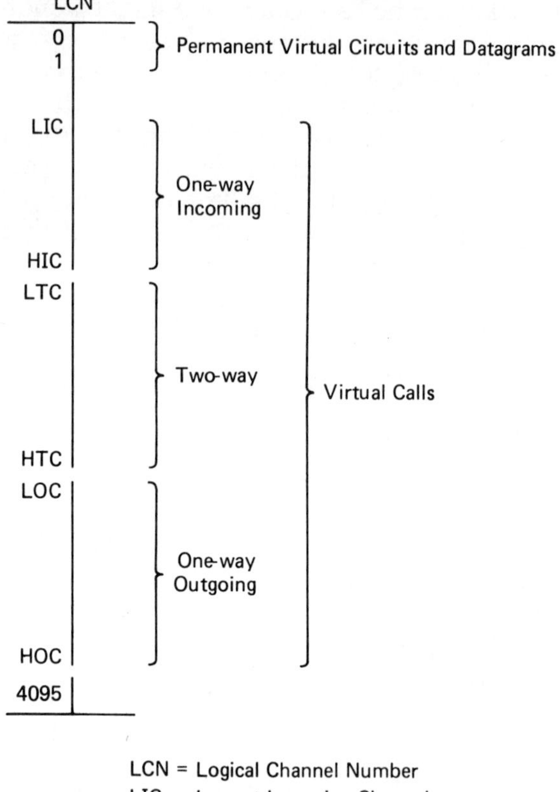

FIGURE 8-7. X.25 Logical Channels.

The address length and address fields contain information on the calling and called DTEs. The facility fields are present when the user site requests facilities (discussed shortly).

The protocol identification field provides for the identification of features such as a datagram. In the event the datagram is used, the call user data field is available for user data. Figure 8-8 is an example of a control packet. An actual data packet also contains fields to maintain sequencing of the packets between the two sites. The fields are the packet receive sequence number P(s) and the packet send sequence number P(s).

X.25 specifies some very useful, optional user facilities. A description of several of these features should give the reader a better idea of X.25 functions.

- Extended packet sequence numbering changes the P(s) and P(r) parameters from 8 to 128.
- Throughput classes assignment provides for a definition of

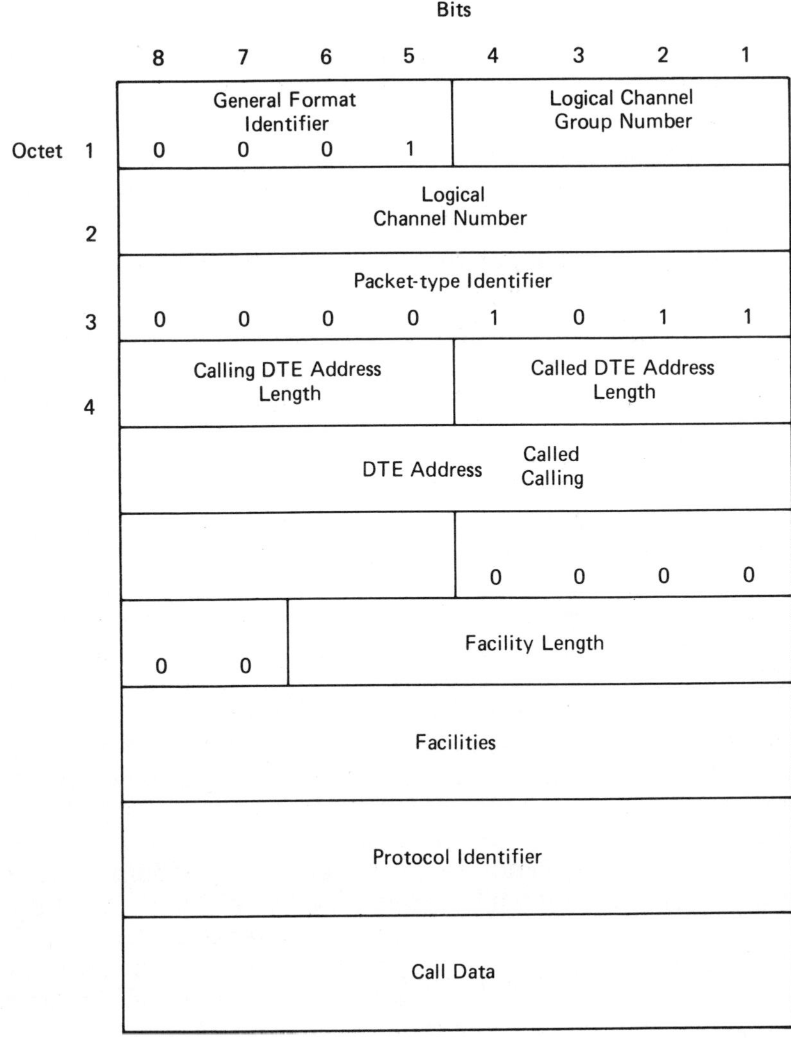

FIGURE 8-8. Packet Format.

the bit/s rate from the calling DTE. Rates vary from 75 bit/s to 48 Kbit/s.

- Packet retransmitting sets rules for retransmission using P(s) and P(r) as well as reject packets.
- Incoming calls barred prevents incoming virtual calls and datagrams from being presented to the DTE.
- Outgoing calls barred prevents the DCE from accepting outgoing virtual calls and datagrams.
- One-way logical channels restricts the logical channel to either outgoing or incoming calls.
- Closed user group permits preestablished DTEs to communicate with each other but precludes communication with all other DTE s.
- Several other user group facilities provide for group incoming access, outgoing access, incoming calls barred, and outgoing calls barred.
- Reverse charging charges the receiving DTE for the virtual call.
- Nonstandard packet sizes allows use of packets other than 128 bytes (octets).
- Datagram and fast select (discussed earlier) provide service to transaction-type applications.

The flow control principles of X.25 are similar to the ideas discussed on DLC windows and flow control in Chapter 7. The control is established at the DTE/DCE interface for each logical channel in each direction. The control is established by the receiver. The P(s) and P(r) fields are used to regulate the windows. Notice that both layer 2 (SDLC/HDLC) and layer 3 (X.25) use windows for flow control. X.25 permits windows (w) of 2 for each direction, or other windows established by a facility. When the P(s) of the next packet to be transmitted by the DCE is within the window, the DCE can transmit. Likewise, the receiving DCE can receive, assuming the packet is in sequence. The transmission of P(r) back to the originator serves to reopen or widen w; when transmitted, it becomes the lower window edge. The value of P(r) must be within the range from the last P(r) received by the DCE and up to and including the P(s) of the next data packet to be transmitted by the DCE. Again, these concepts are quite similar to the windows discussed in Chapter 7.

Transport Layer

The transport layer is not completely defined. General agreement exists that its major function is to relieve the network users of the details of quality and cost-effective service. Subcommittee 16 will soon publish a draft recommendation. The major functions provided by this layer are:

- Mapping transport addresses onto network addresses.
- Multiplexing transport connections onto network layer connections to increase user throughput across the layer.
- Error detection and monitoring of service quality.
- Error recovery.
- Segmentation and blocking.
- Flow control of individual connections of transport layer to network and session layers.
- Expedited data transfer.

The transport layer establishes a transport connection between two users by obtaining a network connection that best matches the user requirements for costs, quality of service, multiplexing needs, data unit size, and address mapping. During the data transfer the layer provides for sequencing, blocking, segmenting, multiplexing, flow control, identification, error control, and error recovery. Error detection and recovery are important considerations since the transport layer is the first layer at the end user site. Although the lower layers may be physically located in an end user location, they are considered part of the network and are not in the user's domain.

The ISO has adopted a five class approach for the transport layer. The classes are established to accommodate different lower layer (N-1, N-2, N-3) entity functions and are based primarily on the amount of error checking and recover furnished by the lower layers. The classes are summarized as follows:

Class 0: Provides very little error recovery. Oriented toward text transmission.

Class 1: Provides for some error recovery. Oriented toward the use of X.25 at the network layer.

Class 2: Provides for more error recovery. Oriented toward a reliable network that provides error recovery but not much error notification to higher levels. This class also provides for the multiplexing capability.

Class 3: Combines the provisions of Classes 1 and 2.

Class 4: Provides for extensive error detection and recovery. Checks for damaged data and lost and out-of-sequence packets. It is supportive of the self-contained datagrams.

Session Layer

A formal description of the session layer is not available yet. The purpose of the layer is to provide the means for the presentation entities to organize and synchronize their dialogue and manage their data exchange. The following services are provided by the layer:

- Quarantine service allows the sending presentation layer to request that one or more data units be held from the receiving presentation layer until the sending entity releases the data. The session layer will also discard all the data if requested and the receiving entity will be unaware of these actions.
- Interaction management establishes a two-way simultaneous interaction (TWS), one-way interaction, or a two-way alternate (TWA) interaction between the peer presentation layers.
- Expedited and normal data exchange provides for methods of prioritizing data traffic.
- Error recovery and exception reporting are also provided, but details are not available.

Presentation Layer

The presentation layer has some of its functions defined. It provides services such as code and character set translation, formating, and syntax resolution. Presently, working groups are developing three protocols for the presentation layer. The first standard, the virtual terminal protocol, will permit a number of different types of terminals to support different applications. Next, the virtual file protocol will provide for code conversion in files, file communication, and file formating. Third, a job transfer and manipulation protocol will provide for control of jobs and record structures.

Application Layer

The application layer is also not complete. Generally, it provides the sole means for a user application, or a person operating a terminal, to access the lower layers. Three categories of service elements (SE) in the application layer are recognized:

- Common application layer SE provides for services from lower layers independently of the nature of the user application.
- Application specific SE provides for capabilities required to satisfy data/information transfers such as data base access, bulk data transmittals, and remote job entry (RJE) transfers.
- User specific SE provides the means to satisfy specific applications such as credit checks and point-of-sales.

OTHER ISO STANDARDS

The ISO layered model holds great promise and, as the top layers are published, many vendors will develop their products in accordance with the specifications. The lower three levels already have seen wide use and X.25 is becoming quite prevalent in many networks. Yet, one is still left with the problem of interfacing existing techniques and protocols to the new methods. For example, the asynchronous, stop-and-wait terminals (discussed on pages 27 through 29) are widespread and not likely to go away in the near future. Consequently, ISO has published a provisional standard for nonpacket mode interface into the X.25 layers. Figure 8-9 shows the packet assembly/disassembly (PAD) function, also known as X.3. Its basic functions are as follows:

- Provides for a start-stop, asynchronous DTE (terminal) access into the X.25 DTE (computer).
- Assembles characters into packets destined for packet-mode DTE.
- Disassembles packets destined for start-stop DTE.
- Handles virtual call setup and clearing procedures.
- Provides a protocol conversion for the start-stop DTE into the X.25 layer.

The specifications actually encompass three protocols. X.3 defines the PAD; X.28 defines the protocol between the terminal

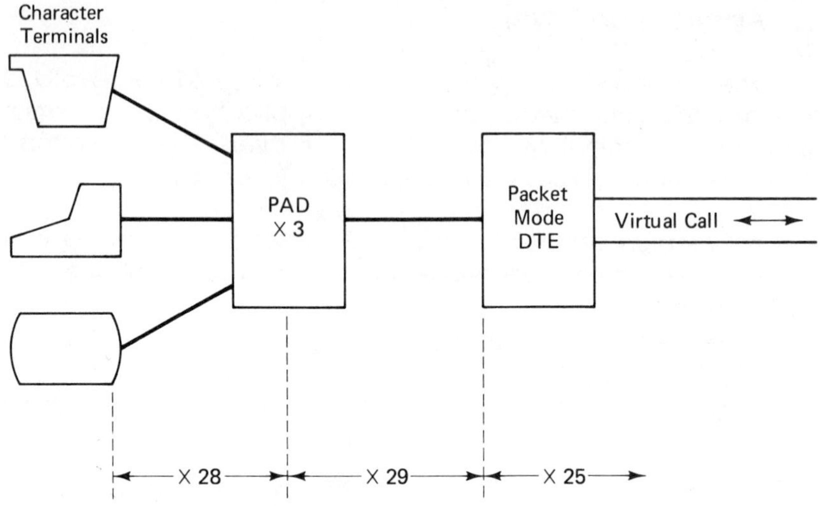

FIGURE 8-9. Packet Assembly/Disassembly (PAD) Function.

and the PAD; X.29 defines the protocol between the PAD and the packet-mode DTE. X.25 then provides the interface between the packet-mode DTE and the DCE. Controversy still exists on certain details of the three protocols and some vendors are implementing PAD-like functions, but not exactly as described in the PAD specifications. Nonetheless, the specifications are valuable and useful; they should be studied before an organization chooses to develop its own interface. A few additional comments will give the reader a better understanding of this important interface protocol.

The PAD facility keeps a profile of each terminal it services. The terminal user selects certain functions that the PAD is to perform. The possible functions include the terminal's ability to escape from a data transfer state and a character echo back to terminal. The terminal also has the ability to signal (data forwarding) that all data has been transferred by sending the PAD a carriage return, a DELETE, a timeout, or some other ASCII control character. In addition, an idle time delay service gives the PAD permission to send data if no data forwarding signal has been transmitted. The PAD also provides the terminal with an option to discard all traffic sent to it. The terminal has carriage return and line folding options, as well as the ability to select input/output speeds ranging from 50 bit/s to 64 Kbit/s.

The start-stop mode DTE gains access to the PAD through a switched or leased telephone circuit using V.21 (up to 300 bit/s) or V.24/V.28 (RS232-C).

X.28 provides procedures for the terminal and PAD interface. The terminal sends signals to the PAD requesting certain functions such as setting up a virtual call. In turn, the PAD notifies the terminal of its activity in fulfilling the terminal commands and its servicing of the preestablished profile. Table 8-2 shows the terminal command

TABLE 8-2. Terminal Command Signals.

DTE → PAD		DTE ← PAD
PAD command signal format	Function	PAD service signal sent in response
STAT	To request status information regarding a virtual call connected to the DTE	FREE or ENGAGED
CLR	To clear down a virtual call	CLR CONF or CLR ERR (in the case of local procedure error)
PAR? List of parameter references	To request the current values of specified parameters	PAR (list of parameter references with their current values)
SET? List of parameter references and corresponding values	To request changing or setting of the current values of the specified parameters	PAR (list of parameter references with their current values)
PROF (identifier)	To give to PAD parameters a standard set of values	Acknowledgment
RESET	To reset the virtual call or permanent virtual circuit	Acknowledgment
INT or INTD	To transmit an *interrupt packet* to the packet mode DTE	Acknowledgment
SET List of parameters with requested values	To set or change parameter values	Acknowledgment
Selection PAD command signal	To set-up a virtual call	Acknowledgment

From: The *Data Communications* Seminar on Standard Architecture, Interfaces, and Protocols, December 9–11, 1980, Dallas, Texas.

signals; Table 8-3 contains the PAD service signals; Table 8-4 contains the clear indication PAD service signals.

X.29 specifies the protocol between the PAD and the packet-mode DTE. The X.25 packet (see Figure 8-8) contains a field to identify the call request from a start-stop DTE. The first octet (bits 1-4) contains codes to determine the use of the PAD message. Subsequent octets contain parameters values, status flags, and error codes.

POTENTIAL PROBLEMS OF MERGING TWO TECHNOLOGIES

Many communications systems were designed using the polling/selection concept (see Chapter 6) which uses overhead messages to poll and select the remote sites and to ACK or NAK the

TABLE 8-3. PAD Service Signals.

Format of the PAD service signal		Explanation
RESET	DTE	Indication that the remote DTE has reset the virtual call or permanent virtual circuit
	ERR	Indication of a reset of a virtual call or permanent virtual circuit due to a local procedure error
	NC	Indication of a reset of a virtual call or permanent virtual circuit due to network congestion
COM	—	Indication of call connected
CLR	See Table 8-2	Indication of clearing
PAD identification PAD service signal	The characters to be sent are network dependent and are for further study	
ERROR	ERR	Identification that a *PAD command signal* is in error

From: The *Data Communications* Seminar on Standard Architecture, Interfaces, and Protocols, December 9–11, 1980, Dallas, Texas.

TABLE 8-4. PAD Clear Indication Service Signals.

Clear indication PAD service signal	Possible mnemonics	Explanation
Number busy	OCC	The called number is fully engaged in other calls
Network congestion	NC	Congestion conditions within the network temporarily prevent the requested virtual call from being established
Invalid call	INV	Facility invalid requested
Access barred	NA	The calling DTE is not permitted to obtain the connection to the called DTE. Incompatible closed user group would be an example
Local procedure error	ERR	The call is cleared because of a local procedure error
Remote procedure error	RPE	The call is cleared because of a remote procedure error
Not obtainable	NP	The called number is not assigned or is no longer assigned
Out of order	DER	The called number is out of order
Clearing after invitation	PAD	The PAD has cleared the call following the receipt of an invitation to clear from the packet mode DTE

From: The *Data Communications* Seminar on Standard Architecture, Interfaces, and Protocols, December 9–11, 1980, Dallas, Texas.

receipt of data. Polling entails a solicitation for data. On the other hand, packet-switching technology entails data exchange only on demand (with a Call Request, a datagram, or a fast select). In addition to the PAD-like facilities of X.3, X.28, and X.29, a public network vendor often supports a protocol-conversion function of polling/selection to packet switching for its customers. The user should examine carefully the function from the standpoint of (a) end-to-end integrity and (b) costs to transmit data.

Figure 6-2 provides a review of the polling/selection protocol. Figures 8-10, 8-11, and 8-12 depict three options in implement-

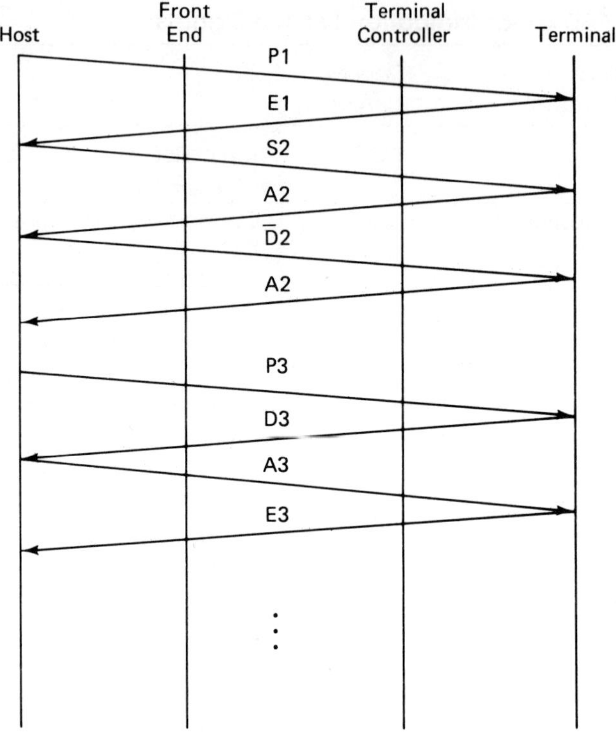

SYMBOLS

Pn = Poll for Terminal n
Sn = Select for Terminal n
En = EOT for Terminal n
An = ACK
Nn = NAK
Dn = Data from Terminal n to Host
$\overline{D}$n = Data from Host to Terminal n

FIGURE 8-10. Option 1 of Polling/Selection Packet Switching Interface.

ing the interface.[3] In the first option, *all* messages are transported from end-to-end, including polls, ACKs, NAKs, selects, and negative responses to polls. This option is quite attractive from the

---

[3]Analysis conducted by Charles E. Bading, Patrick J. Nichols of Western Union, McLean Virginia and Walter I. Landauer of Computer Sciences Corp., Falls Church, Virginia. Study was conducted for the Department of Defense AUTODIN II packet network, September 1979.

standpoint of simplicity and end-to-end accountability (two features that should be given high priority). However, the network uses much of its capacity carrying packets containing the overhead messages of the polling/selection protocol.

Option 2 terminates the polling/selection protocol locally at both ends of the network. The host and front end establish their own polling/selection routine, as do the terminals and the terminal controller. Polls, selects, and other commands are issued to buffers at the local sites. The transfer of data across the packet network occurs when either the front end or terminal controller ACK a block of data sent from the host or a terminal. Due to this approach, the buffers at the sites might overflow. Consequently, the interface logic must provide for closing the receive windows of the front end or controller when their buffers fill. Figure 8-11 shows the window closing at point 1 after the D2 packet is received. The window is open at point 2 upon a buffer being released later in the process.

Option 2 certainly cuts down much of the overhead found in option 1. Only user data is packetized for transfer through the network. However, the end user loses end-to-end accountability with this option. The user must assume that everything is working correctly and that all data sent is received and forwarded to the user application or terminal. Another disadvantage is the software complexity to manage the local protocols and control the flow of messages into and out of the buffers.

The third option represents a compromise between the first two choices. This approach continues with a local protocol termination but transmits end-to-end ACKs/NAKs and selects through the network. Option 3 must provide for the potential cross-flow through the networks of conflicting packets. For example, at point 1 in Figure 8-12 the host has issued a select to terminal 2. However, at the same time data from terminal 2 has been transmitted to the front end buffer. A typical polling/selection protocol requires an ACK to terminal 2 in this situation. However, the terminal controller must transmit a NAK that forces the host to poll the data from terminal 2; after the poll, a select is then issued and data is finally transmitted to the host.

Point 2 in Figure 8-12 illustrates other software complexities. The front end responds negatively to the select 4 because data from terminal 3 is awaiting service. Although two different terminals are involved, the terminal controller expects an ACK for *any* outstanding data block before accepting any other commands.

The advantages of option 3 are end-to-end accountability and the absence of polls across the packet network. The software complexity is the main disadvantage of option 3.

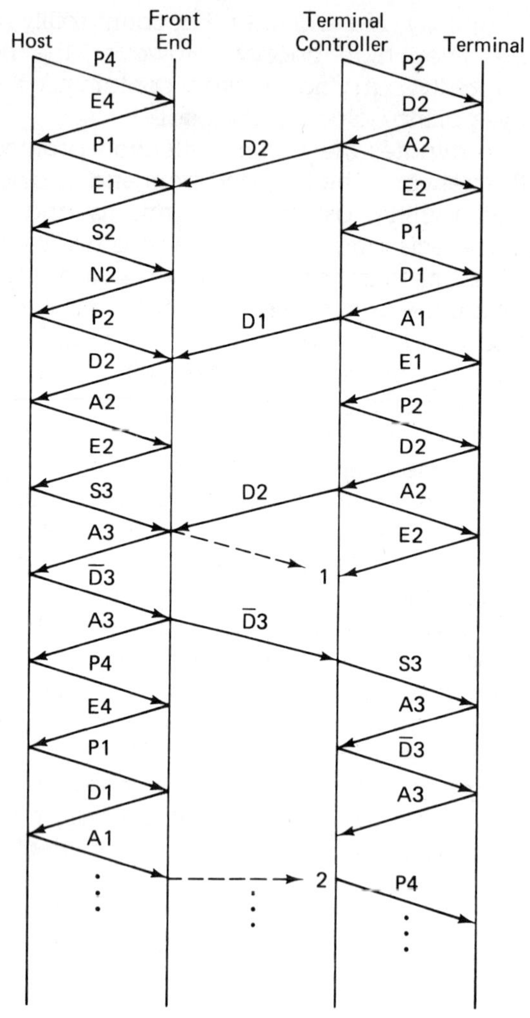

SYMBOLS

Pn = Poll for Terminal n
Sn = Select for Terminal n
En = EOT for Terminal n
An = ACK
Nn = NAK
Dn = Data from Terminal n to Host
$\overline{D}$n = Data from Host to Terminal n

FIGURE 8-11. Option 2 of Polling/Selection Packet Switching Interface.

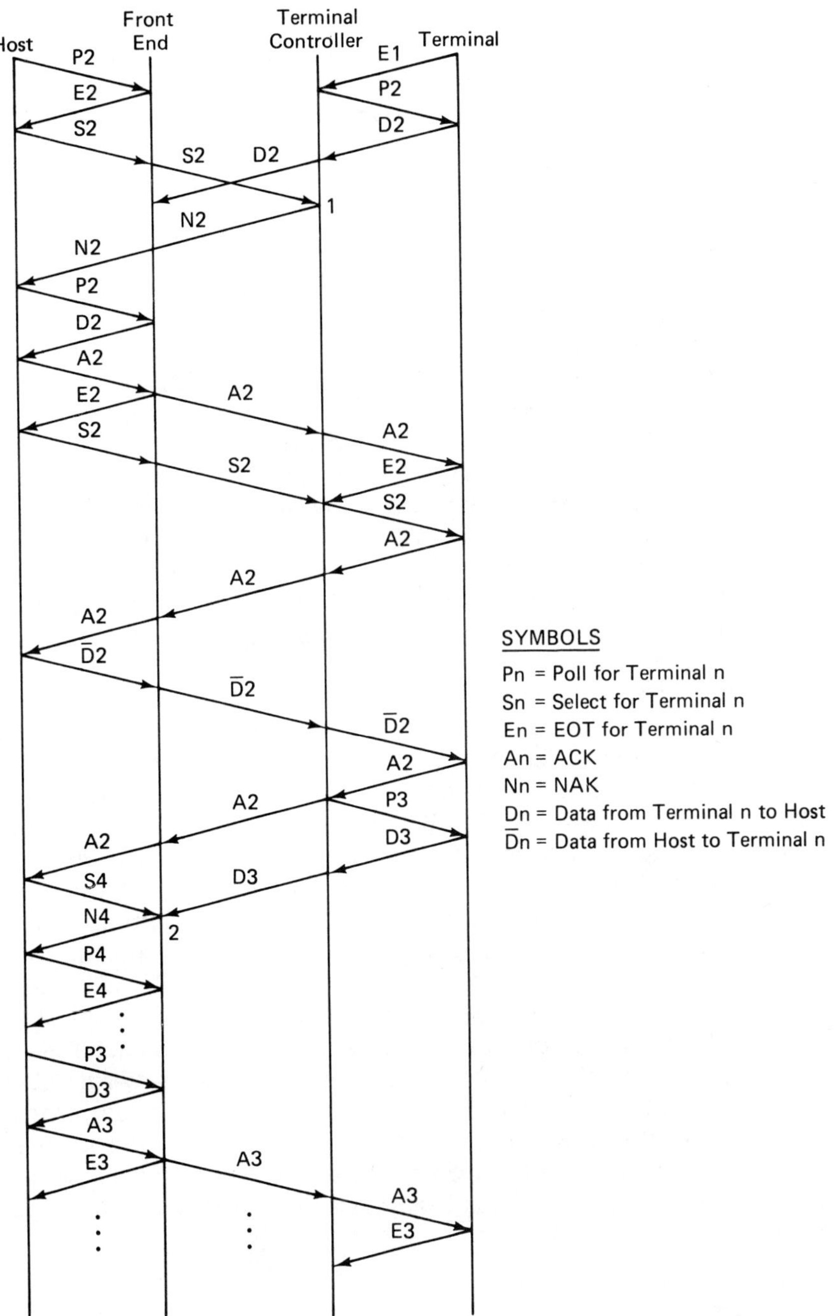

FIGURE 8-12. Option 3 of Polling/Selection Packet Switching Interface.

Public networks should provide for a variation of these options. Since users ordinarily pay for usage based on packet volume, the vendor-specific approach is well worth a close examination. Moreover, the vendor's method of providing data integrity and end-to-end accountability is often of paramount importance for certain user applications.

SYSTEMS NETWORK ARCHITECTURE (SNA)

IBM introduced Systems Network Architecture (SNA) in September 1973 as its major commitment to communications systems and networks. After a slow start, 350 SNA sites existed by 1976. At year-end 1978, the number had increased to 1,250 and, in 1980, over 2,500 installations had SNA. Prior to SNA, IBM had more than 35 major different communications software (CSW) products and 15 different data link controls (DLC). In addition to providing a cohesive communications system, SNA is intended to eliminate many of these incompatible products. Our discussion will provide an introduction to SNA with a description of its concepts and unique terms.

SNA is a specification describing the architecture for a distributed data processing network.[4] It defines the rules and protocols for the interaction of the components (computers, terminals, software) in the network. SNA is organized around the concept of a domain (see Figure 8-13). An IBM host node contains VTAM (see Chapter 4) and the System Services Control Point (SSCP), which is the focal point in the network for managing the configuration, operation, and sessions of components within the domain. Each SSCP in a network has its own domain. The communications controller node (CUCN) is IBM's 3705 front end with NCP. (See Chapter 4 for a description of these facilities.) The cluster controller node (CCN) is a peripheral device (much like a terminal controller) that controls a variety of devices. IBM has developed several CCNs that perform special applications functions, such as banking transactions and retail point-of-sale. The terminal node is the farthest point out in SNA's distributed path. Unlike the other nodes, it is not user-programmable and has less processing capability than the cluster node.

---

[4]*Systems Network Architecture, Concepts and Products*, IBM document No. GC30-3072-0. IBM, Dept. E02, Box 12195, Research Triangle Park NC 27709. For additional information refer to Chapter 4, which contains examples of VTAM and NCP code used to configure SNA components such as LUs, PUs, and links.

End users in an SNA network are individuals or applications programs. An end user is not considered part of SNA, so a logical unit (LU) acts as an access point into the network. The logical unit is software or microcode. An end user-to-end user session, as depicted in Figure 8-13, requires an LU-LU session to acquire the resources for the end users. The LUs provide for any buffers, processor capacity, and software required to satisfy the end user requirements. Each LU has a network name associated with it. The name is used by SNA to determine a network address and the actual location of the needed resources. The end user is not concerned with the physical aspects of the network.

SNA provides the same kinds of functions available in ISO's Open Systems Interconnection (OSI) model, but the actual implementation differs. For example, SNA does not use the packet switching ideas found in ISO's network layer. However, it does provide for extensive user session services such as pacing, expedited routing, and dialogue control.

SNA is organized further into network addressable units (NAUs). The system provides for three kinds of NAUs: logical units (LU), System Services Control Point (SSCP), and physical units (PU). The physical unit represents the terminal, controller, or computer to the SNA network. Every node has one PU address and one or more LUs. In addition to LU-LU sessions, the other network addressable units establish sessions to provide user services. For example, SNA establishes an SSCP-PU session to activate a communications line. Also, an SSCP-SSCP session is used to distribute and share functions across domains in the network.

In addition to the four nodes previously described, SNA distinguishes between subarea or peripheral nodes (see Figure 8-13[b]). A subarea node can route user data (messages) throughout the entire network. It uses the network address and a route number to determine a transmission line to the next node in the network. A peripheral node is more locally oriented. It does not route between subarea nodes. The nodes are connected by common carrier links and the links are controlled by synchronous data link control (SDLC). Messages are routed across the links to/from the nodes based on the address contained in the message.

Every SNA domain component is identified by a subarea address and an element address. The subarea address is the same for all NAUs in the subarea and the element address is unique to each NAU in the subarea. In Figure 8-13(b), three subareas are identified as part of the SSCP domain. The elements within the subareas are NAUs identified with the element address. Each subarea node has a routing table containing the components and

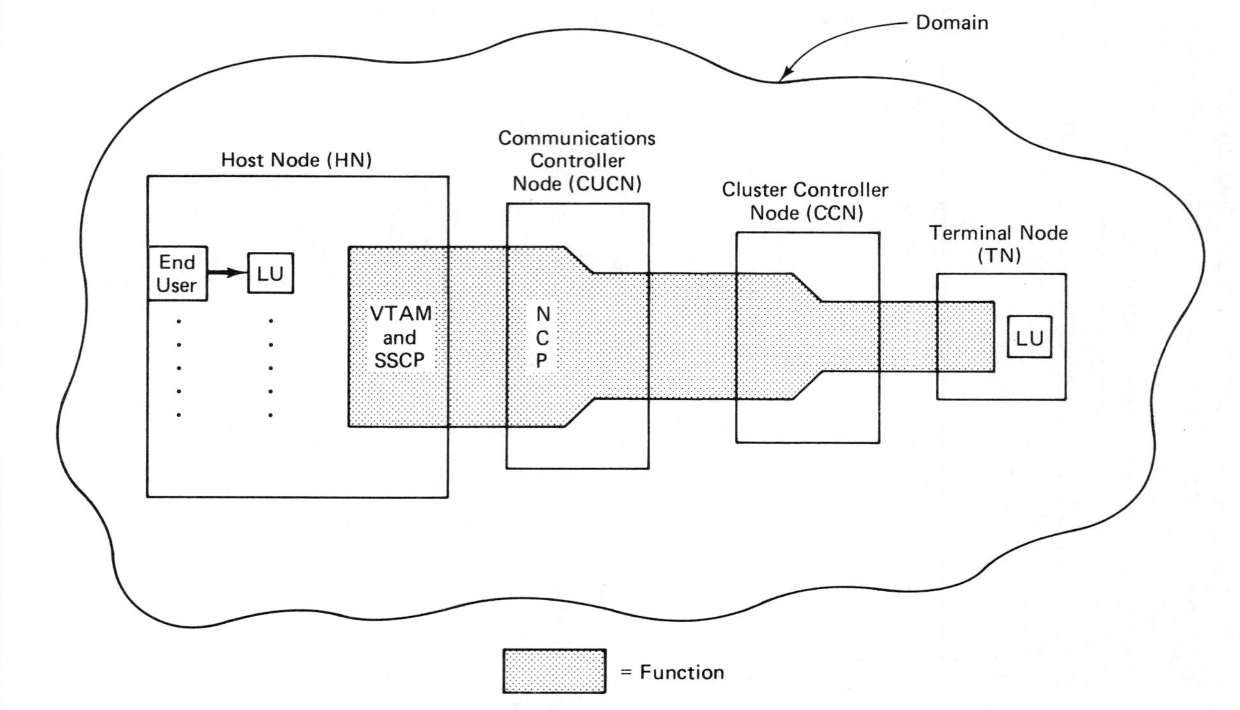

(a) SNA Domain and Nodes

FIGURE 8-13. SNA (From *SNA for Managers*, IBM Manual #SR20- 4517-0.)

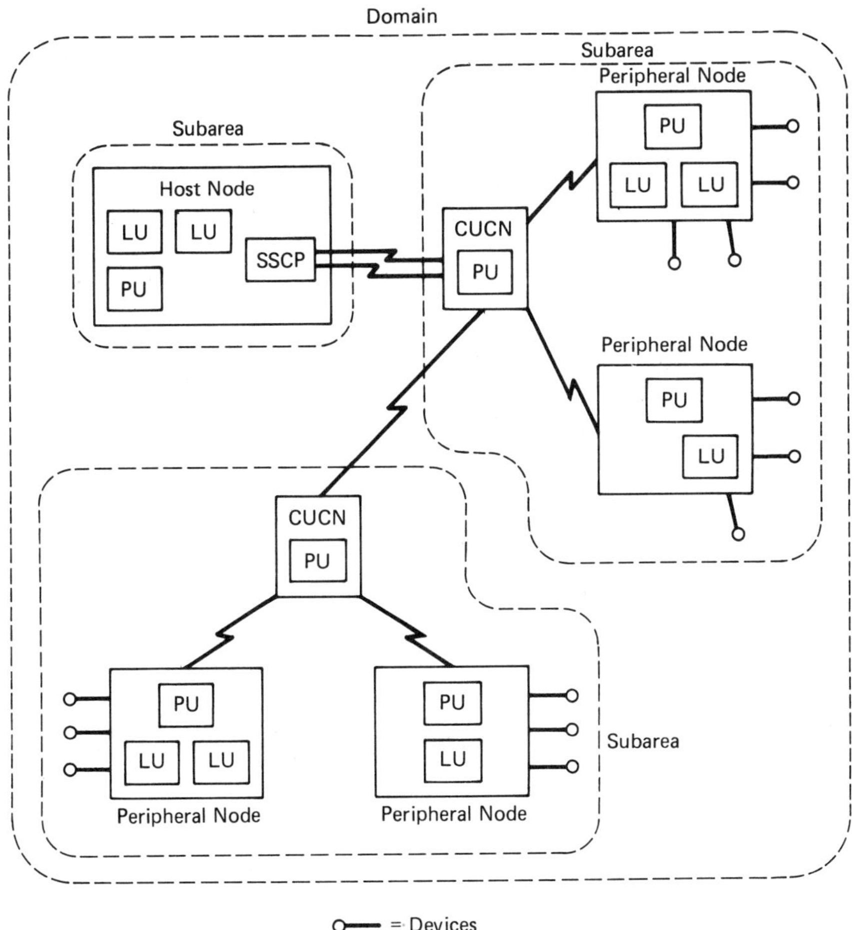

(b) Subarea and Peripheral Nodes

FIGURE 8-13 (Continued)

links attached to it. A message is passed from one subarea node to the next until it reaches its destination subarea. The element address is then used to pass the message to its destination NAU.

Like packet switching, SNA provides for more than one route between any two subarea nodes. This technique, called *multiple routing*, allows traffic loads for different sessions to be distributed over several routes.

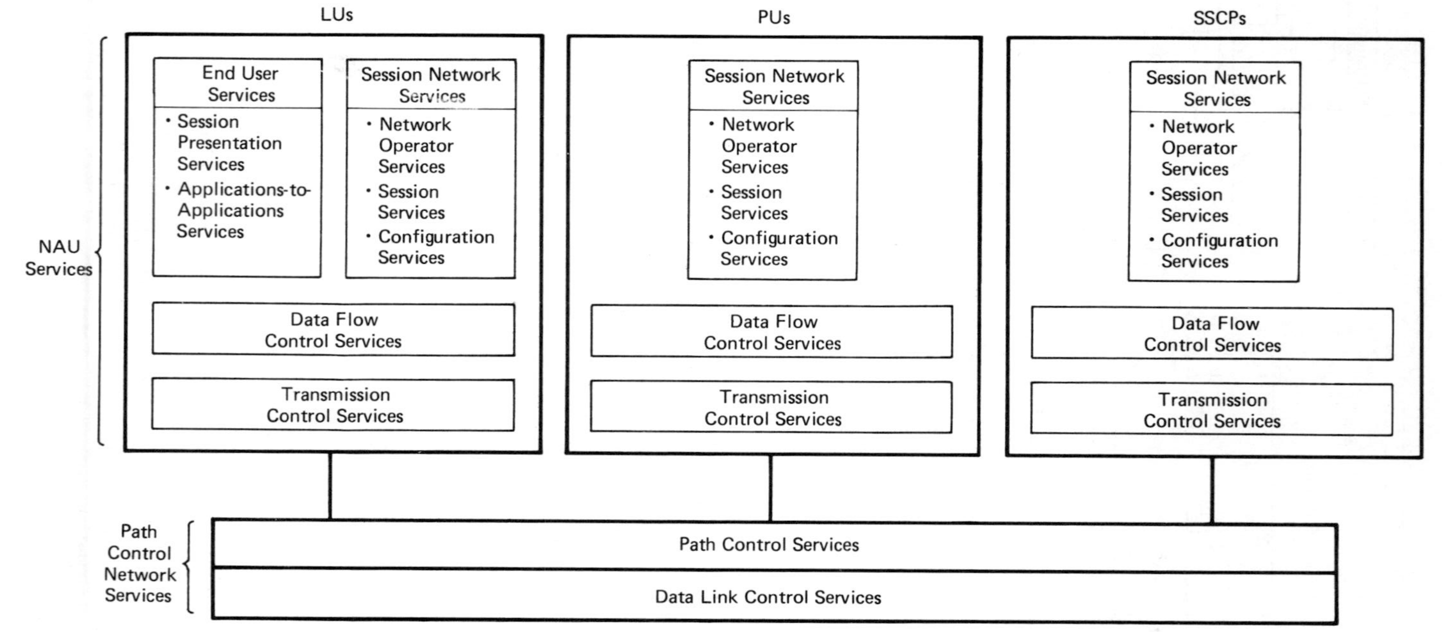

FIGURE 8-14. SNA Services.

SNA Services and Layers

SNA networks use the layering concept discussed earlier in the chapter. These layers provide for two broad categories of service, NAU service and path control network services (see Figures 8-14 and 8-15).

Most of the layers and services have a one-to-one relationship. The two top layers are the one exception: Together the NAU Services Manager and FMD Services Layers provide for End User Services and Session Network Services.

NAU Services Manager Layer and FMD Services Layer. End User Services assist in the exchange of data between Logical Units (LUs). Two categories of End User Services exist. First, Session Presentation Services provides for common formats between end users and device control characters between dissimilar devices. Session Presentation Services provides compression and compaction facilities. It also provides for a common set of display screen

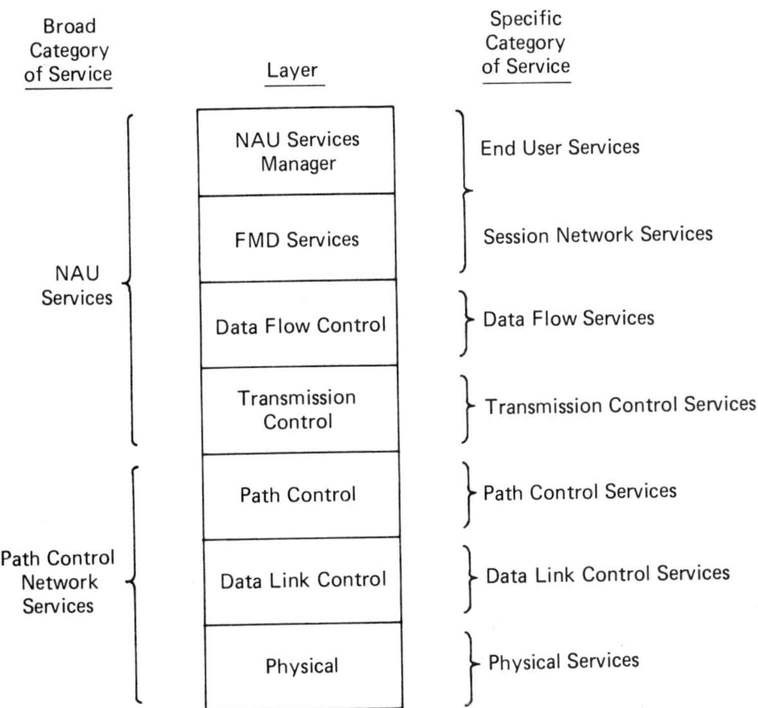

FIGURE 8-15. SNA Layers.

formats between applications. The application program need only transmit the name of the screen format. Second, Application-to-Application Services provides support for joining two transaction processing programs (for example IMS/DC or CICS/VS) in the network. It also provides for remote data base access across nodes and provides for synchronization protocols for updating distributed data bases. These layers and services are quite similar to the ISO applications and presentation layers.

Session Network Services is divided into three categories. First, Network Operator Services facilitates communication among network operators and SSCPs. Second, Configuration Services is responsible for activating/deactivating links, loading programs into SNA nodes, and maintaining tables within SSCP containing the name, address, and status of each link and NAU in the domain in the network. Configuration services are invoked by a network operator using an SSCP-PU session. Third, Session Services is responsible for conversion of the logical network names provided by LUs into corresponding network addresses. This service is similar to the address mapping feature found in ISO's transport layer.

Data Flow Control Layer. This layer also provides several of the functions of the ISO transport layer. For example, support is given for full-duplex dialogue between NAUs. Another is half-duplex, flip-flop mode, commonly found in inquiry-response applications such as order entry systems. In this mode the sending NAU can change the direction of data flow by requesting the receiving NAU to begin sending. Half-duplex contention dialogue is also supported. In this mode, either Logical Unit can begin sending data, in which case the LU that first gains access continues sending a chain of data until it is complete. The chaining allows the layer to group undirectional, related messages together and treat them as a whole. An error in any message of a chain will cause all units in the chain to be ignored. This is a useful feature when downline loading files or jobs to other sites in the network. The data flow control layer also provides for brackets. This permits grouping bidirectional, related messages together that move between two logical units. This keeps messages logically together that are related to one transaction.

Messages transmitted through an SNA network are acknowledged by the LUs that receive them. Data flow control allows the receiving LU to provide a response for every message (Definite Response), a response only under error conditions (Exception Response), or no response at all (No Response). This capability allows the network user to tailor the quality of service and data integrity in accordance with the individual applications requirement.

Transmission Control Layer. The transmission control layer keeps the status of an active session, provides for sequencing of the data messages, and paces the flow of data into and out of the sessions. It also routes data up through the N+1 layer to appropriate points within a NAU. Transmission control provides session window management by session-level pacing, which allows the LUs to control the number messages processed. This prevents the overrun of LUs that have limited processing and buffering capabilities. Transmission control also provides headers (Request/Response Header, or RH) to the message for the chaining, bracket, and pacing functions. Encryption can also be obtained in this layer.

Path Control Layer. The path control layer has two primary responsibilities: routing and flow control. Routing is accomplished by path control examining network names in the message and determining the appropriate line (or perhaps a group of lines) in order for the message to move to its destination. The layer resolves addressing for both subarea and peripheral nodes. Path control also performs message segmenting, which is similiar to the idea of packetizing messages discussed earlier. Transmission efficiency often can be improved by dividing a message into smaller pieces. SNA allows different segment sizes for each link or group of links. In addition, path control also blocks messages together. This is a useful function for reducing channel input/output interrupts and operations. Path control provides for various classes of service such as fast response, secure routes, or more reliable connections. Three transmission priorities (high, medium, and low) are provided.

The layer also contains a flow-control mechanism, called *virtual-route pacing*, to limit the flow of data from a transmitting subarea node. SNA assigns virtual routes to the two subareas involved in a session. The virtual route is assigned to an explicit route (see Figure 8-16). The concept is similiar to session-level pacing found in transmission control, but virtual-route pacing affects the flow of data in all sessions assigned to a virtual route between two subareas. Session-level pacing applies only to individual sessions.

Data Link Control and Physical Layers. These layers are discussed elsewhere in the book. The data link layer is implemented with SDLC (see Chapter 6) and the physical layer is available with RS232-C (Chapter 3) and X.21 (Chapter 8).

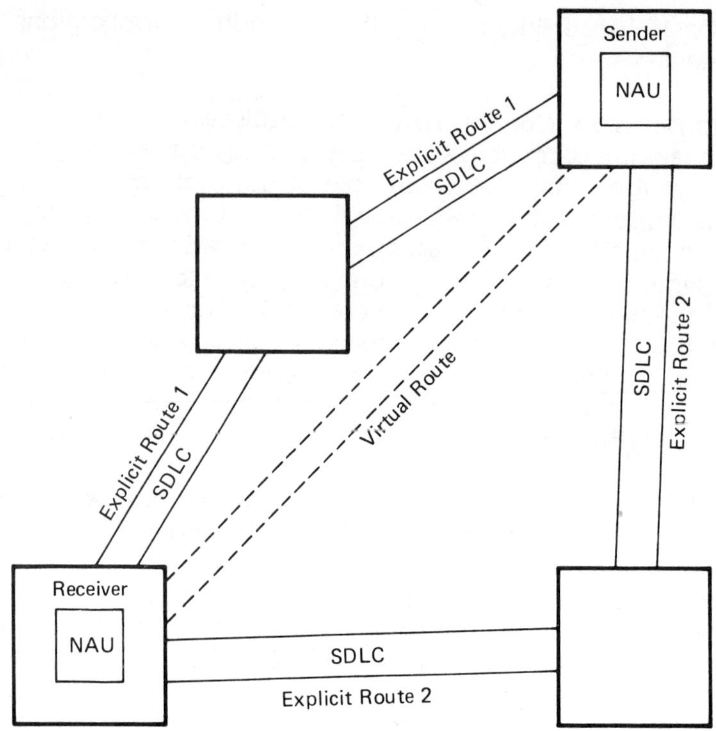

NOTE: Virtual Route Can Take One of Two Explicit Routes.

FIGURE 8-16. Virtual and Explicit Routes.

SNA Messages and Routing

SNA uses its layers to route the user data through the network. The message format consists of the Request/Response Unit (RU), the Request/Response Header (RH), the Transmission Header (TH), and the Data Link Header/Trailer (DH/DT) (see Figure 8-17). The RU contains user data and control information. It is passed from the top three layers to the transmission control layer, where the RH is attached. The RH contains control parameters on chaining, bracketing, and pacing. Path control adds the TH by converting network names to network addresses. Last, SDLC adds the five fields to complete the DH and DT.

Figure 8-18 illustrates the relationship of the layers and message across the origin node, an intermediate node, and the destination node. Notice that the path control layer at the intermediate node determines that the message is destined for another site in the network.

Multisystem Networking Facility (MSNF)

MSNF is one possible implementation of SNA. It permits the interconnection of many single-host SNAs and their respective domains into a larger multiple-host network. MSNF provides an extensive distributed processing framework with the ability to share resources across sites. Terminals can access any MSNF controlled application in any host by going directly through the local and remote front ends. Additionally, two applications programs can communicate with each other across domains by using MSNF facilities.

MSNF is established with tables containing the resources that are owned by domains and shared by other domains. The tables are part of the System Services Control Point (SSCP) and the Cross-Domain Resource Manager (CDRM). A cross-domain session is initiated by two CDRMs (a) exchanging handshaking messages, (b) determining the validity of the session request, (c) determining if the requested resources are available, and (d) binding the session users together.

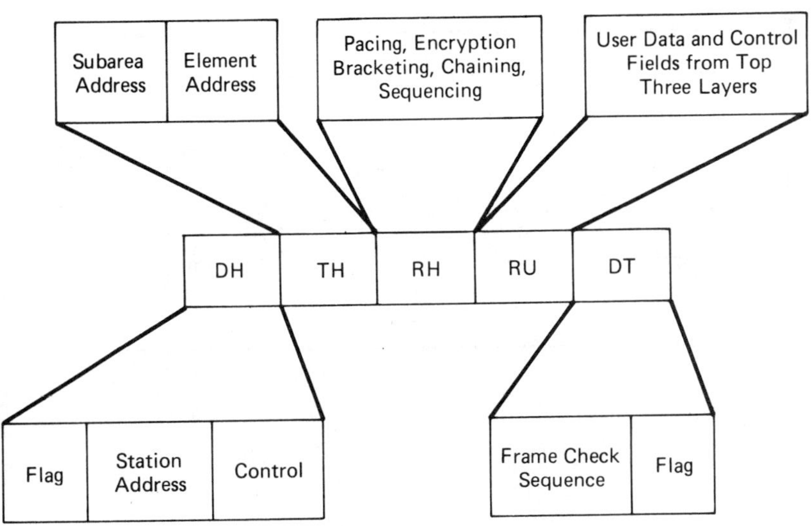

FIGURE 8-17. SNA Message Format.

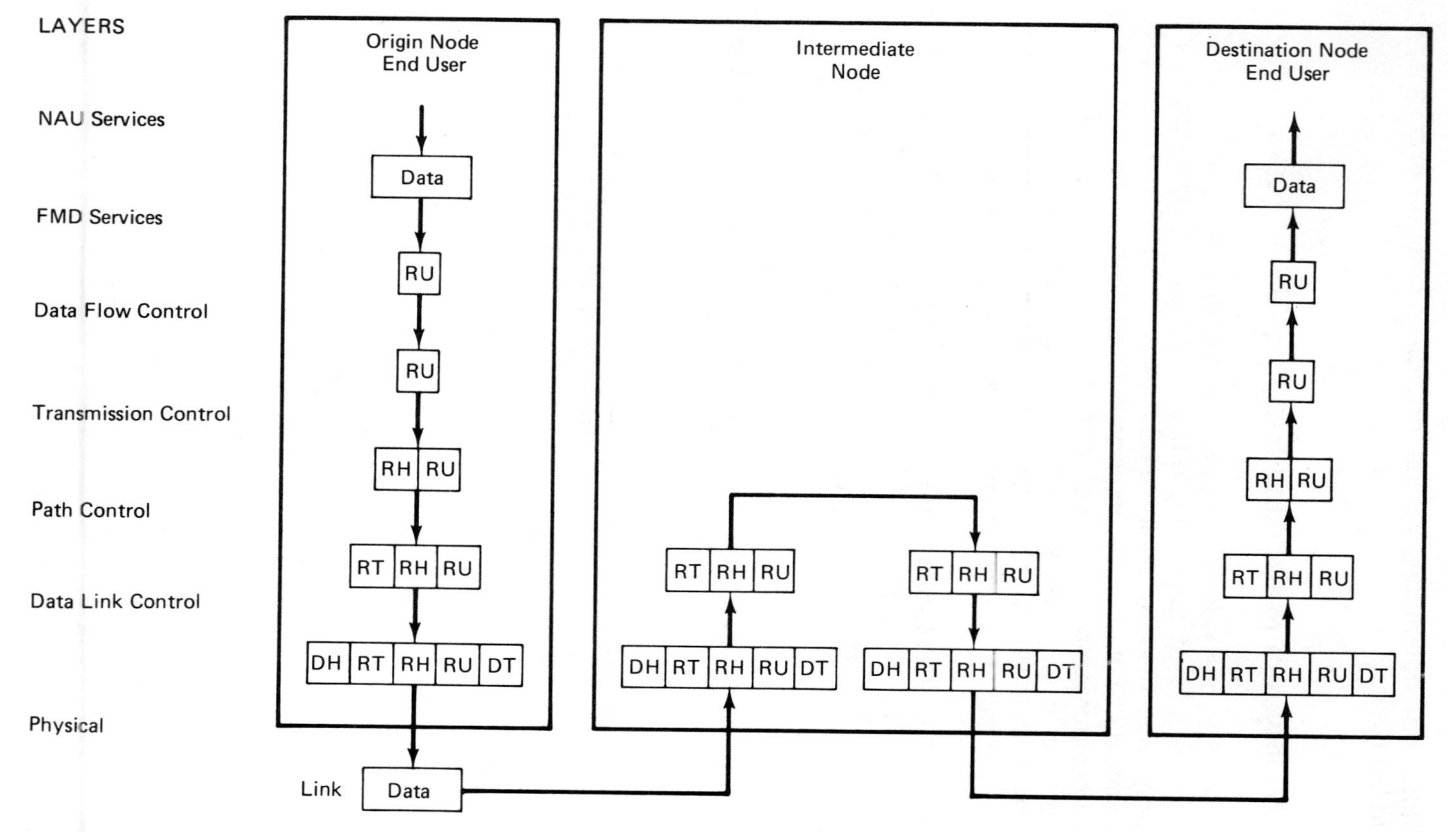

FIGURE 8-18. Message Routing.

The cross-domain resources are stored in Cross Domain Resources Tables (CDRT). Refer to Figure 8-19. Each entry contains the name of an LU or an application program and the name of the CDRM that owns the resource. Let us examine how the distribution and sharing of SNA resources is accomplished with MSNF.

Assume Terminal B in the Host A domain requests a logon to application 5 in the Host B domain. The NCP receives the message and forwards the request to the owner of the terminal, SSCP A. The SSCP logic examines its tables and determines that the requested program is not resident within its domain. SSCP A then forwards the request to its CDRM to see if the resource has been identified as a cross-domain application.

CDRM A examines its Cross-Domain Resource Table (CDRT) and finds an entry for application 5. It also sees that CDRM B owns the requested application. CDRM A then sends a message to CDRM B requesting service. CDRM B receives the request and examines its own CDRTs to verify that terminal B is a valid cross-domain resource and that program 5 is accepting messages.

When both CDRMs have given their approval, SSCP B binds the application-terminal session. Thereafter, data flows directly from Host B, through its front end, through the SDLC link, through the Host A front end, and to the terminal. Host A owns the terminal but does not receive or transmit any further messages until the session terminates.

The OSI and SNA Layers

From the standpoint of functions provided to an end user, the OSI model and SNA have many similarities, but the manner in which the functions are implemented are quite different. For example, the OSI Network layer provides for packet switching; SNA does not use this concept in any layer but its Path Control layer does perform a similar function with message segmenting. Table 8-5 compares the two approaches and shows the similarities and differences existing in each layer.

In 1981, IBM announced products that allow SNA components to interface into X.25 packet switching networks. The support functions reside in the front end NCP (X.25 program product) or in a remote device (5973 X.25 Network Interface Adapter [NIA]). The products connect between an SDLC link and the X.25 virtual circuit and the NIA terminates polling locally. The IBM interface supports both permanent virtual circuits and switched virtual circuits. In addition, IBM also has products available to support the X.21 switched and nonswitched features.

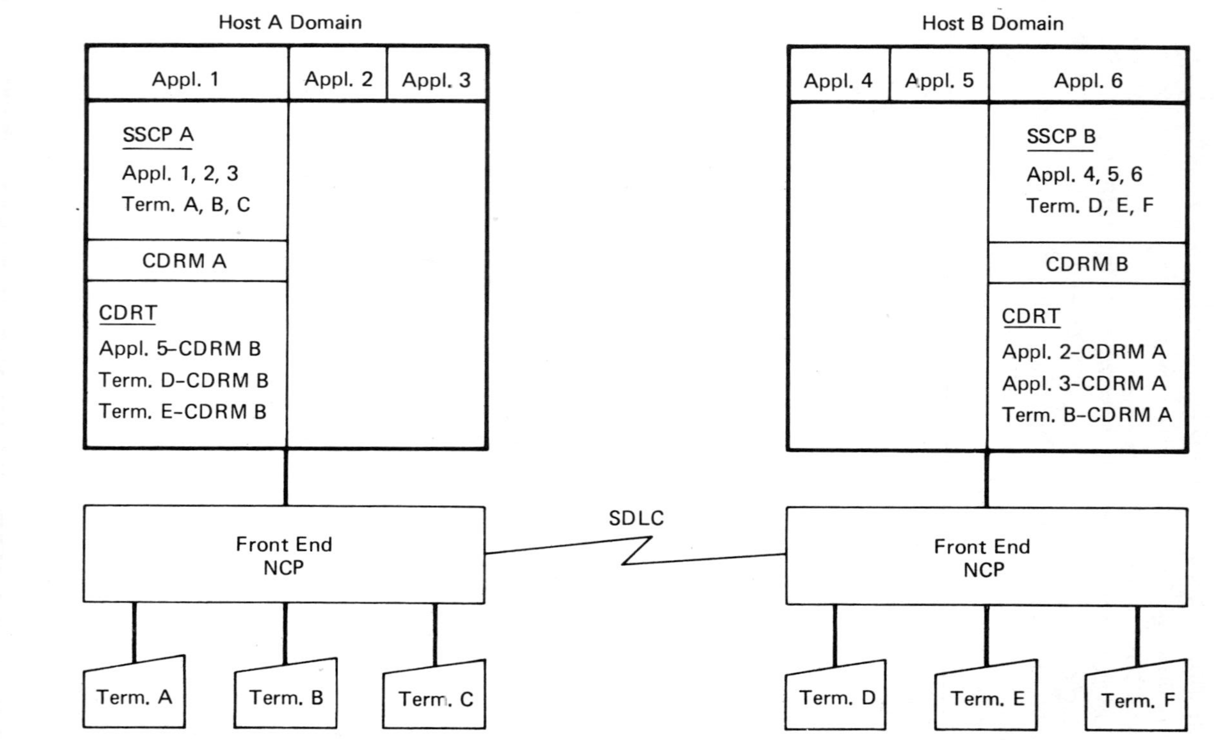

FIGURE 8-19. Cross-Domain Resource Tables.

TABLE 8-5. OSI and SNA Layers.

Layer Name		Layer Function	
OSI	SNA	OSI	SNA
Application	NAU Services Manager	User bridge to layers. File management. Service element management.	Data exchange between LUs: format, and device management; compression and compaction; common syntax and screens; file management; address mapping
Presentation	FMD Services	Syntax, code, format management. Some file management.	
Session	Data Flow Control	User dialogue synchronization, manage data exchange, guarantine service	Dialogue synchronization; chaining and bracketing; response management
Transport	Transmission Control	Error control, address mapping, segmentation, blocking, priorities, service quality	Data pacing; encryption; session status management
Network	Path Control	DCE-DTE packet interface (X.25)	Routing and flow control; address mapping; message segmenting and blocking.
Data Link Control	Data Link Control	Manage data flow across a link (HDLC)	Manage data flow across a link (SDLC)
Physical	Physical	Physical, electrical interface to network (X.21)	Physical electrical interface to network (X.21, RS232-C)

9

LOCAL AREA NETWORKS

INTRODUCTION

During the past few years, local area networks (LANs) have become one of the most publicized and controversial topics in the data communications industry. The publicity stems from *what* the LANs are purported to do for an organization; the controversy comes from *how* they are to do it. Due to the publicity, and in spite of the controversy, LANs will play a prominent role in data communications, networks, and distributed processing in the 1980s.

The LAN is distinguished by the area it encompasses; it is geographically limited from a distance of several thousand feet to a few miles and is usually confined to a building or a plant housing a group of buildings. In addition to its local nature, the LAN has substantially higher transmission rates than networks covering large areas. Typical transmission speeds range from 1 mbit/s to 30 mbit/s. LANs do not ordinarily include the services of a common carrier. Most LANs are privately owned and operated, thus avoiding the regulations of the FCC or the State Public Utility Commission. LANs are usually designed to transport data between computers, terminals, and other devices. Some LANs are capable of voice and video signaling as well. The LANs employ many of the techniques discussed in this book to manage data flow; for example, switching, digitizing schemes, data link controls, modulation, and multiplexing are often found in local area networks.

Local area networks have become popular for a number of reasons, the primary one being that most businesses transmit over 80% of their data and information locally, that is, within the local office or branch. This locality of data flow requires a transport system to move the data between the local machines. Moreover, many local applications (such as computer-to-computer traffic) require high transmission rates—certainly higher than the voice-grade technology (300 — 56,000 bit/s) of the local common carrier. LANs are seen by some organizations as a means to bypass local loops and all the problems inherent in the common carrier's end office connection.

Initially, LANs were developed as a means to tie together expensive resources for backup and sharing. For example, in the early 1960s vendors built channel adaptors to join CPUs and others developed interface boxes to allow smaller computers to act as "peripheral devices" to large mainframes. In the early 1970s, LANs were used to share memory and printers in order to expand the life of an organization's systems. Lately, LANs have been used to tie together components that have outgrown the centralized computer room.

LANs are also coming to the forefront as a means to implement distributed data processing (DDP) in an organization. As discussed in Chapter 10, DDP is becoming a prevalent technology in the industry and local networks are one way to implement it.

Last but not least, LANs are seen as a path for increased office automation. (The latest buzz phrase is "office of the future.") LAN vendors are pushing their networks to sell their work stations such as word processors, printers, electronic files, and calendars. It is estimated that the vendor's revenue from office peripherals attached to the network will be 8 to 10 times that of the network itself. There is little debate that the white collar office worker's environment can be made substantially more productive. The LAN-automated office is seen as the solution to the productivity problem.[1]

[1] Many office automation products on the market today are very poorly designed for the human interface. It is the author's experience that some systems are more of a hinderance than a help. Nonetheless, properly designed, the automated office has great promise and potential for increasing productivity.

MAJOR COMPONENTS OF A LOCAL AREA NETWORK

A LAN usually contains four major components (see Figure 9-1), which serve to transport data between end users.

(1) LANs' path most often consists of coaxial TV cable or a coax baseband cable. Cable TV (CATV) coax is used on many networks because it has a high capacity, a very good signal-to-noise ratio, low signal radiation, and low error rates (1:10^7 to 10^{11}). Twisted pair cable and microwave are also found in many LANs. Baseband coax is another widely used transmission path, giving high capacity as well as low error rates and low noise distortion.

Thus far, optic fiber paths have seen limited application, but their positive attributes (see Chapter 3) virtually assure their use in the future. The immediate use of lightwave transmissions on local networks is point-to-point, high speed connections of up to 10 miles. A transfer rate of over 44 mbit/s can be achieved on this type of path. Infrared schemes using line of sight transmission are also used on the LAN path. Several vendors offer infrared equipment for modem and local loop replacement. Up to 100 Kbit/s over one mile distances are possible with infrared schemes.

(2) The interface between the path and the protocol logic can take several forms. It may be a single CATV tap, infrared diodes for infrared paths, microwave antennas, or complex laser-emitting semiconductors for optic fibers. Some LANs provide regenerative repeaters at the interface; others use the interface as buffers for data flow and/or simple connections, like that of RS232-C.

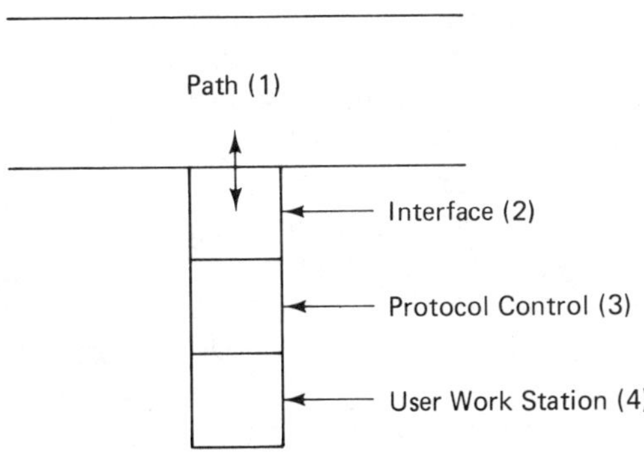

FIGURE 9-1. Major Components in a Local Area Network.

(3) The protocol control logic component controls the LAN and provides for the end user's access onto the network. Most LAN protocols employ methods and techniques discussed in Chapters 6 and 8. Other widely used LAN protocols are discussed later in this chapter.

(4) Last of the four major components is the user work station. It can be anything from a word processor to a mainframe computer. Several LAN vendors provide support for other vendor's products and certain layers of the ISO model.

LAN PROTOCOLS

Chapter 6 discusses data link controls (DLCs), which are used to manage the flow of data on a communications path (link). Local networks employ most of these concepts. For example, polling/selection, hub polling, contention, and time slots are all used in one form or another. Sliding windows and cyclic redundancy checks are also employed. Two other approaches are also popular in the LAN industry. Consequently, they are discussed in detail here.

Contention: CSMA/CD

A popular form of contention protocol is CSMA/CD (carrier sense multiple access/collision detect). Its origin is the University of Hawaii's Aloha network. The Aloha network used a radio-based packet scheme in which the secondary stations independently transmitted to the master without regard to the other station's signals. The master station broadcasted at one band and all secondary stations at another. Since the secondary stations transmitted at random, packets often "collided" when transmitted from different stations at the same time. After such collisions, the stations waited a random time before retransmitting. The Aloha scheme yielded only 18.4% maximum channel utilization. The slotted Aloha scheme provided for more effective use of the channel. With this approach, each station was synchronized on a master clock and any transmitted packet began on a specific clock interval. Collisions still occurred, but slotted Aloha provided for 36.8% maximum channel utilization.

CSMA/CD uses some of the Aloha concepts. However, before transmitting a packet, a station "listens" for a signal on the path and does not transmit until the signal (i.e., another station's packet) has passed through the cable. The sender then transmits its packet. Collisions can still occur when two or more stations sense an idle channel and begin transmission. However, the CSMA/CD protocol monitors the channel for a collision during transmission. If a station's output does not match the signal on the channel, it knows a collision has occurred. The protocol then ensures that all other stations know of the collision. Some carrier sense protocols (like Aloha) provide for a central site to transmit a busy signal on a separate subchannel when it is receiving data. The CSMA/CD does not work this way because it has no master station.

After deferring to a packet, the station transmits its signal in one of two methods. Under the persistent carrier sense approach, stations transmit immediately after the busy signal goes off. Most CSMA/CD protocols use a nonpersistent carrier sense where stations transmit with a random delay to avoid repeated collisions. Since each station generates its own randomizing variable, retransmissions among the stations rarely occur at the same time. Under heavy workloads, CSMA/CD increases this delay (with an exponential backoff algorithm) to prevent channel satuation. A central site carrier sense protocol can achieve 80% effective channel utilization; CSMA/CD achieves better than 90% utilization rates.

CSMA/CD networks work best on a bus, multipoint topology. All stations are attached to one path and monitor the signals on the channel through transceivers attached to the cable. Figure 9-2(a) shows a typical CSMA/CD bus structure.

Token Passing: Empty Slot

Token passing or empty slot protocols usually reside on a ring topology (see Figure 9-2(b)). A *token* is a time slot or packet that is passed to the next station on the ring network. The packet may contain the address of the station or may simply be an "empty slot" available to any station that has traffic to place in the packet. Upon using the packet, the station sets a flag and places a destination address in the packet header to indicate the slot is full. The token moves around the physical ring, is checked at each intermediate station for a relevant address, and is eventually passed to the destination node. The receiving station checks for errors and relays an ACK or NAK back to the originator, which then sets the token to "empty."

As with any protocol, some form of window control is needed with the token passing scheme. (Windows are discussed in Chap-

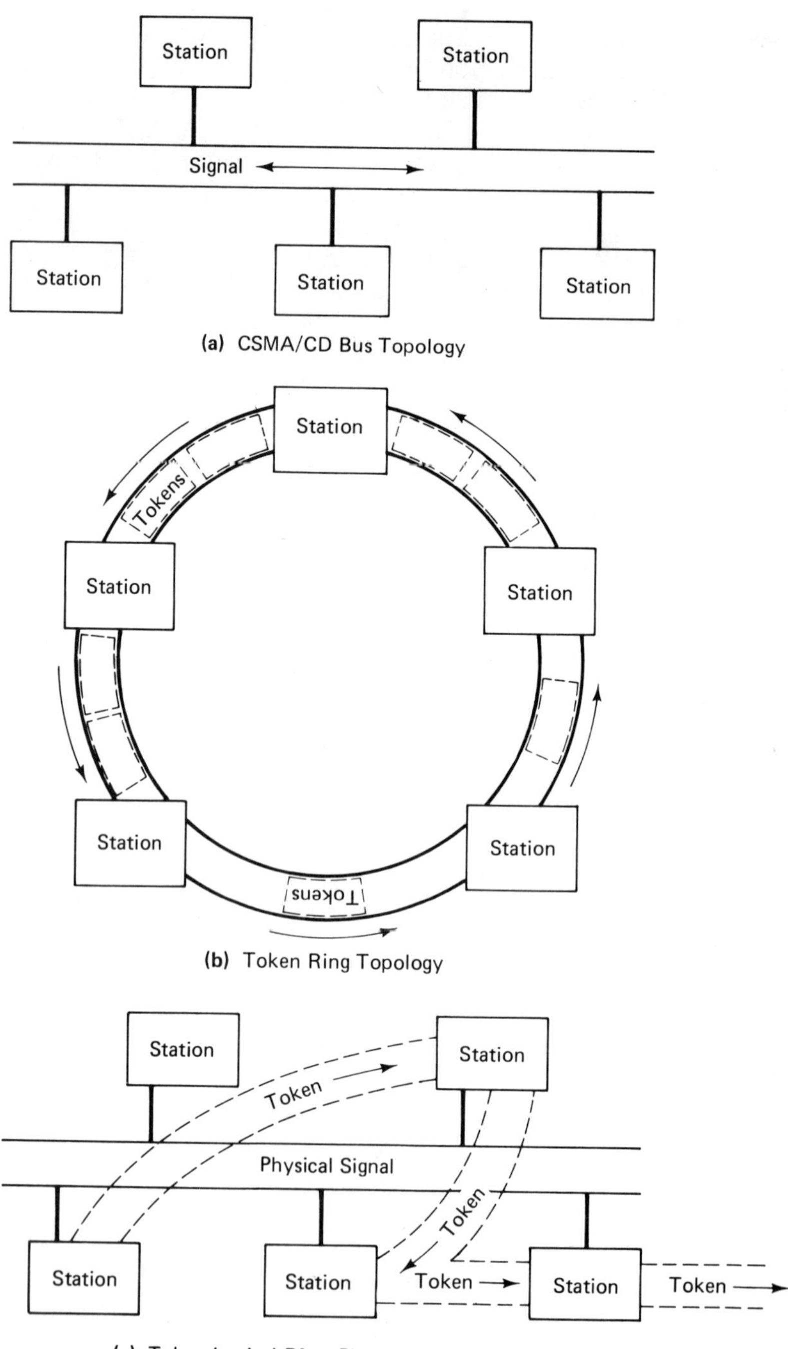

FIGURE 9-2. Local Area Network Topologies.

ter 6.) For example, multiple tokens may not be allowed; a station is not allowed to use the same token twice; and stations may be limited in the number of sessions that can exist between end users at any one time. Duplicate tokens are usually not allowed due to the complexity of the ACK/NAK logic.

An interesting variation to the token physical-ring approach is the token logical-ring, physical-bus topology illustrated in Figure 9-2(c). Using a bus path, the stations pass the tokens by placing the address of the next logical recipient in the header of the packet. The signal passes along the bus and is physically monitored by all stations, but the token is made available to a receiving station based on the sending stations' placement of a station address in the destination header. In the event a token is passed to a failed node, the originator will time out, retransmit a given number of times, and eventually "patch around" the defective station.

The IEEE 802 Standard supports both CSMA/CD and token passing. Bus and ring topologies are included in the recommended standard. The preliminary 802 Standard also includes specifications for token bus and token ring access. As of this writing, the standard is not complete and will likely see revision before final publication.

THE BROADBAND-BASEBAND ISSUE

Perhaps the most confusing and controversial aspect of LANs is the broadband-baseband issue. Vendors are literally going at each other's professional throats on this question. Therefore, at the onset of this discussion, this author wishes to make clear that he considers both approaches viable. Indeed, they can compliment each other. Each technology has advantages and disadvantages; neither is clearly preferable to the other.

The primary difference between broadband and baseband is the method of signal generation. Broadband uses a radio frequency (RF) modem to generate an AC signal. The signal lies within a preassigned bandwidth for the LAN path; data is used to modulate the RF carrier at that fixed frequency. Broadband uses frequency division multiplexing (FDM) to provide multiple channels on the network. In contrast, baseband generates digital waveforms and uses time division multiplexing (TDM) schemes. The data are propagated on the path as voltage differences. Many of the dializing schemes discussed in Chapter 7 are used in baseband networks (for example, Manchester encoding). A baseband LAN uses one channel.

Coaxial cable is the most widely used media for both broadband and baseband networks. The baseband cable is 3/8 of an inch in diameter and is surrounded by a copper cover. The broadband cable is slightly larger, is covered with aluminum, and costs approximately 50% more than the baseband cable. Typically, baseband nets use digital repeaters for extended cable lengths; broadband uses amplifiers. The protocols vary widely. For example, CSMA/CD is used as both broadband and baseband networks. Token passing is also implemented using both technologies.

Broadband employs a central retransmission facility (CRF) to provide frequency allocation for the transmit and receive signals on the network. Figure 9-3 shows how the CRF operates. A signal is transmitted to the CRF from a station on the network at a frequency within the low band of an RF bandwidth. The CRF receives the signal and transmits it on the forward channel in the low band. The return channel band is usually in the 5-110 MHZ band and the forward channel from the CRF is in the 160-300 MHZ band. Each channel is undirectional. The baseband LAN uses no CRF since it has no subchannels.

Figure 9-3 illustrates the recommended broadband standard from the IEEE 802 working group. In addition, the standard establishes the band of 108-162 MHz as a guardband between the two major channels and provides for TV channels within the 180-216 MHz range. The 802 committee also recommends a single cable, which is in variance with some recent vendor announcements.

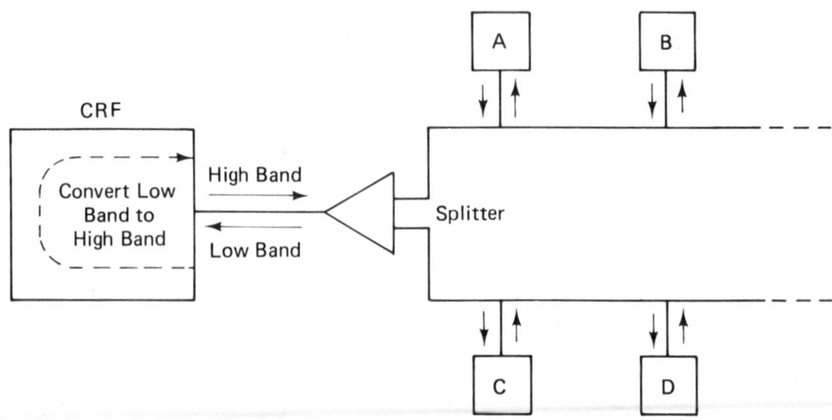

FIGURE 9-3. Central Retransmission Facility.

Various claims to the pros and cons of the two approaches are as follows: Broadband nets have greater capacity; baseband LANs are capable of 1-10 mbit/s rate, whereas a broadband LAN can operate at speeds well over 150 mbit/s. However, the broadband subchannels operate at substantially slower rates, so the individual user is getting only the RF modem speed supporting the user work station. Broadband permits multiple channels, providing for multiple types of protocols and subnetworks on one system. While baseband does not have this feature and is restricted to one protocol, nothing precludes the baseband net from supporting subnetworks and multiple applications through TDM techniques. Broadband can accommodate more users than baseband; RF modems and amplifiers can provide for more extended coverage. Moreover, its technology is based on CATV and is highly reliable. For example, mean time between failure (MTBF) of its coupler taps is 30 to 40 years and 18 years for its amplifiers.

Some broadband proponents say the costs are comparable. However, be aware that a baseband operation is simple and has small start-up costs. All that is needed (excluding software) is a cable, two baseband transceivers, and two interface units. The broadband components can be considerably more expensive. Each RF modem ranges in price between $500 and $2,000; amplifiers can cost as much as $1,500 and the CRF runs from $4,000 to $12,000.

In summary, the vendor's claims should be carefully examined (especially the costs). Keep in mind the current cliche about the issue, "Broadband, baseband: The vendors want us on their own Bandwagon." As an example of using baseband and broadband, Ungerman-Bass has taken an approach with its Net/One LAN that permits the interconnection of both techniques. The RF modems and the baseband transceiver can be exchanged; the resolution is made by altering an "encoder/decoder" board in Net/One's protocol and network logic boards. It is likely the Net/One approach will see wider use by other LAN vendors.

LANs: VENDOR OFFERINGS

Many local networks are available today. All use either baseband, broadband, or both technologies, and the majority of the networks use some form of CSMA/CD, token passing, or the conventional polling/selection protocol. Several of the vendors have announced support for selected layers of the ISO Open

Systems Interconnection model such as X.25 and X.21. (See Chapter 8 for an explanation of the model.) Table 9-1 is a partial list of current vendor offerings. The networks discussed in this section are representative of what is available in the market today. These products have also been selected because they illustrate the major features of LANs that were discussed earlier in the chapter.

Ethernet

Ethernet is one of the better known LANs and has one of the larger user bases of the LANs. It was developed by the Xerox Corporation at its Palo Alto Laboratories in the 1970s and was modeled after the Aloha network. In 1980 the Digital Equipment Corporation (DEC) and the Intel Corporation jointly published a local area network specification based on the Ethernet concepts.[2]

Ethernet's primary characteristics include the CSMA/CD protocol and baseband signaling on a shielded coaxial cable. The data rate is 10 mbit/s, with a provision of up to 1024 stations on the path. It uses a layering concept, somewhat similar to the low levels of the ISO model, and transmits user data in frames or packets. Ethernet uses Manchester encoding (see Chapter 7); at a 10 mbit/s rate, each bit cell is 100 ns long (1 sec/10,000,000 bit/s = .0000001). Since the Manchester code provides for a signal transition within each bit cell, the Ethernet signal is self-clocking.

The network architecture is shown in Figure 9-4. It consists of the coaxial cable, terminators, transceivers, controllers, and work stations. The cable can be as long as 550 feet with extensions of 1,650 feet using one or two repeaters. The terminators complete the electrical circuit. The transceivers transmit and receive signals, detect packet collisions, and maintain signal quality on the bus. The controllers provide for collision management, translation of signals (encoding/decoding), and other CSMA/CD tasks. The work stations are the end user devices such as word processors, computers, and terminals.

The Ethernet packet contains six fields (see Figure 9-5). The preamble is sent before the data to provide for channel stabilization and synchronization. It performs similar functions to the SYN bytes described in Chapter 2. The preamble is a 64-bit pattern of

[2] *The Ethernet, A Local Area Network*, Version 1.0, September 30, 1980. Published by Digital Equipment Corporation, Maynard MA; Intel Corporation, Santa Clara CA, and Xerox Corporation, Stanford, CT.

recurring 10s, with the last two bits coded as 11. The last bits indicate the end of the preamble and the beginning of the data. Upon reception of the double-1, successive bits are then passed into the station.

The destination and source address identify the receiving and sending stations respectively. Each address is 48 bits in length. The destination address can also be a multicast-group address (a group of logically related stations) or a broadcast address (all stations on the Ethernet).

The type field of 16 bits is used by end users (and not the Ethernet protocol) at a higher level in the local network. It is defined at the Ethernet level to provide for a uniform convention between higher levels. The data field contains user data with any arbitrary bit sequence allowed. The frame check sequence field provides for a cyclic redundancy value (CRC, see Chapter 6). The CRC field is 32 bits in length.

Ethernet is designed around the three layers depicted in Figure 9-6. The User or Client Layer is the work station; the Data Link Layer contains the data encapsulation and link management functions; the Physical Layer provides for data encoding/decoding and channel access. These layers will be used to illustrate how Ethernet transports packets and manages data flow.

Transmission of Packet. The client layer passes data to data encapsulation, which constructs the frame, calculates the frame check field, and appends it to the frame for error detection. The frame is passed to link management. This sublayer monitors the carrier signal on the bus and defers to passing traffic. When the channel is free, link management waits a brief period and sends a stream of bits to the Physical Layer. Data encoding first sets up and sends out the preamble to allow receivers and repeaters to synchronize clocks and other circuitry. It then translates the binary bit stream into Manchester-phase encoding. The encoder drives the transmit part of the transceiver cable. The channel access sublayer (transceiver) actually generates the electrical signals for the coaxial cable. In addition, it simultaneously monitors the bus for any collision.

Receiving of Packet. The receiving channel access detects the incoming signal and turns on a carrier sense signal for use by the Data Link Layer. It synchronizes with the incoming preamble and passes the bits up to the next sublayer. Data decoding translates the Manchester code back to binary data, discards the preamble, and sends the data to the Data Link Layer. Link management has

TABLE 9-1. LAN Vendors.

Company	Name	Maximum Distance	Path & Type	Protocol	Speed	Maximum in Connections
AMDAX Corp. Bohemia NY	CABLENET	>75 mi.	Broadband Coax; Bus	TDMA & FDM	7 or 14 mbit/s	16,000
Apollo Computer Chelmsford MA	DOMAIN	3,250 ft.	Baseband Coax, Ring	Token Passing	10 mbit/s	>100
Corvus Systems San Jose CA	OMNINET	4,000 ft.	Twisted-pair Wire; Bus	CSMA	1 mbit/s	64
Datapoint Corp. San Antonio TX	ARC	4 mi.	Coax; Bus	Token Passing	2.5 mbit/s	255
Digital Equip. Maynard MA	DECDATAWAY	15,000 ft.	Twisted-pair wire; Bus	HDLC-like & Others	56 Kbit/s	31
Digital Comm. Germantown MD	INFOBUS	>75 mi.	Broadband Coax; Bus	CSMA/CD & FDM	1 mbit/s @ Channel	256
Gould, Inc. Andover MA	MODWAY	15,000 ft.	Coax; Bus	Token Passing	1.544 mbit/s	250
IBM (GS) Atlanta GA	Series 1/Ring	5,000 ft.	Coax; Ring	CSMA/CD Variation	2 mbit/s	16
Interactive Sys. Ann Arbor MI	Videodata	40 mi.	Broadband Coax; Bus	TDM	100 Kbits/ @ Channel	248 @ Channel
LOGKA Ltd. London, England	POLYNET	NA	Baseband Coax; Ring	Empty Slot	10 mbit/s	NA

NESTAR Sys. Palo Alto CA	CLUSTER/One	1,000 ft.	16-wire cable	CSMA/CD	240 Kbit/s	65 Apples
Network Sys. Minneapolis MN	HYPERCHANNEL & HYPERBUS	3,000 ft.	Coax; Bus	Priority CSMA	50 mbit/s & 6.3 mbit/s	256 256
Prime Framingham MA	PRIMENET	750 ft.	Coax; Ring	Token Passing	8 mbit/s	>200
Stratus Computer Natick MA	STRATALINK	750 or 1,500	Baseband Coax; Ring	Token Passing	2.8 mbit/s	32
SYTECK, Inc. Sunnyvale, CA	LOCALNET	5–20 mi.	Broadband Coax; Bus	CSMA/CD	128 Kbit/s– 2 mbit/s	16,000 to 64,000
Ungerman-Bass Santa Clara CA	NET/ONE	4,000 ft.	Baseband Coax; Bus	CSMA/CD	4–10 mbit/s	200
WANG, Inc. Lowell MA	WANGNET	2 mi.	Broadband Coax; Bus	CSMA/CD & FDM	9.6 Kbit/s to 12 mbit/s	512 to 1000's
XEROX Corp. El Segundo CA	ETHERNET	1.5 mi.	Baseband Coax; Bus	CSMA/CD	10 mbit/s	100 to 1000
ZEDA Ltd. Provo UT	INFINET	3.1 mi.	Twisted-pair wire, Bus	CSMA/CD	25 Kbit/s	30 micros
ZILOG, Inc. Cupertino CA	Z-NET	1.2 mi.	Baseband Coax; Bus	CSMA/CD	800 Kbit/s	255

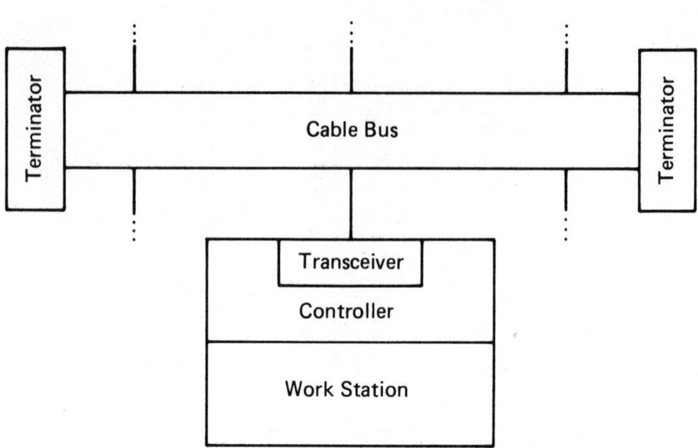

FIGURE 9-4. Ethernet Architecture.

detected the carrier sense from the transceiver (channel access sublayer). It receives the bits from the Physical Layer. Upon the carrier sense signal going off, it passes the frame to data decapsulation. This function examines the destination address field and, if appropriate, passes the frame to the Client Layer. It also checks the frame for damage and sends an error status code to the user station.

Handling Collisions. A station's signal can collide with another station. For example, if Station A's signal does not reach Station B before Station B transmits a packet, then Station B's link management sublayer has not activated its deferral logic. The signals collide. This period of vulnerability is called the *collision window* and is determined by the total propagation time on the channel.

A collision is noticed by the transmit side of the channel access sublayer. The transceiver "listens to itself" and, upon detecting the collision, turns on the collision detect signal. The collision detect signal is sensed by the transmit side of link management, which

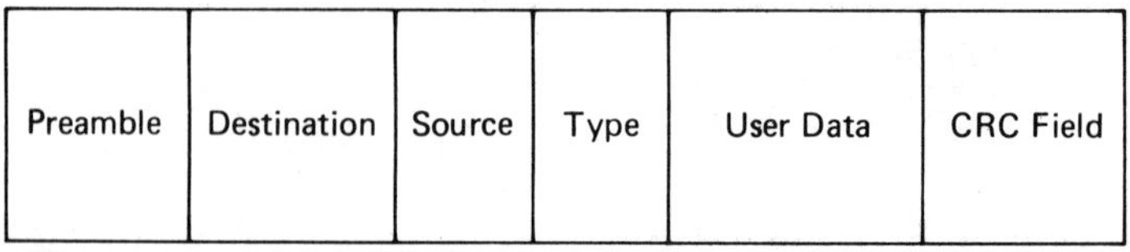

Preamble	Destination	Source	Type	User Data	CRC Field

FIGURE 9-5. Ethernet Packet.

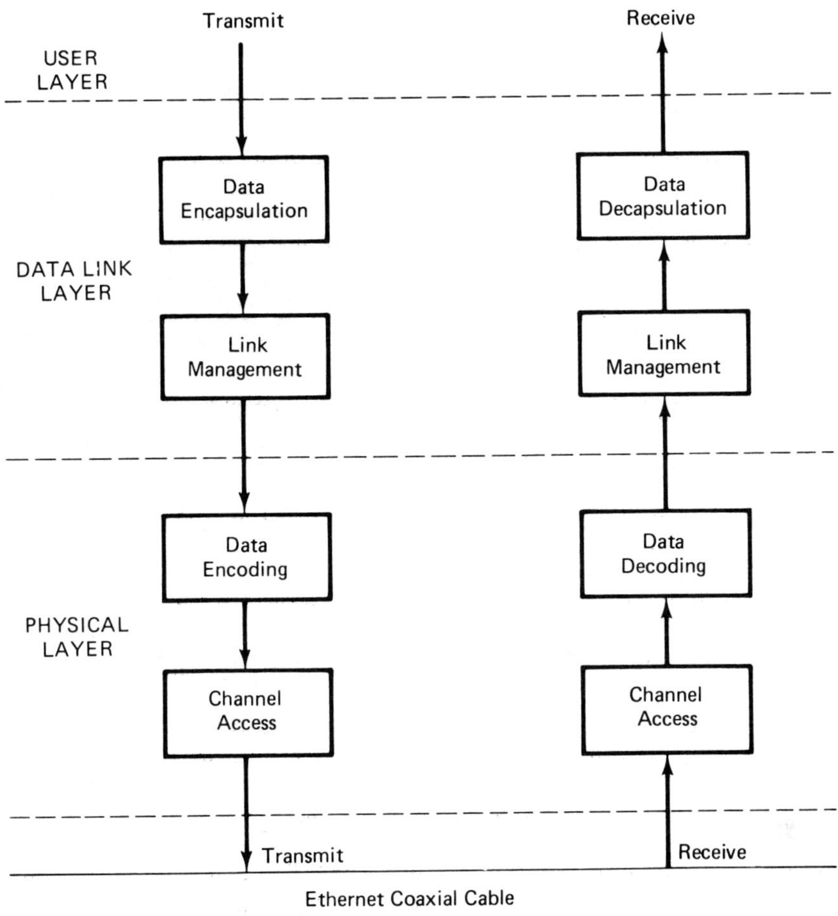

FIGURE 9-6. Ethernet Layers.

transmits a bit sequence called a *jam*. The jam ensures that all transmitting stations are aware of the collision (in case the collision is very short). Transmit link management then terminates the transmission and randomly schedules a retransmission. Excessively busy conditions will trigger exponential backoffs from the channel by the link management logic.

Receiving stations do not have their collision detection signal turned on by the collision. Rather, receive link management notes the fragmentary frames that resulted from the collision and simply discards these frames. A collision fragment is designed to be shorter than the shortest valid frame.

Ethernet personnel have indicated the network will soon support gateway interfaces into X.25 networks as well as IBM's SNA. DEC has announced that it will integrate Ethernet into phase IV of its Digital Network Architecture (DNA) before 1985. Bell systems widely used Unix operating system has been made available to Ethernet users by the 3Com Corporation of Mountain View, California. Finally, the IEEE LAN standards committee has endorsed almost all aspects of the Ethernet specification.[3]

WangNet

Wang Laboratories, Inc., announced its entry into the LAN arena in June 1981. Named WangNet, it is significantly different from Ethernet, primarily because it is based on broadband technology. The system uses a 340 megahertz cable bus. A portion of WangNet uses modems, which are attached to the bus to divide the bandwidth into subchannels. User work stations are then attached to the modems. Figure 9-7 depicts the Wangnet topology. It uses two cables, one for transmit and one for receive. A transmit-receive cross-over point at the mid-point translates the transmit signal to the receive signal.

WangNet currently allots approximately 35% of the 340 megahertz bandwidth into three bands. (The remaining 65% is reserved for future capabilities.) The bands are Wang Band, Interconnect Band, and Utility Band (see Figure 9-8).

Wang band. This subnet is reserved for use by Wang computer systems. The channel is controlled by the CSMA/CD protocol. Packets are transmitted between the Wang processors at a 12 mbit/s rate. The work stations are attached to the bus by a Cable Interface Unit (CIU), which manages the packet assembly/disassembly process as well as the CSMA/CD tasks. The CIU permits addressing up to 65,535 stations. Obviously, traffic load throughput and response time would determine the practical limit.

Interconnect band. This band is broken into three groups. Sixteen dedicated point-to-point or multipoint channels comprise the first group. These channels emulate leased lines operating at speeds of up to 64 Kbit/s. Attachment to the cable is through the

[3]The IEEE group also endorsed a token ring and a token bus approach. The token ring standards closely resemble IBM's position, discussed later in this chapter.

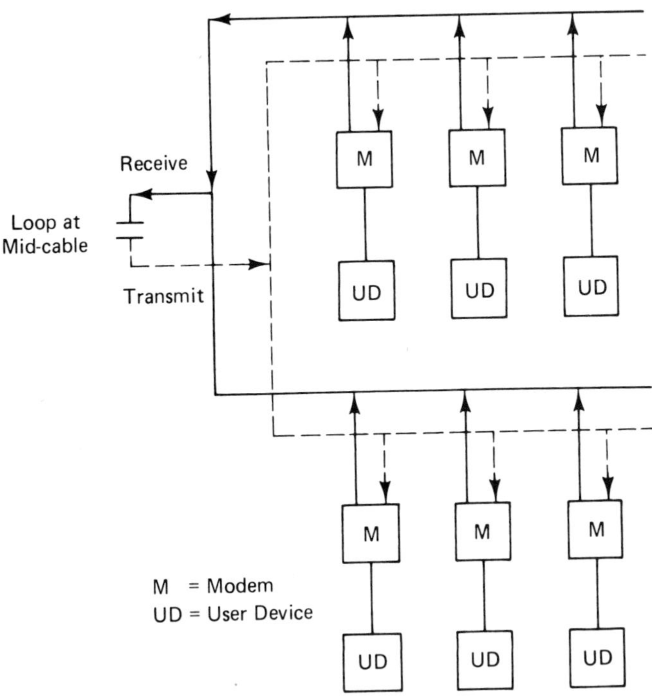

FIGURE 9-7. WangNet Topology.

RS449 standard (see Chapter 3). Thirty-two point-to-point or multipoint channels make up the second group. These "leased" lines operate at rates up to 9.6 Kbit/s and use the RS232-C interface standards. The dedicated channels on both groups use fixed-frequency modems. The third group, emulating a switched line facility, provides for 256 channels using frequency agile modems. The 256 channels provide for 512 station connections. The modems can be set to one of the 256 channel frequencies and are assigned to the channels by the Wang Dataswitch. The switch controls the attached modems in the following manner:

- A transmitting site issues a request for a bus connection to the Dataswitch. The request is transmitted through a control channel.
- The switch receives the signal, establishes the handshake with the transmitting modem, and also receives the station number of the receiving modem.

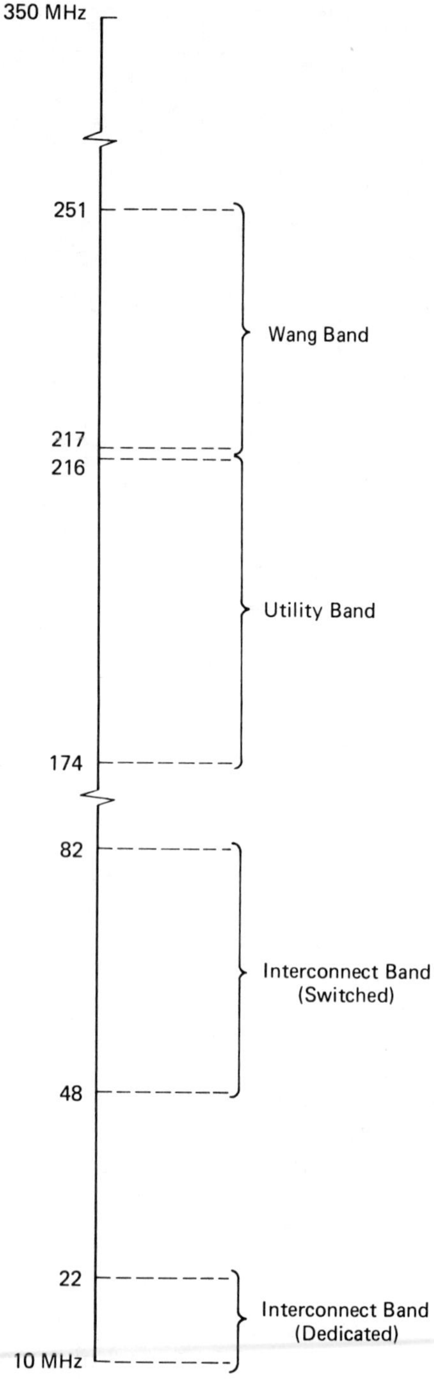

FIGURE 9-8. WangNet Frequency Allocation.

- Dataswitch relays the signal through a manual operater or an RS366 call unit.
- If the receiving station is free, the switch assigns on unoccupied frequency to the two modems.
- The two stations then have a point-to-point connection through a RS232-C interface.

This group's 256 channels operate at a 9.6 Kbit/s data rate. The point-to-point connections are not to exceed two cable miles and operate in full-duplex.

Utility band. Seven 6 megahertz channels are allocated for television. The band is designed for closed-circuit TV and videoconferencing. It operates in the 174 MHz to 216 MHz range commonly used by CATV vendors.

WangNet has a limited user base since it went operational as late as 1982. Nonetheless, it is an attractive and interesting approach. Wang's previous success and current presence in office products should bode well for WangNet use.

IBM Token Ring

The IEEE 802 Committee has endorsed IBM's token ring LAN architecture. The major characteristics of the network are as follows:

- The topology is a combination of a star and ring hierarchy.
- The transmission technology is baseband.
- The transmission path is twisted-pair copper wire.
- The protocol uses a single-token access.

Figure 9-9 shows the IBM approach. Rings, largely self-contained, are connected through bridges. The bridge provides switching, buffering, protocol, and speed changes between the local rings. The bridges are provided for fault isolation. They are also intended to interface broadband networks using CATV technology and (eventually) optic fiber LANs.

The IBM announcement has not yet seen actual supporting commercial products. However, an examination of IBM's LAN frame format (Figure 9-10) reveals that the network will provide interface capabilities into IBM's major communications systems and

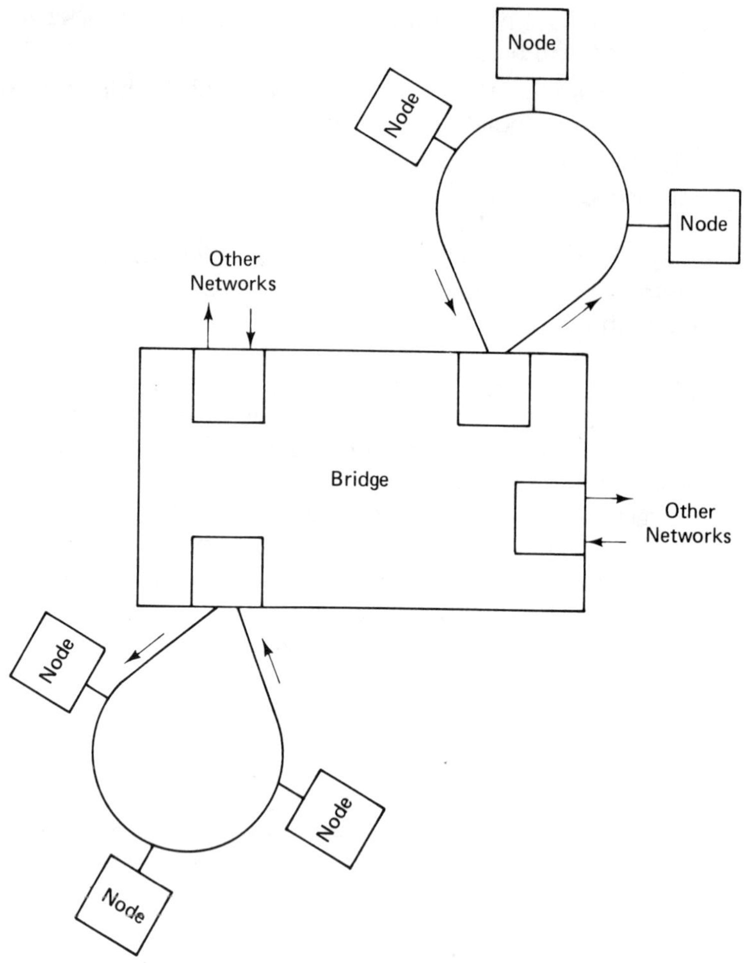

FIGURE 9-9. The IBM Token Ring LAN.

networks such as SDLC and SNA. Moreover, the format could provide for an X.25 interface.

For example, the information field could contain SNA's Path Information Unit (PIU) and SDLC's frame. Next, the data packet could incorporate ISO's bottom three layers and X.25. The PIU, SDLC frame, and X.25 packet could then be wrapped into the IBM LAN token ring frame. It is probable that IBM's product announcements will include features along this line of thought.

QUESTIONS FOR LAN EVALUATION

Ethernet, WangNet, and the IBM token ring are illustrative of the offerings available today. Certainly, one should not restrict oneself to analyzing only these three products. Again, Table 9-1 provides a sampling of other local networks. Several questions should be posed to the vendors and their responses should be used in evaluating their networks:

1. What is the user base (how many networks have been installed)? Can users be contacted?
2. Given that many LANs are new, with limited users, what are the results of beta test sites (i.e., customer prototyping the LAN)?
3. How many stations are supported?
4. What type of stations are supported (word processors, printers, etc.)?
5. What types of software packages or functions are available?
6. Are security features such as encryption and password logons available?
7. What are the performance statistics, given varying stations and varying traffic loads?
8. What is the cost and effort of adding stations to the network?
9. Does the LAN use layered protocols?

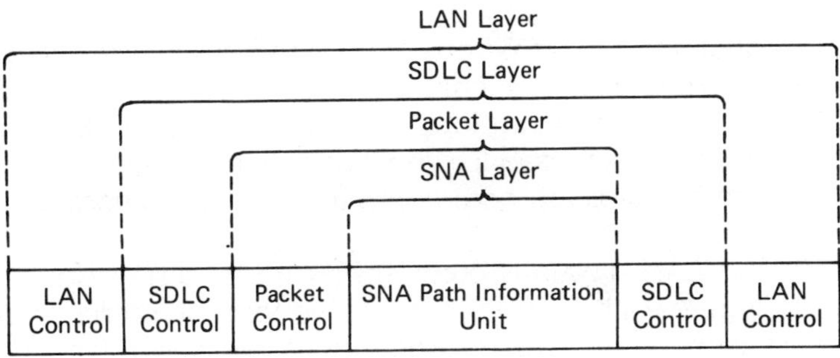

FIGURE 9-10. Possible IBM Token Ring Format.

10. Are accepted U.S. standards used in the architecture (RS232-C, RS449)?
11. Does the vendor have products or plans to support the ISO seven-layer model?
12. What are the failsafe and backup capabilities?
13. Is the net subject to a single-point failure?
14. What is the maximum distance allowed and the costs involved in extending cable distances?
15. What is the maximum drop length of work stations to the cable and does this fit with the user's building or plant layout?
16. Is the network transparent to other vendor's products? If not, which vendor products are supported and at what cost?
17. What is the network protocol?
18. Is the network broadband or baseband?
19. What is the service/maintenance plan?
20. Is documentation adequate?
21. Does the vendor have a realistic and attractive plan to use the LAN for office automation?
22. Does the vendor plan to integrate its LAN with other communications products? If so, will it be transparent to the user?
23. What is the cost of the network?

PBX AND PORT CONTENTION LANs

The attention on cable bus local networks has resulted in a very viable and reliable technology being pushed to the background—the PBX. (See Chapter 3 for a description of PBXs.) These instruments have been used for years as switching devices and can now provide facilities for local computer devices to communicate with each other.

The PBX LAN is usually configured around a star topology as shown in Figure 9-11. Most PBXs use some form of matrix or digital switching technology. The technology is attractive for an LAN because it is reliable, readily available, and relatively inexpensive. Moreover, unlike a bus architecture, a malfunctioning user device

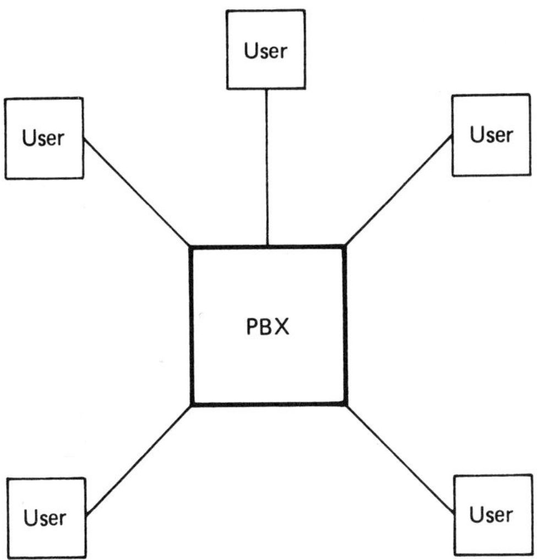

FIGURE 9-11. PBX LAN.

is isolated from other user devices since the PBX is between the devices.

The PBXs can provide for analog or digital transmission with data transmission speeds as high as 56 Kbit/s. Connections to the machine are through simple direct-wire or RS232-C facilities. Some offerings provide for a distance as great as 5,000 feet between user devices and the PBX.

10

DISTRIBUTED SYSTEMS

INTRODUCTION

Distributed systems have become commonplace today. Vendors regularly announce products that support distributed processing and user applications are rapidly adopting distributed techniques. Several years ago, the idea was viewed with considerable skepticism, considered by many to be too complex for practical implementation. However, the changing data processing environment has created attractive costs and benefits for distributed processing and has provided the impetus for both user and vendor to support the concept. In today's environment, the question is becoming less of "Should we distribute?" and more of "How do we distribute?"

Distributed data processing (DDP) is defined as follows:

- In contrast to the conventional single, centralized site, processing is organized around multiple processing elements (PEs). A PE is a computer or other device capable of performing automated, intelligent functions.
- PEs are organized on a functional and/or geographical basis.
- Distributed elements cooperate in the support of user requirements.

- Connection of processing elements is through common carriers or private links.
- Implication is the dispersion of hardware and/or software and/or data to multiple PEs.

Distribution is *not* decentralization. The latter connotes dispersion with no integration of the parts to the whole. Distributed processing provides functional or geographical dispersion, yet the dispersed parts are integrated into a whole coherent system. One of the greatest pitfalls an organization faces in dealing with DDP is the mistaken belief that decentralization is the same as distribution. Decentralized systems make good sense for certain organizations; for others, the approach can be very disruptive and expensive. Both are viable options as long as it is recognized which approach is being implemented. This point will be emphasized throughout this chapter.

WHY GO DISTRIBUTED?

Given the definition of a distributed system, one might question why an organization would choose the distributed approach. After all, its description seems to imply increased complexity. The seeming anomaly can be answered by an examination of certain events that have taken place in the industry during the last decade.

Several years ago, the idea of economies of scale encouraged the industry to develop large, centralized computers. All processing for user departments was performed by one or a few large mainframe machines. Grosch's Law held that the cost of an executed machine instruction was inversely proportional to the square of the size of the machine. Hence, the larger the system, the better the economies of scale.

Grosch's Law seemed to hold for some time but other factors were at work. For example, the increased complexity of the large centralized computers resulted in diminishing returns in using the larger systems. It is quite easy to see the diminishing returns problems in operation today. The large scale centralized systems are indeed powerful; some machines execute more than ten million instructions per second (MIPS). These systems have been generalized to accommodate a wide variety of user needs and applications. Yet, the machines are often overkill for many user systems. The large-scale systems have elaborate and complex operating systems designed to perform many functions, but not many of them in an optimum manner. A simple user enquiry (transaction) may require the execution of tens of thousands of instructions on these machines. The resulting overhead negates or diminishes much of their raw processing power. According to some proponents of DDP, the cost per computation now favors the smaller computers.[1]

The large centralized systems seem to suffer from over generalization—they do many things but none of them very well. The revolutionary microelectronic age has given genesis to the notion that specialized machines can perform fewer functions in a cost-effective manner and perform them very well. These machines are being used increasingly for dedicated applications. The issue of generalization versus specialization was partially resolved by the extraordinary cost and performance benefits of small, inexpensive computers.

Most of the components in an information system have been declining in cost and improving in performance. A significant aspect of these decreasing costs is the fact that communications costs (for example, leased lines) have been decreasing at a lesser rate than computing costs. Organizations have realized that local computing and local storage of data can lead to overall savings by decreasing the message flow across the more expensive communications lines, thereby decreasing the use of communications facilities. Moreover, auxiliary storage costs such as disks have also been decreasing significantly. Redundant data and duplicate data bases are not so much a cost issue today as they are an issue of coordination and control. Consequently, local storage of data is a very viable option today, especially if large scale disk units are used.

Another major reason for the growth of distributed processing has been pressure from users to gain control of some of the computing power in the organization. In many organizations, the users are not satisfied with the services received from the applications development staff. Projects are often late. Typically there is a backlog of applications to be developed and, in many instances, the backlog is increasing. The maintenance costs for applications running on large-scale systems is truly extraordinary. One would be amazed at the amount of time a programmer must spend in simply keeping the applications system synchronized with the

---

[1] "Moving Away From Mainframes," *Business Week*, February 18, 1982, p. 78.

changing mainframe environment. Weekly and sometimes daily changes are made to job control language, data base control blocks, job streams, and software libraries. Users do not understand the environment and do not understand why a programmer must spend so much time in system maintenance. (The smaller and simpler mini- and microcomputers require less of this kind of maintenance.)

Users have become disenchanted with the closed-shop approach. The *closed shop* means that users are given limited personal access to the computer. (It is too complex for them to operate anyway.) More important, in a closed shop the user has little input to or control of the data processing department. It is believed by many that the centralized data processing department is not appreciative of user needs nor responsive to their requirements. Consequently, many users believe the large centralized environment militates against taking full advantage of the real power of the computer.

Financial factors are also at work within the organization. There are increasing cost-related and profit-related pressures to gain more control of information systems management by user departments. The users of today are much more knowledgeable and sophisticated than they were several years ago. They understand that information is power and that control of the information resource is important to their performance and even their bottom-line profit. Many users believe their professional destiny is greatly dependent upon the computer and they want more control over their destiny.

Distributed processing's growth is also tied to the intelligent or programmable "terminal" a.k.a. personal computer (PC). However, this "terminal" is now a full computer with disk, memory, tape, communications interfaces, and powerful software—all for a price of as low as a few thousand dollars. The intelligent terminal market, growing at a rate of over 30% annually, is finding its way into the hands of the user and line manager. An increasing number of people consider the intelligent "terminal", or PC, the answer to many of their data processing problems.

Perhaps it appears as if I am stating that centralization and the supercomputer have been failures, but that is not the case. The large scale centralized computer is absolutely essential for many applications today. In fact, the industry needs even more powerful computers to solve many of its problems and they are needed to increase productivity of many of our offices. Large mainframes will also play a major role in the DDP arena by managing large complex data bases. The point to be made is that the supercomputer and large scale mainframes are not needed nor suited for many applications that are currently running on them.

Moreover, common sense dictates giving more responsibility and control to those people who use and provide input (data, for example) into a system and receive the reports of the system. Regardless of whether the information resource is distributed or centralized, a higher quality product results from more involvement of users and their staff. Most individuals identify with projects over which they have a vested interest and responsibility. In turn, those projects will receive more care and attention and the resulting products will be of higher quality. For example, it has been demonstrated many times that data bases are more accurate, more timely, and of generally higher quality when users participate in establishing edit criteria, monitor the maintenance of the data base, input the source data, and review data-related errors. The key ingredient to a successful DDP is to establish control over the distribution process but provide sufficient autonomy not to stifle initiative and diminish identification with and support of the effort.

The data processing industry is in the initial stages of a radical transformation. Distributed data processing will be one end result of this transformation. DDP will affect practically everyone. Just witness the effects of advances in microelectronics and data communications. Hardly a day goes by without our using a computer or a computer network. The use of an electronic game, the telephone request for a hotel reservation, the enquiry for a bank balance, the entering of a manuscript on a word processor—all use computers and some use networks. As the industry matures (as the required high-level software is designed to provide end user access to data), more people will begin to use and acquire the systems. Computers and supporting networks will be distributed into practically every aspect of our lives.

However, certain organizations will not reap the benefits of DDP. These companies' managers will not manage nor control the move to DDP. The companies will experience spiraling applications software costs, redundant and conflicting data bases, and incompatible data communications protocols. This point will be emphasized throughout the chapter: DDP requires unified coordination within the organization if it is to succeed. It requires top management involvement, active user participation, and strategic planning. In essence, problems in DDP usually stem from the lack of top management control and commitment.

PROS AND CONS OF DISTRIBUTED PROCESSING

The Disadvantages

Practically speaking, distribution may not be for everyone. And, if it is, it offers some potential dangers.

Loss of Control. The most serious potential problem with DDP is the loss of control over the organization's automated resources. This includes the inability to provide proper audits among all the distributed sites, the emergence of multiple standards within the organization, and the lack of cost control measures over the distributed departments. The loss of control stems from the problems cited below.

Duplication of Software Resources. The distributed sites are often tempted to develop their own automated systems to fit perceived local needs. Consequently, sites that have the same requirements may develop duplicate software systems that have the same or nearly the same functions. This often results in unduly high software costs.

Duplication of Data, Leading to Data and Report Conflicts. Independent distributed facilities are also prone to develop their own version of a corporate data base. This can occur for several reasons: (a) The corporate data base is designed to optimize efficiency for all users. The individual site will "spin off" a portion of the data base and optimize the data to its specific needs. (b) The distributed site wishes to add data to the corporate data base; again, it duplicates the data base and then appends its unique source data to the duplicate copy. (c) The distributed site wishes to use special software (a different DBMS or data dictionary) that is not suited to the corporate data base.

It should be emphasized that redundancy, per se, is not necessarily a poor approach. However, uncontrolled redundancy creates problems. When the same data is presented in a variety of ways, it is often difficult and expensive to resolve the differences. This topic is covered in more detail later.

Hardware Problems. Independent acquisition of hardware by autonomous managers will likely result in the inability to share the computers among the sites because of incompatibilities among the vendors and models. Different vendors' machines often have different word sizes, varying instruction sets, and conflicting compilers. Moreover, if the resources cannot be shared, some machines may be underutilized and others overutilized because peak loads cannot be shifted to other sites that are not being fully used.

Reversion to Past Inefficiencies. The data processing profession has made great strides during the past several years. Project control, programming techniques, and data management have become more disciplined and more productive. Data processing professionals are usually highly skilled and highly trained. The placement of automated resources under the responsibility of untrained and inexperienced users (or other personnel) can lead to the same inefficiencies that existed in earlier years. The uninitiated user tends to ignore sound programming techniques. Documentation is often not prepared under this environment. At a minimum, budget and cost control problems may become severe because the data processing labor to produce systems may not be accountable when the systems are developed in the user department. The resulting inefficiencies can lead to escalating complexities in the system.

Maintenance of the Remote Sites. Those sites that are located in remote areas may have difficulty in obtaining timely response to technical problems. Vendor support personnel and the organization's technical support staff may not be readily available and maintaining a trained staff at each site is often prohibitively expensive.

Incompatibility of Hardware and Software. Unless the acquisition of automated resources is controlled and coordinated, the distributed sites will likely develop hardware and software systems that are incompatible and that cannot communicate with each other. This problem often surfaces in the communications protocols and data link controls (DLCs) that were discussed in previous chapters. As a consequence of this independence, a considerable amount of time and effort may be spent in developing emulation software or protocol conversion packages. This is not to say one should go sole source (i.e., one vendor for everything), but the organization cannot allow the distributed sites complete independence in choosing the product.

Lack of Sufficient Machines and Program Power. The distributed environment often makes use of the less powerful mini- or microcomputer and certain problems cannot be solved on these systems. The powerful centralized maxicomputer may be essential. It may be necessary to store all data centrally for reasons of performance, simplicity, and auditing, in which case a large-scale computer may be required. Moreover, applications tend to grow in size and complexity. As a system evolves it may outgrow the small machine environment. A distributed system requires a very careful design that (a) allows larger problems to be divided for execution on multiple processing elements or (b) provides a maxi within the distributed system for the "larger" applications.

The Advantages

While the potential dangers of distributed processing are real and can be serious, the offsetting advantages of a *properly implemented* system are quite significant:

Reduction of Costs. In many instances, DDP saves the organization money. Local processing can reduce the amount of traffic across a communications line. Data can be entered and edited for errors at the distributed sites. In several actual cases, the author knows where DDP paid for itself solely by local data entry and editing. Distributed systems can provide opportunities for cost reductions in other areas: (a) use of less expensive and less complex computers; (b) decreased complexity for certain applications (e.g. payroll, accounts receivable) that run in the simpler small computer environment; and (c) better utilization of personnel that use the system (more on this subject later).

Response Time Improvements. The lengthy delays encountered in a centralized system often stem from an overloaded CPU and slow communications lines. Local processing of a transaction eliminates the relatively slow common carrier lines in favor of high-speed channels. It also eliminates the lengthy waiting queues that accumulate at the single-server, centralized host. Local processing can also shorten the time for an application to complete the processing of its cycle. For example, the Otis Elevator Company installed local processors at some of its remote sites for editing its weekly payroll records before they were transmitted to the host central site. The distributed system cut down errors significantly and decreased the payroll cycle by a full day. Detecting errors before they enter a system proved very valuable to this company.[2]

Distributed processing can improve response time for certain applications that can be partitioned (or divided) into pieces for simultaneous execution. Therefore, instead of the serial processing approach, the functions are executed in parallel on multiple functionally distributed computers. This technique decreases the lag that is inherent with a serial process (see Figure 10-1). Parallel processing techniques are increasing in the industry and will likely become a dominant method of DDP implementation. The topic of partitioning is covered in Chapter 11.

User Control. Earlier it was stated that users wish to have more control over the critical data processing resources that largely determine their profit or loss. Distributed processing can provide the user with this control. However, the point should be re-emphasized that unregulated user autonomy and control will surely do more harm than good. Later discussions will provide guidelines on delineating user independence and user control.

Without question, users need to participate more in the development and selection of automated resources. To do so would decrease the tension and political infighting that often occurs between the user community and the data processing department. The impression that the users are getting more tailored support and, to an extent, are controlling their information systems can have very high payoffs, both in terms of morale and productivity.

Backup. If one wishes to provide backup to a large-scale centralized machine, one must acquire another large-scale centralized machine. Some organizations attempt to acquire services from other companies or time sharing facilities to provide backup of critical applications. However, this approach is quite cumbersome in keeping the two organizations' systems compatible with respect to operating systems releases, compiler enhancements, and a myriad of other daily ongoing chores. If the organization can use two maxis, then coupling the machines together properly will allow for one machine to backup the other.

---

[2] "Distributed Processing at Otis Elevator," Joseph Kelly. *Data Management*, May 1981, pp. 29–33.

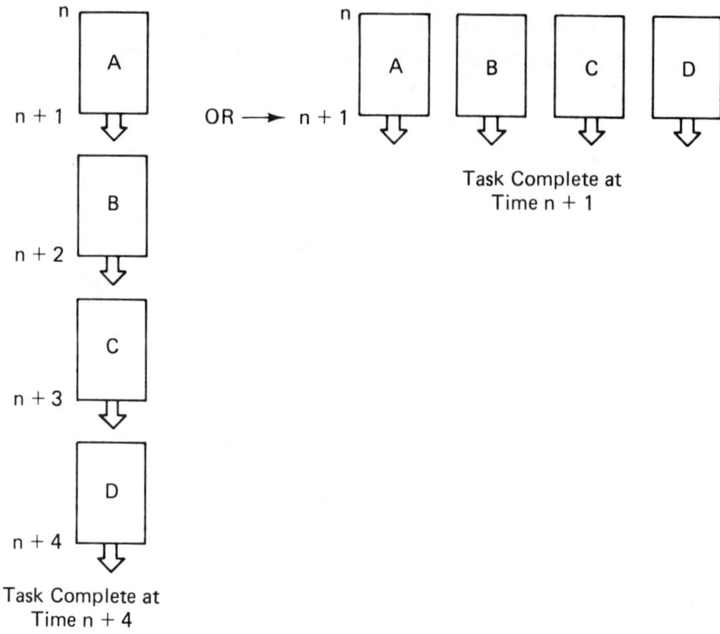

FIGURE 10-1. Serial and Parallel Processing.

On the other hand, distributed systems can provide very flexible and cost-effective backup. Since the system has its parts integrated to a whole, a failed processing element can have its work transferred to another site. We see this technique being refined today; it is available in several vendor offerings at the present time.

Resource Sharing. A distributed system allows the remote sites to use each others' facilities (machines, data, and software). Redundancy can be reduced since a user knows that resources are to be shared within the distributed system.

FORMS OF DISTRIBUTED PROCESSING

Distributed systems can be implemented in a variety of ways. This section discusses the more prevalent forms of DDP and explains the following concepts:[3]

[3]Some of these concepts are explained in more detail in *Design and Strategy for Distributed Processing* by James Martin. Englewood Cliffs NJ: Prentice-Hall, 1982.

- Vertical/Horizontal Systems.
- Functional/Geographical Distribution.
- Heterogeneous/Homogeneous Systems.

Vertical/Horizontal Systems

The choice of the distributed topology (*topology* is the shape of the system) depends upon factors such as cost, response time, throughput, capacity, load sharing, and capacity needs. The topology is also highly dependent upon the amount of distribution to be allowed, the degree of autonomy at the distributed sites, and the type of functional distribution.

The vertical topology usually appears as illustrated in Figure 10-2. The processing occurs at different levels in the network. The higher level processors may be more powerful and larger in capacity. Transactions or batch jobs can be submitted at the lower levels and passed up the hierarchy to the appropriate processing element. The lower level sites perform local functions and edit data before passing the data to another element.

The vertical network is often implemented with the master/slave concept. The sites act as masters to lower-level elements and slaves to higher levels. Control and flow on the communications lines between the distributed sites take place through data link controls (DLCs, see Chapter 6). The initiation of sessions between the sites is the responsibility of higher level protocols.

The midlevel sites in Figure 10-2 are in an interesting position in the hierarchy. For example, site B acts as a slave to Site A and as a master station to sites D, E, and F. Site B controls all processing, resource sharing, load leveling, and data access between sites D, E, and F. If site F wishes to establish a session with site D, then site F must initiate the request through site B. The control functions for the local environment are the responsibility of site B and site A is not tasked with the local operations. (The reader may wish a refresher on sessions; if so refer to Chapter 2.) If site F wishes to communicate with site K, it must go through sites B, A, C, H and, finally, site K. One can see that frequent access across the vertical hierarchy will create significant overhead in session establishments and dis-establishments.

Figure 10-2 also illustrates the idea of vertical distributed environments or domains.[4] Site A's domain is universal, encom-

[4]The reader should not confuse my use of *domain* with that of IBM's SNA (see Chapter 8).

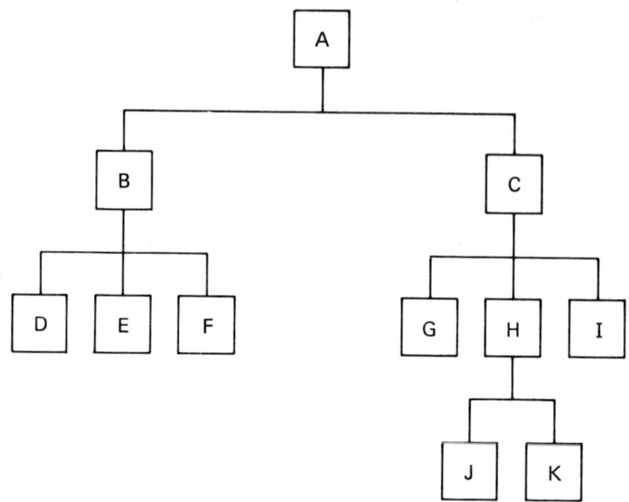

FIGURE 10-2. Vertical Topology.

passing the entire network. Site B controls a subset of the universe, its local environment. Site C also controls an environment and, within the site C environment, site H controls a local environment of sites J and K. The concept of vertical domain is quite attractive for the systems designer because loading of work, partitioning of functions, and isolation of activities are simplified with the vertical approach. The chapter on design amplifies this point.

The horizontal system is a peer-oriented network (see Figure 10-3). No master/slave relationship exists in the system; all sites are of equal status. While message flow across the communications lines is controlled by a DLC, the initial sessions between sites are usually established between the sites involved in the process and not by any intervening master station. For example, if site A wished to access data at site C, site F would act as a conduit between sites A and C. Site F could not deny the request unless it was too busy to handle A's traffic. The horizontal networks are often found in local area networks.

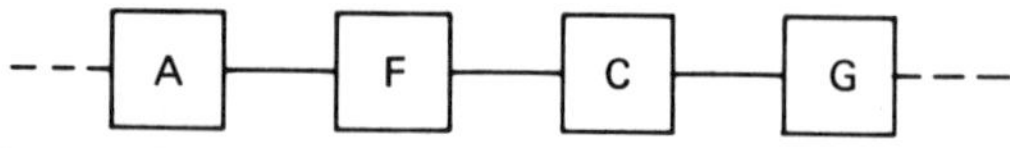

FIGURE 10-3. Horizontal Topology.

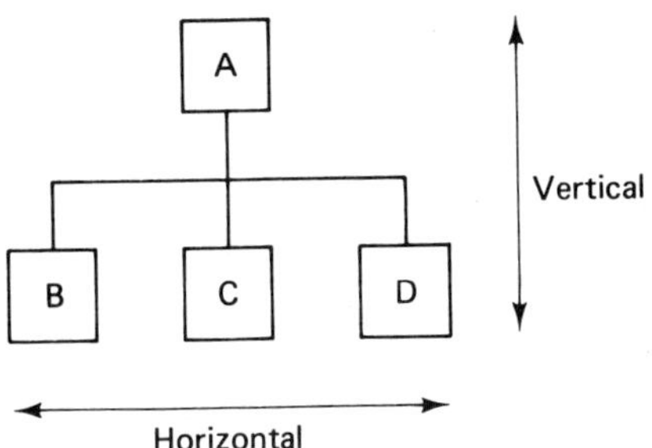

FIGURE 10-4. Vertical and Horizontal Topology.

Networks often exhibit both horizontal and vertical characteristics (see Figure 10-4). The advantages of each are combined in a variety of ways, depending on the specific needs of the organization. The summary in this section provides guidelines and ideas for implementing the distributed topology.

Functional/Geographical Distribution

The processing elements need not be located far apart in order to be considered a distributed system. Indeed, a Local Area Network (LAN) frequently uses distributed techniques and seldom encompasses an area greater than a few thousand feet. DDP may be accomplished within the confines of a building or even within a room. The term *functional distribution* refers to processing elements that are organized to perform specific tasks (or functions).

Figure 10-5 shows an example of functional distribution. This example is the Bank of America System, one of the pioneers in implementing a commercially viable distributed system. Each processing module is organized around four minicomputers, one pair for front end work and one pair for data base management. The two pairs act as backups to each other. The module is tailored to process-specific applications such as commercial loans, savings accounts, and credit cards. Each module also contains the functional data base for the respective application.

Figure 10-5 shows that the Bank of America network also exhibits geographical distribution. The system is organized around two processing centers, one in Los Angeles and the other in San Francisco. Messages may be exchanged between the centers to complete a transaction on a customer account.

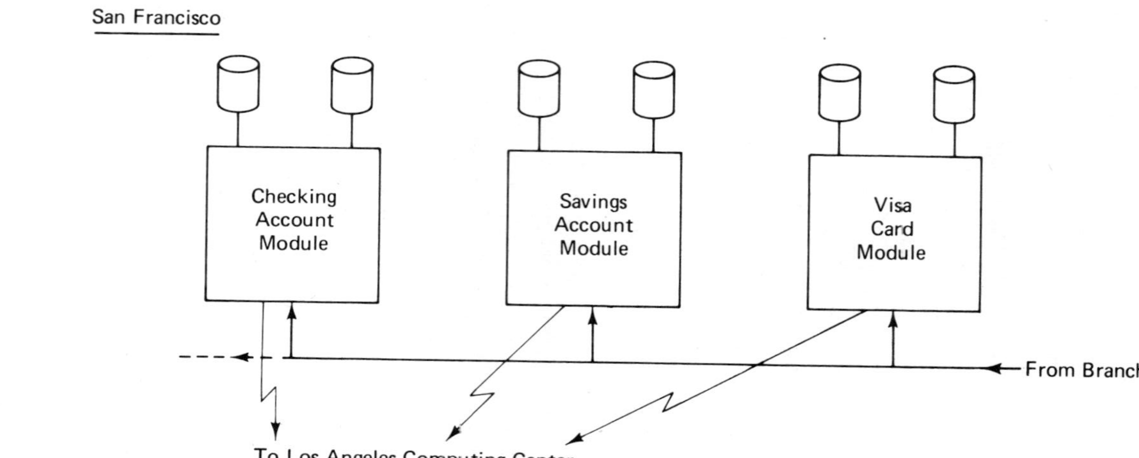

FIGURE 10-5. Bank of America Network.

This network uses leased lines between the two centers; the DLC is SDLC, a Go-Back-N, bit synchronous control; line speed is 9.6 Kbit/s. The lines between each processing module are coaxial cables operating at a speed of 1.5 Mbit/s. The lines between the branches and the processing center are leased lines under SDLC and operate at 2.4 Kbit/s. These lines have an average of 10 branches and 70 terminals multidropped onto them.

Heterogeneous/Homogeneous Systems

The Bank of America network is also an example of a homogeneous system: All processing elements are the same identical machines, protocols, DBMSs, and operating systems. Heterogeneous systems contain varied and dissimilar support components. Common examples of heterogeneous networks are the public packet networks such as Telenet, Tymmet, and the nascent NET/1000 from American Bell.

Summary

The vertical and horizontal networks can exist with many different topologies and can exhibit many forms of functional/geographical distribution as well as different varieties of homogeneity and heterogeneity. The vertical approach allows each environment or domain to be different—thereby permitting a heterogeneous network. For example, in Figure 10-2, site B's domain could be Prime computers and domain C could have Data General machines. If messages were to be exchanged between the Domains of B and C, either site B, C, or A would provide the necessary conversions to interface the heterogeneous domains.

The vertical network is sometimes cumbersome for dynamic resource sharing among the sites. Session establishment requires moving up and down the network hierarchy. The horizontal network decreases some of this overhead since the transaction can move more directly to the cooperating processing elements.

A "pure" vertical network is also subject to reliability problems. A failure at any site above the low level local domain will render useless all processing elements under the failed masters. Practically speaking, few vertical networks are implemented that allow this kind of vulnerability; multiple coupled elements and hybrids of horizontal networking are usually implemented to avoid this problem.

The heterogeneous network is attractive for those organizations that are composed of separate independent divisions or that chose to go with a multivendor network. While conversion routines, emulation packages, and gateway functions present very real problems, the multivendor approach allows the organization to select the best processing elements for each site and application. It gives the organization more choice than does the homogeneous network approach. However, resource sharing under a heterogeneous system is often considerably more complex due to the different products that must be interfaced together.

The choice between geographical or functional distribution is really not one of citing advantages or disadvantages. The user requirements and user locations will dictate the choice.

DISTRIBUTED DATA BASES

Since the purpose of computers is to process data and to manipulate the data into something meaningful (information), the issue of data in a distributed environment is very important. Distributed data represents one of the more interesting and challenging problems of DDP and considerable care must be taken in implementing distributed data bases if the potential benefits of a distributed system are to be realized.

Reasons for Distributing Data

The reasons for distributing data are similar to the rationale for DDP discussed at the beginning of this chapter:

- Placing data at the local sites can reduce the amount of data transferred among the network computers, resulting in reduced communications costs.
- Local access of data can improve the response time to obtain the data since the delays of remote transmission are eliminated.
- Distributed data bases can give increased reliability to a system because the data is located at more than one site. The failure of a node need not close down all operations of data access in the network.

- The provision for local storage gives users more control over the data. As stated earlier, increased quality is likely to result from user involvement and in user control of the data.
- Distributed data bases present a challenging technical problem. While this statement is made somewhat tongue in cheek, the author knows of instances where organizations allowed technicians to move to the distributed data environment because the technicians wanted to "make it work." To be sure, this is not a very good reason and it once again points up to the need for management control of the process.

Types of Distribution

Three basic types of data distribution can be implemented. The first type is centralized data [see Figure 10-6(a)] and is not really distributed, but will be used for purposes of comparison. In this approach, all data reside at one site; all data queries and updates from the remote sites are transmitted to the central site.

Figure 10-6(b) illustrates the partitioned data base approach. In contrast to the centralized scheme, the data are split into pieces or partitions and assigned to selected sites in the network. The partitioned data bases all may be of the same data structure, format, or access method—in which case the system is a homogeneous partitioned data base. If the partitioned data consists of different structures, formats or access methods, the system is a heterogeneous partitioned system.

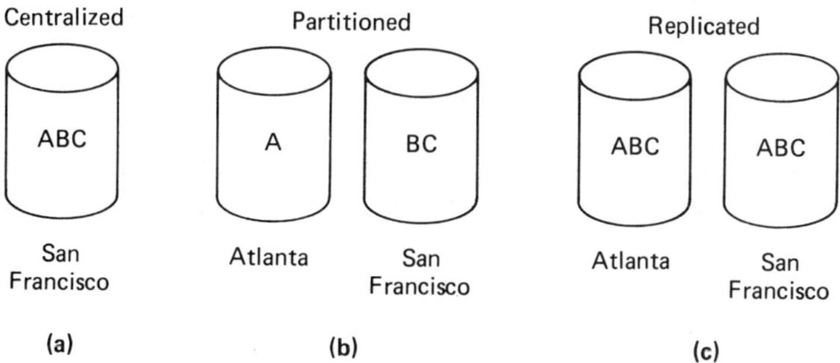

FIGURE 10-6. Types of Data Base Distribution.

Replicated data is the third approach [see Figure 10-6(c)]. This type stores multiple copies of the same data at different sites in the network. The same data structures among the replicated data is called *homogeneous replicated data*. Replicated data bases may be classified as heterogeneous replicated systems if the duplicate data is reformated or placed under a different access method to fit local needs.

Distributed data bases are further classified as being vertically or horizontally distributed. The vertical system exhibits the form of the vertical topology (see Figure 10-2) where data are distributed in a hierarchical manner. Detailed data may be stored locally and general, aggregated data or less detailed data stored farther up in the hierarchy. Horizontally distributed data exhibits the form of the horizontal topology (see Figure 10-3) wherein data are distributed across peer sites in the network.

In summary, data bases can be placed in the network in various combinations of:

- Centralization
- Partitioning
- Replication
- Homogeneity
- Heterogeneity
- Vertical distribution
- Horizontal distribution

The choice depends on factors that will be explained in this chapter and in Chapter 11.

Distributed Data Terms and Concepts

In order to follow the subsequent discussion on distributed data systems, several basic terms and concepts should be understood. Many different names are attached to these concepts and different vendors or designers may use their own term to describe the same concept.[5] The reader may wish to review Chapter 4 on basic data base concepts.

[5]Upon review of many variations, I settled on the terms and definitions found in *ACM Computing Surveys*, Volume 13, Number 2, June 1981. I also am indebted to ACM for their examples in this particular issue.

Consistent State. All data in the network data bases are accurate and correct. The replicated copies contain the same values in the data fields. Two examples illustrate the idea: (a) A customer's total bank deposits reflect the sum of his checking deposits and savings deposits. If check deposits = $100 and savings deposits = $250, total deposits must contain the value $350 if the data are to be consistent. (b) The customers' total deposit value is stored in replicated data bases at the branch bank and the main office. *Both* data bases show $350 for the customer's total deposit, otherwise the data is not in a consistent state.

Transaction. A transaction is a sequence of operations transforming a current consistent state to a new consistent state. For example, transferring funds from a customer's bank account to another customer's bank account involves more than one operation and also involves moving the data base from one consistent state to another consistent state.

Some data base systems permit a transaction to create an undefined number of operations or even another transaction. These systems are usually difficult to control and often create loading problems on the network computers. The transaction sets off a chain reaction to multiple data bases or programs, often resulting in a chaotic environment. This practice, sometimes called a *multifunction transaction*, is to be avoided and will be discussed later in Chapter 11.

Temporary Inconsistency. This term describes the state of the data during the execution of a transaction. The state occurs in between the operations. For example, an applications program that transfers funds between two customers will inherently create a temporary inconsistency because the program's instructions are executed sequentially. For example, transaction 1 (T1) is processed by the following program:

Code to Transfer Funds (Coding Example 1)

```
        Sequence of Operations
        BEGIN
T1 - 1      Read Customer A Account
T1 - 2      Read Customer B Account
T1 - 3      Write Customer A Account −$100 to Customer A Account
T1 - 4      Write Customer B Account +$100 to Customer B Account
        END
```

The program transfers $100 from the customer A account to the customer B account. After the execution of the first WRITE instruction, the data is temporarily inconsistent by $100. Only with the full completion of the transaction does the data move from one consistent state to a new consistent state.

Conflict. Upon the completion of an operation or a series of operations, the resulting state is inconsistent. Conflict usually occurs when two or more transactions are involved in an update of the same data. The following code depicts two transactions executing simultaneously on the computer. The code in Coding Example 1 is used for both transactions, so processing the two transactions interleaves the instructions of the two copies of the same code. As in the first example, transaction 1 transfers $100 from customer A to customer B. Transaction 2 transfers $75 from A to B. The code shows the sequence of instructions involved for the two transactions and the values in the customer's account after the execution of each instruction.

Code to Transfer Funds (Coding Example 2)

```
        BEGIN
T1 - 1          Read Customer A Account
T2 - 1          Read Customer B Account
T1 - 2          Read Customer B Account
T1 - 3          Write Customer A Account -$100 to Customer A Account
T2 - 2          Read Customer B Account
T1 - 4          Write Customer B Account +$100 to Customer B Account
T2 - 3          Write Customer A Account -$75 to Customer A Account
T2 - 4          Write Customer B Account +$75 to Customer B Account
        END
```

The result is a conflict. The data is inconsistent. The total deposits of the two customers should be $600, with $125 in customer A's account and $475 in customer B's account. Due to the overlapping of operations between the two transactions, customer A has $225 and customer B has only $375 for a total deposit of $600. The initial $100 transaction was undone. A careful analysis of the program and the workspace of transactions 1 and 2 shows how this problem occurs (see table on facing page):

Execution of	Results in					
	Transaction 1 Workspace		Transaction 2 Workspace		Data Base	
	A Account	B Account	A Account	B Account	A Account	B Account
T1 - 1	300	—	—	—	300	300
T2 - 1	300	—	300	—	300	300
T1 - 2	300	300	300	—	300	300
T1 - 3	200	300	300	—	200	300
T2 - 2	200	300	300	300	200	300
T1 - 4	200	400	300	300	200	400
T2 - 3	200	400	225	300	225	400
T2 - 4	200	400	225	375	225	375

The transactions are using separate workspaces for the reading and writing of the data base. Notice that, upon execution of T1 - 3, the transaction 1 A Account workspace is reduced by $100 and the value written to the data base. *However*, transaction 2 A Account workspace remains at 300. Later, with the execution of T2 - 3, transaction 2's A Account workspace is reduced by $75 and the value is written to the data base. The conflict has occurred with one copy of the data base. Later discussions show how distributed copies further complicate the problem. Chapter 11 describes methods to analyze and prevent conflict.

Schedule. A *schedule* is an ordering of events or instructions within multiple transactions. The events in Coding Example 2 represent a schedule but not a very good one. Conflict can be avoided by running each transaction individually until it is completed. However, this approach is not effective due to resulting performance problems.

Serializable Schedule. The effect of interleaving the operations of multiple transactions is the same as running the transactions serially—that is, the result is a consistent state. A serializable schedule can be obtained in Coding Example 2 by issuing both WRITEs for transaction 1 before issuing the READs for transaction 2. However, this then becomes a straight serial process, so other methods (discussed in Chapter 11) must be used to achieve the effect of serial processing.

Locking. The locking of data ensures the inaccessibility of that data while it is in a temporarily inconsistent state. Locking is used to prevent multiple transactions from creating a conflict in the data base. The data base management system (DBMS) is usually responsible for locking the affected data, typically at a record or segment level. Each application that has WRITE code to the data base must be defined to the DBMS prior to execution. Thereafter, when the application code is executing, a READ instruction will signal the DBMS to disallow other applications from obtaining the data that was accessed by the READ. The lock is released after the application completes the transaction, and the other waiting applications are given access to the data. For example, in Coding Example 2, the initiation of transaction 1 locks the data from transaction 2.

Resiliency. A dictionary definition of *resiliency* is the ability to return to the original form. The term aptly describes a critical characteristic of a distributed data base: The failure of a site should not affect the operations of another site. A down site should not evoke a "sympathetic failure" at other sites. An incomplete transaction (for example, a hardware failure during the execution of the update code) should be completely backed out at all sites and the data restored to its original form.

THE CHALLENGE OF DISTRIBUTED DATA

A distributed data system is deceptively complex. This section will describe the complexities and problems, as well as techniques to avoid some of the possible pitfalls. Figure 10-7 is a guideline for decisions on distributing data. Generally speaking, the figure shows that as an organization moves to (a) larger data bases with (b) multiple copies of data and (c) shorter response time on updates, the organization also moves to an increasingly complex distributed environment. If possible, complexities should be avoided because of the increased risk of system failures, data inconsistencies, and increased costs. If the organization needs this type of environment, it should recognize the trade-offs.

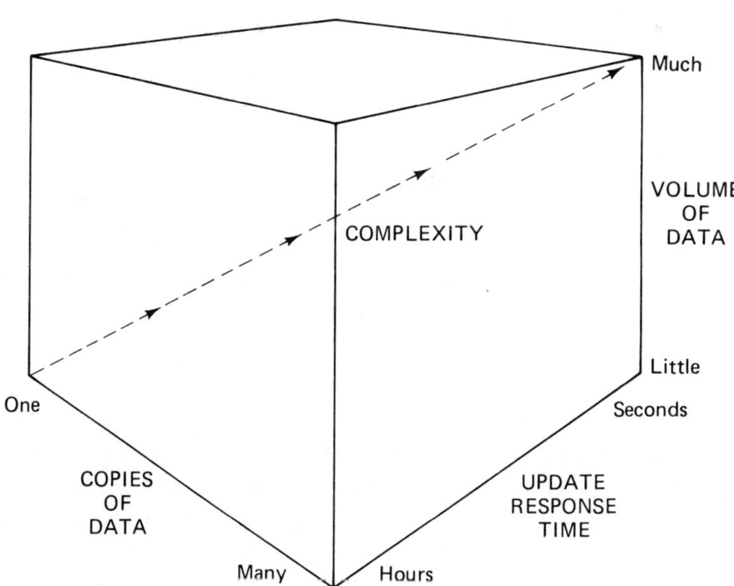

FIGURE 10-7. Distribution Complexity.

An example of the complexities and problems is provided in Figure 10-8(a). Sites A and B simultaneously update an item in the data base. In the absence of control mechanisms, the data base reflects only the one update; the other update is lost. This happens when both sites retrieve the data item, change it (add or subtract from the value), and write their revised value back into the data base. In the illustration, the revised value should have been 250 instead of 150. This example is similar to the problem encountered in Coding Example 2.

The problem is compounded in Figure 10-8(b). At a later time, site C retrieves the data and stores it in a replicated local data base. The organization now has a rather serious consistency problem and it is not a temporary inconsistency. The data bases are in conflict. This is only one example of many possibilities. Coding Example 2 represents the problem with only one copy of the data; as multiple copies are distributed the problem becomes more complex.

Lockouts and the Deadly Embrace. The most common solution to this problem is preventing sites A and B from simultaneous executions on the same data. Through the use of lockouts, for example, site B would not be allowed to execute until site A had completed its transaction (see Figure 10-8[c]).

Lockouts work reasonably well with a centralized data base. However, in a distributed environment, the sites may possibly lock each other out and prevent either transaction from completing its task. *Mutual lockout*, often called *deadly embrace*, is shown in Figure 10-9. Sites A and B wish to update data base items Y and Z, respectively; consequently, site A locks data Y from site B and site B locks data Z from site A. In order to complete their transactions, both sites need data from the other *locked* data bases. Hence, neither can execute further and the two sites are locked in a deadly embrace. Clearly, the deadly embrace is an unacceptable situation and the system must be able to detect, analyze, and resolve the problem.

Update and Retrieval Overhead. The use of locks is a widely accepted method for achieving consistency. When properly implemented with serialized scheduling, locks provide a very valuable method for maintaining data base integrity. However, consistency of the distributed data does not come without cost. For example, in Figure 10-10 the site in San Francisco issues a transaction to update two replicated data bases in New York and Atlanta. The following

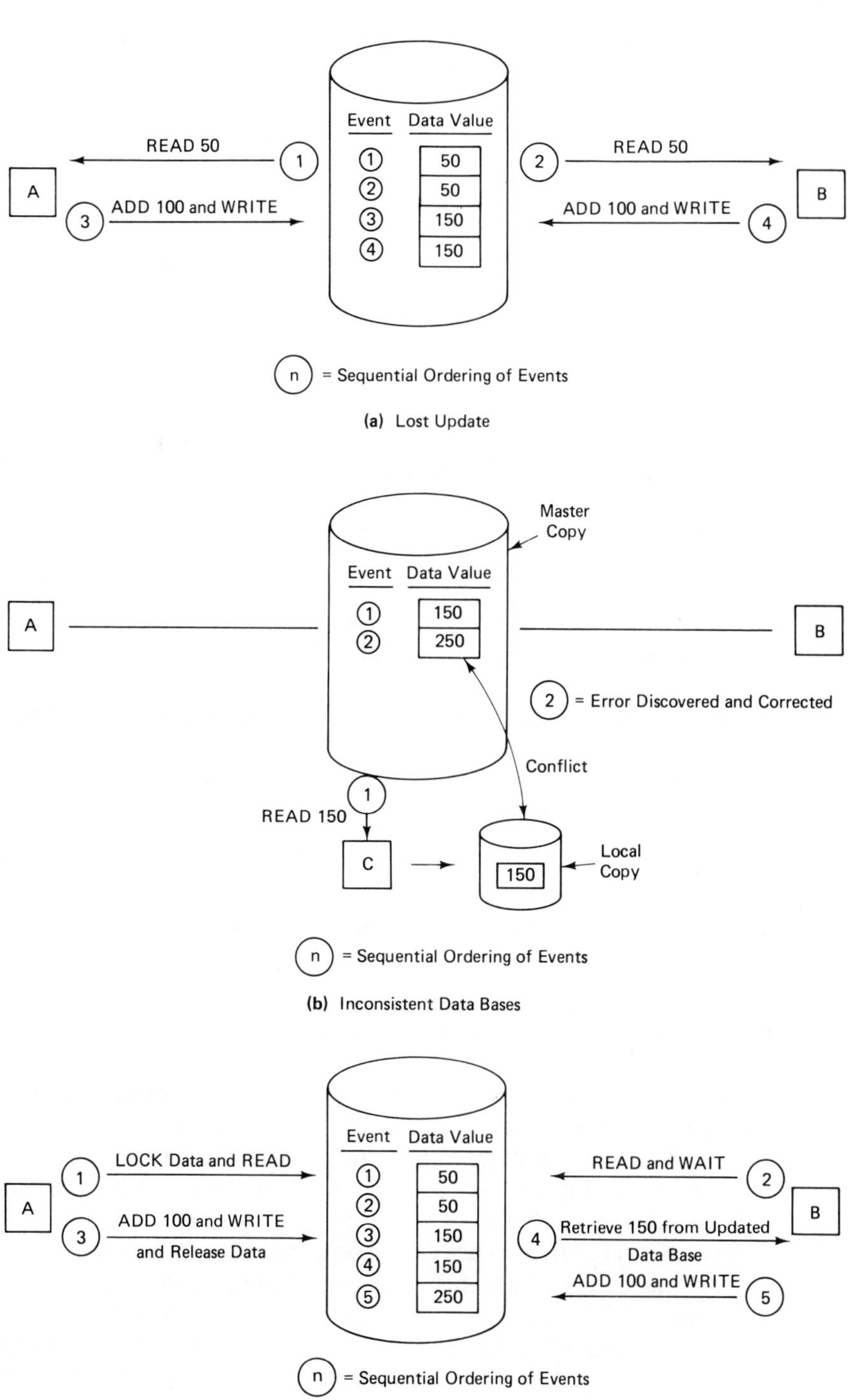

FIGURE 10-8. Network Data Base Problems.

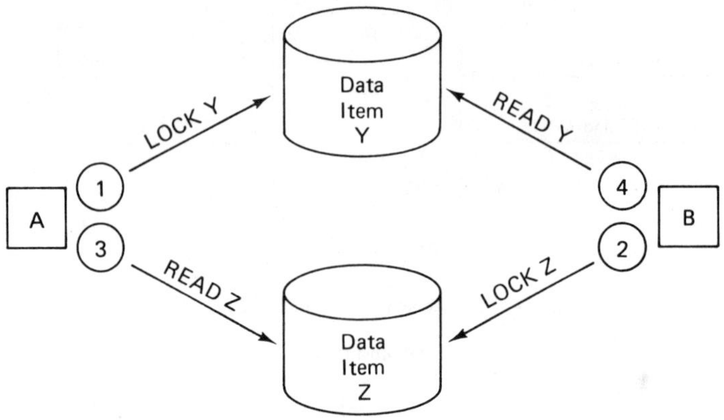

FIGURE 10-9. Deadly Embrace.

communications messages must be exchanged over the network between San Francisco and the two other sites:

Event 1: San Francisco sends lock request messages to New York and Atlanta.

Event 2: New York and Atlanta send lock grant messages to San Francisco (if the request is acceptable).

Event 3: San Francisco transmits the update transaction.

Event 4: The receiving sites update the data bases and transmit to San Francisco the acknowledgment of an update completion.

Event 5: San Francisco receives the acknowledgment and transmits messages to release the lock.

A typical locking algorithm requires 5 (n-1) intersite messages to manage an update transaction among n distributed sites. This could be considered an extreme example. The messages in events 1 and 3 can be piggybacked onto each other. The update portion of the message would only be used after an established time had elapsed and San Francisco had not been sent a revoking message indicating one of the sites had sent San Francisco a lock denial. However, the 5 (n-1) algorithm is simpler and more reliable.

This overhead is the tip of the iceberg; each of the messages requires data link control (DLC) messages to ensure that the data base messages are properly received. As discussed in Chapter 6, a DLC might require four DLC control messages to every user data message. The number varies from three to nine, so the number four is conservative. Consequently, the locking algorithm is actually D (5 [n-1]), where D is the number of overhead DLC messages needed to manage one user message flow. In the example in Figure 10-10 (assuming a DLC overhead factor of 4) the calculation reveals that 40 messages are transported through the network to accomplish *one* update at two sites 4 (5 [3-1]). Finally, the example did not calculate the additional costs of environments that have heterogeneous topologies and/or data systems. Protocol conversions and data structure translators consume additional overhead. It takes little imagination to recognize that the proper placement of replicated data in a distributed network is a very important task.

Retrieval overhead presents problems also, especially in a partitioned data base. A data retrieval request may require the

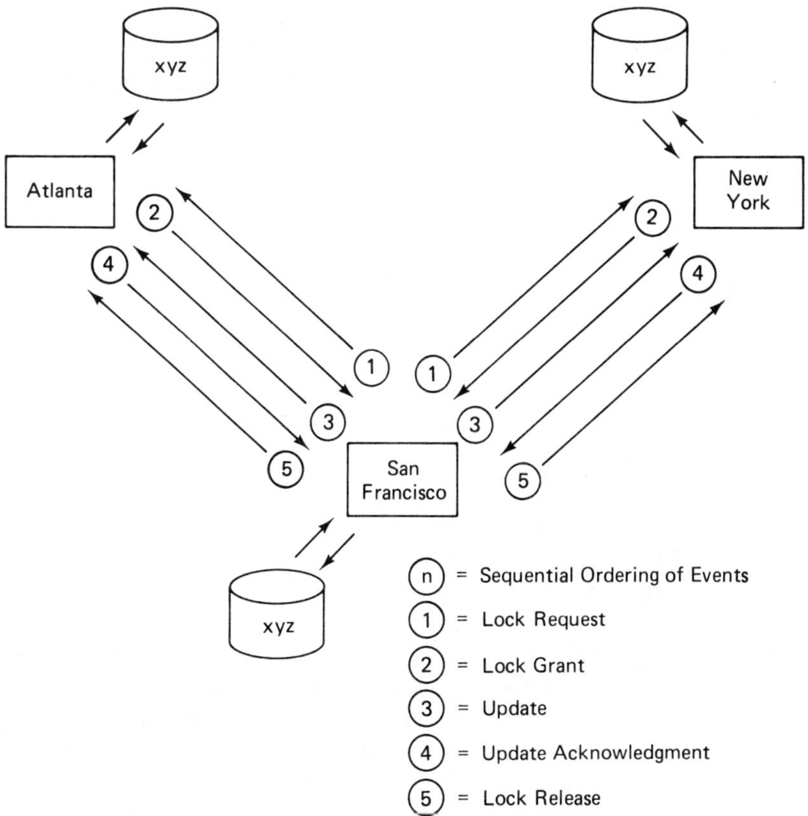

FIGURE 10-10. Update Overhead.

"rounding up" of the needed data at several sites in the network since the data has been partitioned into subsets and assigned to different locations. The DLC, protocol conversion, and data structure translation overhead are important considerations in the decisions of how to partition the data bases.

Failure and Recovery. The efforts to achieve resiliency in a distributed system are quite different than in a conventional centralized environment. The centralized approach assumes the availability of much information about a problem or failure. The operating system can suspend the execution of the problem program and store and query registers and control blocks—during which the problem component does not change. Moreover, the events are local and time delays in the problem analysis are very short.

In a DDP system, the time delay in gathering data for analysis may be significant—in some cases, the data may be outdated upon receipt by the component tasked with the analysis and resolution. The problem may not be suspended as in a centralized system, since some distributed systems have horizontal topologies and autonomous or near autonomous components.

Referring again to Figure 10-10, one can gain an appreciation of the situation. Let us assume the update executed successfully at New York but the Atlanta site experienced problems due to a hardware or software failure. The update must be reversed in the New York copy. All items must be restored, transactions eventually reapplied, backup tapes made, and log files restored to the pre-update image. In the meantime, other transactions must be examined to determine if they were dependent upon the suspended transaction. Of course, one has the option of maintaining the New York update, continuing subsequent updates, and bringing the Atlanta data up to date at a later time. Nonetheless, the affected nodes cannot independently make these decisions—all must be aware of each other if the data bases are to be properly synchronized.

A key question must be answered: Does the organization need timely data at the expense of consistency? Stated another way, must all copies be concurrent with each other? Practically speaking, certain classes of data (such as historical data) may be allowed to exhibit weak consistency: data are not kept concurrent. On the other hand, other classes of data (real time data, for example) may need strong consistency. The designers must examine the user requirements very carefully. The benefits of strong consistency must be weighed against the increased costs of additional complexity and overhead.

Class of Data. The picture may not be as grim as it seems. If an organization's data is relatively stable and seldom updated, the multiple copies approach can reap significant benefits in decreased communications costs and faster response time. The designer must know the ratio of updates to retrievals in order to make intelligent decisions on the location of the data in the distributed network.

Martin provides a useful classification of data based on the frequency of change, the need for fast updates, and the type of updates:[6]

Class 0: Data is never or infrequently changed.

Class 1: Data changed with a single replacement; if data is replaced more than once, no harm is done.

Class 2: Data that must not be updated with a transaction more than once, but updates can take place in any sequence.

Class 3: Data that must be updated in a predetermined order and in a real-time mode.

Class 4: Data is updated and the update affects other data bases and/or computers in the network.

The different classes of data are subject to different data base management schemes. The classes of data can be analyzed with other factors to determine the best approach. For example, if a large data base is Class 3 and exhibits a high ratio of updates to retrieval, one should be very careful of creating multiple copies. To do so would move the complexity arrow (see Figure 10-7) to the extreme corner in the matrix.

Another factor to be examined with data class is frequency of use by each site. For example, if a data base exhibits the characteristics of Class 0, and several sites in the distributed system have equal needs for frequency of access, it makes sense to replicate the data at those sites.

---

[6] James Martin, *Design Strategy for Distributed Systems*. Englewood Cliffs NJ: Prentice-Hall, 1981, pp. 282–283.

Major Factors in Distributed Data Decisions. In Chapter 11, a section is devoted to distributed data base design. The discussion in this chapter has focused on more general guidelines for deciding how to distribute the data in the network. The determining factors are summarized as follows:

1. Frequency of use at each site.
2. User control of data.
3. Real time update requirements.
4. Backup requirements.
5. Class of data.
6. Cost to store locally vs. cost to transmit remotely.
7. Security considerations.
8. Time of use of the data.
9. Volume of data accessed.
10. Retrieval response time requirements.
11. Location of data users.
12. Retrieval access vs. update accesses.

Loading. The term *loading* refers to the assignment of resources to a distributed site. The resources can be data or software. In most cases, loading is accomplished in a nonreal time mode. This approach can simplify the distributed environment considerably. Loading can be accomplished within the vertical or horizontal topology in three ways: downline loading, upline loading, and crossline loading (see Figure 10-11).

Downline loading. Resources are transmitted from the primary or master site to the secondary or slave site(s). Downline loading is often used to transport copies of software developed and tested at the centralized site. The programs are then placed at the secondary sites for execution. This approach avoids staffing the distributed sites with programming personnel. Data is frequently downline loaded as well. For example, data is updated at one site; the one copy greatly simplifies the update process. At selected times, the data is sent to the secondary sites and copied into the files. This approach works well for small data bases, but a large data base copy transmission creates considerable traffic on the network. An alternative approach is transmitting only the segments of the data base that have been altered since the last downline load. Another very useful technique is the downline loading of tasks, in which the

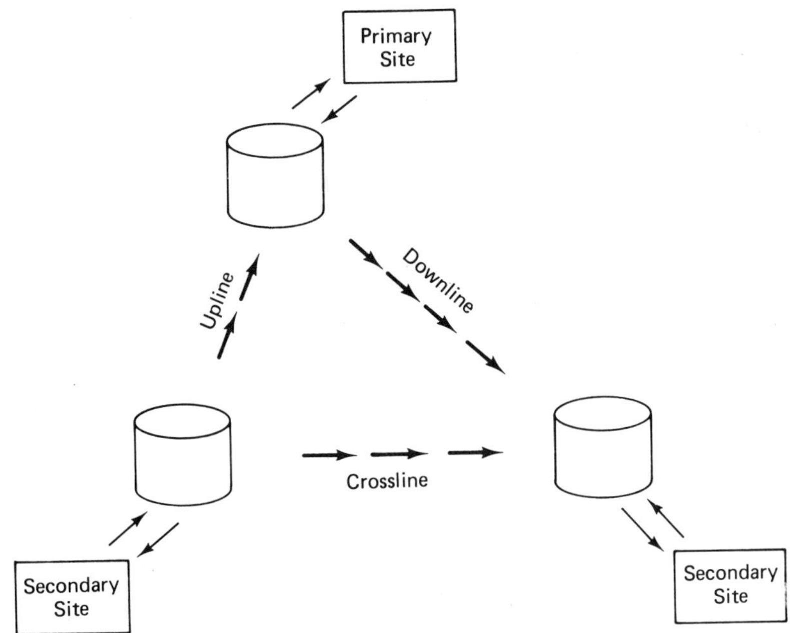

FIGURE 10-11. Loading.

master station receives transactions and decides which processing elements are available and/or have the proper resources for the particular transaction. This concept is also called *workload partitioning* or *transaction downloading*.

Upline loading. This technique is used when the secondary site needs a larger computer to accomplish its task. It transmits work "up stream" to the master site for processing. Upline loading is also used to transmit data. The data can be detailed items, but is frequently summarized or aggregated data. Upline loading is also used to accomplish distributed system development. Each secondary site is responsible for the development of a part of the system (i.e., the development is partitioned). At prearranged times, the subsystems are transmitted to one or two sites (usually the master site) for integration testing. The completed system may then be downline loaded for execution at the other modes. Distributed system development has several advantages, but must be approached carefully if its full benefits are to be realized. Pages 332 through 345 discuss this concept in more detail.

Crossline loading. The horizontal network components can accomplish the same functions discussed above. The term to describe peer to peer exchange is *crossline loading*. As in downline and upline loading, the technique of crossline loading is often used to share resources, partition work, and provide backup for the peer processing elements.

What are some candidates for offloading? The following list provides some examples. It is not all inclusive but serves to show the potential value of offloading:

- *Edits.* Data editing is performed locally. Edit criteria can be developed at the central site and downline loaded.
- *Formats.* Screen and print formats stored locally.
- *Error Processing.* Any error condition software is loaded to the local elements.
- *Protocols.* Certain processing elements are designated as protocol handlers. The protocol overhead is offloaded from central host.
- *Audit Trails.* Approach this area with caution; distributed audit trails can become complex.
- *Data Bases.* Selected data is assigned to the remote sites.
- *Addressing.* Address resolution and message routing is offloaded from the primary station.
- *Processing Programs.* The applications themselves are offloaded to remote sites.

The values of offloading are resource sharing, distribution of peak load, relieving the primary site from certain specialized activities, and backup. The reader is encouraged to examine this idea further; it can be very effective. Table 10-1 provides other points that should be considered when establishing an offloading system.

Summary. The tradeoffs of centralized, replicated, and partitioned data bases are summarized in Figure 10-12. The comparison of partitioned data to centralized data is illustrated in Figure 10-12(a); Figure 10-12(b) compares replicated data to centralized data. Also, Figure 10-7 should be reviewed again. Complexity increases as the organization moves to multiple copies of large data volumes with real time update requirements. Certain points in the matrix provide the appropriate environment for each organization. Do not be deterred from using DDP because of the challenge of managing distributed data bases; however, approach this

TABLE 10-1. Loading Factors.

Determine Functions, Processes, Data, Programs To Be Off-Line Loaded
How Often
Time of Day
Volume of Data to be Transmitted
Rotation of Line Load Sequences
Procedures for Use if Operator Functions Changed
Verification of Transmission and Load
Controls Over What Happens at Up/Down/Cross Line Site
Line Load and Analysis
Transmission Method
Candidates for Offloading:
Forms
Formats
Edit Criteria
Security/Privacy Screens
Master Data Bases
Processing Queues
Address Capability
Error Processing
Protocol Handling
Diagnostics
Processing Programs
Routing Controls
Audit Trail Storage

portion of DDP with care and deliberation. Chapter 11 provides specific ideas for designing a distributed data base system.

PLANNING THE DISTRIBUTED SYSTEM

Distributed system planning involves the following major efforts:

- Determining distributed tendencies.
- Assessing political climate.
- Analyzing current data processing environment problems.
- Determining extent of current distribution.
- Weighing cost-performance-functions trade-offs.

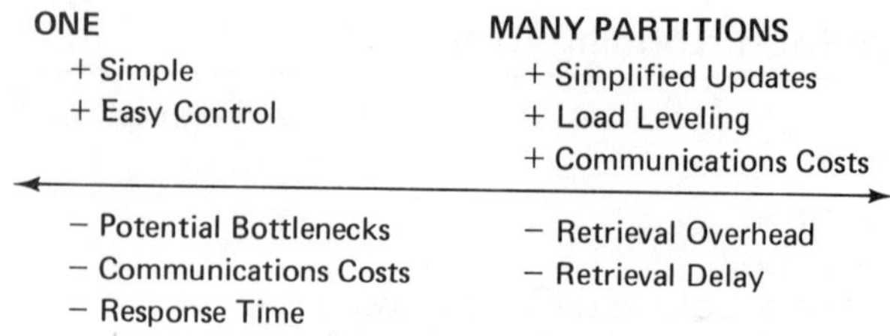

(a) Partitioned Copies Issue

ONE
+ Simple
+ Easy Control

MANY
+ Retrieval Response Time
+ Communications Costs
+ Fewer Bottlenecks

– Potential Bottlenecks
– Communications Costs
– Response Time

– Overhead
– Complexity

(b) Multiple Copies Issue

FIGURE 10-12. Distribution Trade-offs.

Determining Distributed Tendencies

The implementation of a successful distributed system depends on an understanding that information systems should match a company's structure and strategy. Further, the basic characteristics and nature of the organization must be understood. The move toward distribution is greatly simplified if management recognizes which parts of the organization should be distributed, which parts should be centralized, and which should be decentralized.

As a general rule, an organization will most likely benefit from the distributed approach if its departments operate with a high degree of autonomy. Companies organized with profit centers, such as a holding company, or independent subsidiaries are examples of highly autonomous structures. The diversity of business functions creates a demand to split up centralized systems and move them to the autonomous departments or subsidiaries. At the extreme end, highly autonomous subsidiaries often point to a decentralized approach with little or no integration of the parts to a whole. Organizations dispersed throughout a wide geographical region are also oriented toward distributed or local processing in order to reduce the communications costs of sending data to a centralized host.

The distribution or decentralization tendency also increases with organizations that offer multiple products or have varied missions. The situation creates pressures for differentiation within the organization and provides an impetus to develop disparate information systems for supporting the varied products. This tendency is reinforced by product managers wanting to have more control over their information systems by placing automation decisions within their department.

The distributed tendency is summarized in Figure 10-13.[7] The tendency or desirability to distribute is usually low if an organization is single-product oriented, its departments are geographically co-located, and it is operating under one hierarchical structure. The distributed tendency increases for those organizations that offer multiple products, are geographically dispersed, and have departmental independence.

The organization's structure and product determine how data is used. It must be recognized that data is the driving force of an

[7]This concept and Figures 10-13 and 10-15 are explained further by Jack R. Buchanan and Richard G. Linower in "Understanding Distributed Data Processing." *Harvard Business Review*, July–August 1980.

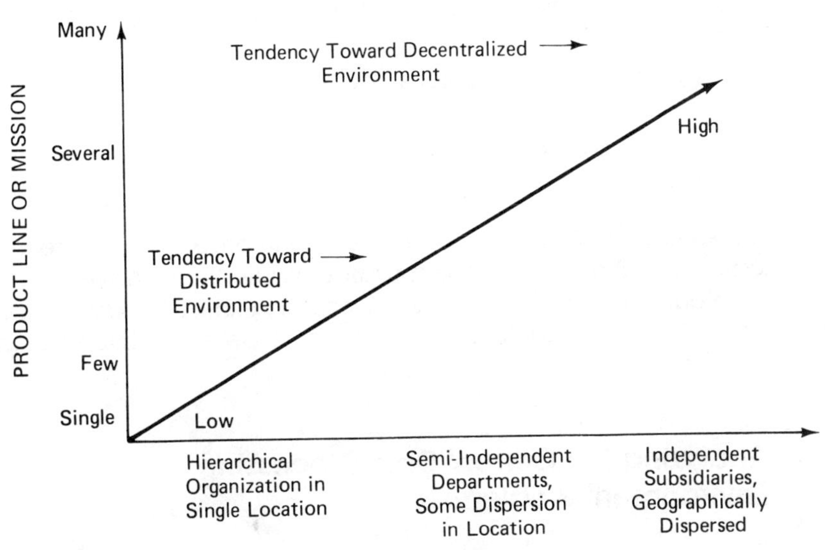

FIGURE 10-13. Tendency Toward Distribution/Decentralization.

automated system—the programs exist to process the organization's data. Consequently, the organization must have a clear understanding of how the data flows support the organizational structure and product.

Assessing the Political Climate

While recognition of distributed tendencies is important, it should also be recognized that the actual implementation of a new system will most likely disrupt the company. Organizational changes and realignments may be required and some managers may be reluctant to accept additional responsibilities or, more seriously, a divestment of responsibilities. A familiar management saying, "I don't mind progress, as long as nothing changes," should be kept in mind at all times. The following questions should be addressed before formally developing a plan for a distributed environment:

- Does high-level management have the time and ability to focus on the issues?
- Is the power structure in the organization too diverse to allow centralized planning and control?
- Can vested interests be melded to the overall corporate plan?
- Are the products/missions too diverse for integrated information support?
- Is the organization structured around autonomous and independent departments or subsidiaries?

These kinds of questions are necessary in order to measure the political atmosphere of the organization. While their answers may not be pleasant or easy to accept, organizational behavior and politics are a fundamental ingredient in large scale information systems planning.

Analyzing the Current Data Processing Environment Problems

Once the distributed tendencies have been studied and the organizational politics analyzed, the focus should turn to questions on the services provided by the data processing environment:

- Is the backlog of user applications increasing?
- Does the organization have redundant applications development and maintenance?
- Are bottlenecks in turnaround time and performance increasing at certain computer facilities?
- Is data of local interest only?
- Are communications costs increasing at a rate higher than tariff increases?
- Are some applications candidates for partitioning and parallel processing?
- Is data entry and editing centralized?
- Are users interested in assuming more responsibility in the data processing area?

The list of questions could cover several pages. The important point is for *each* organization to comprise questions relevant to its environment and then attempt to answer them as a method to point to specific action items in the planning process.

Determining Extent of Current Distribution

The organization should also be analyzed to determine the extent of distributed data processing that currently exists. This effort is important for two reasons:

1. Many decisions will be ill-conceived if it is not known where the organization is with respect to its use of distributed systems.
2. The study must establish a base from which design decisions can be made.

Simply stated, it will be very difficult for the organization to determine which direction to follow if it cannot identify where it is starting from. Figure 10-14 illustrates the importance of the analysis.

The decisions for an organization at point A on the distributed/centralized spectrum will be quite different from those for an organization at point B. For example, point A decisions require the development of software to interface different computers and communications protocols that currently exist. Point B decisions

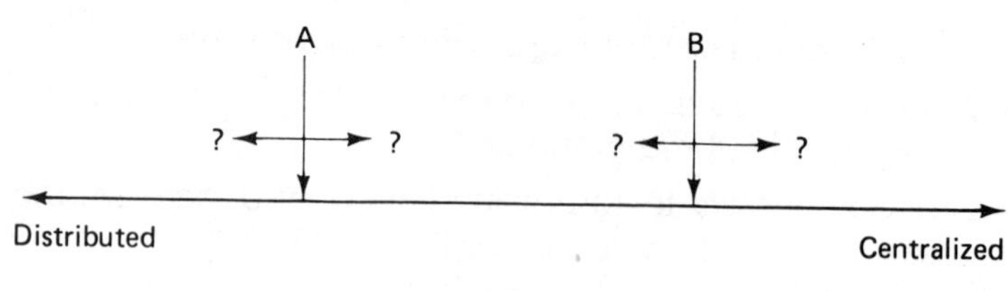

FIGURE 10-14. Distributed/Centralized Spectrum.

focus toward a design for compatible hardware and one communications protocol.

The second reason for conducting a distribution analysis of the current environment is to identify existing conflicts and make assessments of potential problems in future systems. The analysis can be conducted in the following support areas for each business activity (e.g., inventory, marketing) or information system that supports the business activity:

1. Applications software development.
2. Hardware management.
3. Data base design.
4. Data administration.
5. Systems programming functions.
6. Data access.
7. Communications equipment control and use.
8. Communications software development.
9. Budget control for applications development.
10. Standards for software development.

In most instances, the results of this analysis will reveal conflicts among the departments. For example, Figure 10-15 reveals an obvious conflict in support areas 1 and 10. The software development standards are highly decentralized within the user department and the applications software development is also performed largely by distributed user departments. This situation will likely lead to conflicts among the users and the data processing staff. It will certainly result in incompatible systems within an organization that has need for integrated systems and data bases.

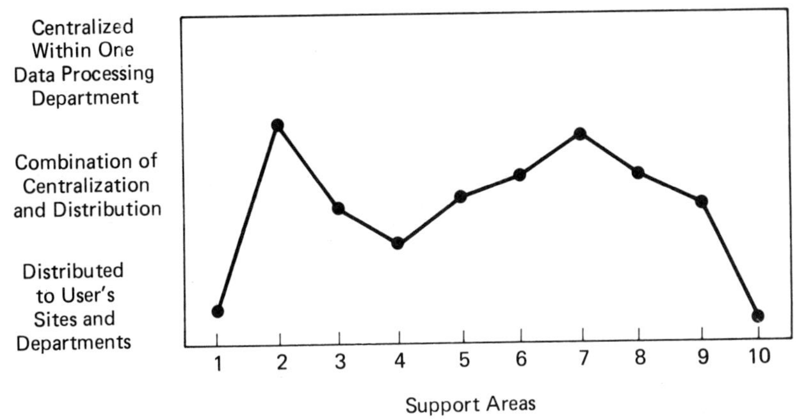

FIGURE 10-15. Current Distribution.

Weighing Cost-Performance-Functions Trade-Offs

A primary objective of an organizational analysis is the provision of a plan that meets the user requirements while meeting cost constraints. To achieve this objective, data processing management must work closely with the organization's user management. All parties must actively participate in balancing the requirements against the costs to satisfy the requirements.

Upper-level management must be involved in these efforts as well, because the operating goals of the organization must affect the information systems. Figure 10-16 exemplifies how the managerial decisions affect the design and costs of the distributed system. Each management decision affects the design, which in turn increases or decreases the system costs.

The decisions to improve local response time (#1) and decrease communications costs (#4) provide for the same design decision to install local computers. However, the decision to reduce communications costs must be weighed against the costs of additional processors. Consequently, the decision to acquire additional processors for reducing communications costs may increase personnel costs to the point of offsettng any potential communications costs reductions. The additional staffs required to operate the computers at the distributed sites can become a significant element in the budget.

MANAGEMENT DECISION	DESIGN IMPLICATIONS	COST IMPLICATIONS
1. Improve Response Time to Local Sites	Install Local Computers	Additional Hardware Purchases
2. Provide Local Control of Software Resource	Staff for Local Program Development	Possible Systems Duplication
3. Move Data Responsibility to Local Level	Distribute Data Base	Complex Data Base Software
4. Decrease Communications Costs	Install Local Computers	Additional Hardware Purchases
⋮	⋮	⋮

FIGURE 10-16. Cost and Design Implications of Management Decisions.

The distributed systems design involves the continuous trade-off of costs, performance, and functions provided. The design cannot provide for the optimum performance of all three components, since the emphasis on one will diminish the other two. This problem is illustrated in Figure 10-17.

The three components are located on the three points of the triangle. If the organization's emphasis is on performance (such as response time or throughput), costs will increase in order to provide a greater bandwidth for the communications lines and more powerful computers at the distributed sites.

The network support functions will affect both costs and performance. Certain support functions (security provisions, providing for different terminal types, code translation) will extend response time and increase overhead costs. Other support functions will reduce throughput performance (for example, adding department, data, terminal, or user identifiers to the message header). Yet, these functions may be essential in meeting the user requirements. All affected parties must recognize that all three points on the triangle cannot be made optimal. Compromise is required and, where compromise cannot be reached, top management must arbitrate and provide firm decisions.

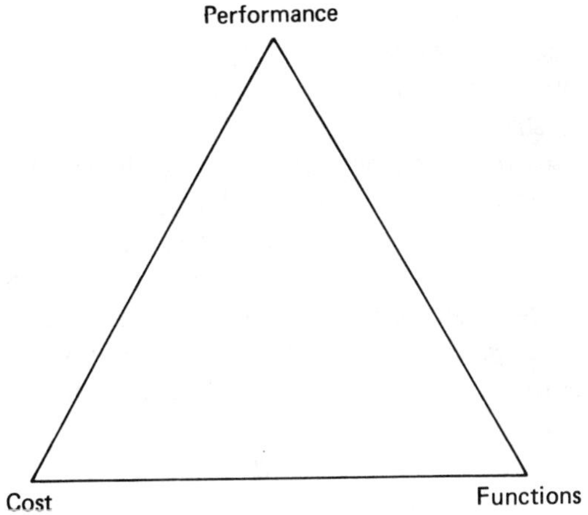

FIGURE 10-17. Decision Trade-offs.

Summary

Many people in the industry believe the widespread movement to distributed systems is inevitable. As stated previously, it is less a matter of deciding whether to distribute or not and more of a matter of deciding how to distribute. The critical point and key ingredient is to establish a plan that supports the organization's goals. The plan may lead to a hybrid system of distributed, centralized, and decentralized components. As long as the decisions are based on some form of rationale and judgments of the technical and business risks, the organization should succeed in its efforts. The major efforts listed earlier and described thereafter are tools to help formulate the plan. The plan, however, is not an end into itself but a means to the end goal of truly effective and supportive distributed systems. The tools and plan will not achieve this goal by themselves—*only* firm, aggressive, knowledge commitment from top management will lead to success.

MANAGING THE DISTRIBUTED RESOURCES[8]

The advancements in distributed data processing (DDP) during the past few years have been remarkable. Many fully tested hardware and software components are now available commercially. Distributed networking concepts have been refined and strides are being made in managing dispersed data bases. If computing costs continue to decline relative to transmission costs, the number of DDP networks will likely increase.

Yet, while the technical matters are being solved, many management issues are not. Ironically, some of the more active features of DDP present the most vexing management problems. For example, distributed network architecture is conducive to load leveling, or resource sharing, among the sites. Although this can lead to more effective use of the network's computers, resources dispersed at these sites are not easily managed. Distributed data processing appeals to some organizations because it lets them move computerized resources to various departments and divisions throughout the organizational structure. But this movement may lead to uncontrolled proliferation of conflicting or redundant software and data.

---

[8]Reprinted from "An Automatic Pilot for the Growing Distributed Network" by Uyless D. Black, *Data Communications*, November 1980. Copyright © by McGraw-Hill, Inc. All rights reserved.

The primary challenge to DDP management is to effectively administer the hardware, software, and data resources that exist throughout the network. This task is especially difficult if the organization has acquired or merged with other companies that have different computer architectures and network protocols. Moreover, many large organizations have separate departmental budgets that often allow managers to develop unique information systems. The resulting incompatibilities can be costly and difficult to manage.

If one fundamental principle were to guide management in the use of DDP, it would be that the organization requires strong unified management control and guidance. At first this might seem contradictory but, in fact, DDP requires a stronger management structure than the traditional centralized approach.

A distributed system may have millions of source statements (such as computer program instructions), hundreds of programs, thousands of data items, and scores of data bases residing in many geographically dispersed sites. Because of performance considerations and communications costs, many of these resources will be duplicated at various network sites.

Typically, the different sites develop, maintain, and update their own data files and programs. It is not uncommon for a site to develop programs or create data files for its own needs and informally pass these resources on to other sites. Some organizations permit sites to modify the code or data files that were originally developed for general use, resulting perhaps in several versions of the same program or data file. Companies do not always track and record these changes. Even worse, they may allow different sites to develop redundant programs that perform similar functions, which usually occurs when departments or subsidiaries of a company have enough political clout to establish their own versions of an information system.

Figure 10-18 illustrates the problem caused by lack of control in an evolving distributed environment. Site I develops a program or data base [Figure 10-18(a)] and, to share resources, passes it to site II. To meet its specific needs, site II modifies the core program [Figure 10-18(b)]. At a later date, site I changes its version of the program [Figure 10-18(c)], then passes the modified program to site II.

However, the altered program does not fulfill site II's needs and, in some instances, may not function in II's environment [Figure 10-18(d)]. Site II is then faced with the difficult choice of modifying its version of the program to accept the change or rejecting the change and expanding its staff to maintain permanently both the core program and the enhanced program.

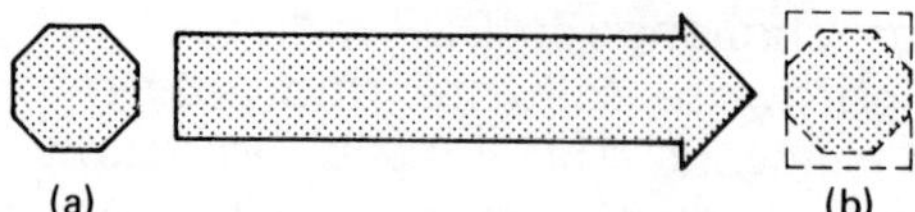

Site I Develops a Program and Passes it to Site II, Which Makes Changes.

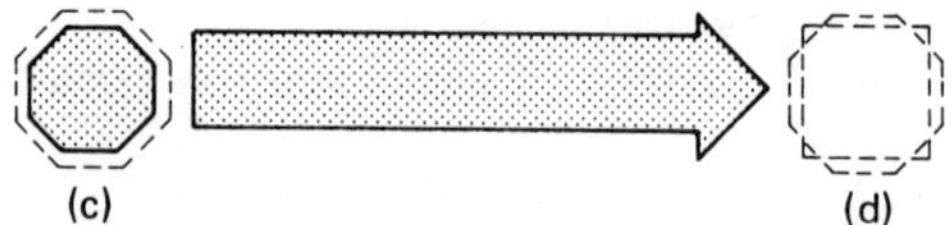

Site I Later Changes the Program and Passes it to Site II, Which Cannot Use it.

FIGURE 10-18. Control Problems.

It is not difficult to see how an organization's distributed information system can become unsynchronized and inaccurate. Disorganized distributed systems have cost companies considerably in dollars and credibility. In addition to paying an inordinate overhead for supporting this kind of environment, these companies experienced system failures and data errors when changes were made to supposedly identical copies of code or data files.

As a company's DDP network grows in size and complexity, the impact of a poorly conceived coordination and control policy becomes more severe. Unfortunately, many organizations begin the initial move to DDP without an awareness of the pitfalls of having no centralized control. The ideal solution is to give the distributed sites the freedom to manage their resources and develop information systems under the general supervision of a companywide entity. This benefits staff morale and productivity and ensures the best use of a distributed network.

Management Model

Distributed automation management (DAM) is a management and control model for a distributed network. Essentially, DAM is an architecture containing detailed information about the automated (computerized) resources existing throughout the network. It can be applied to an organization with distributed departments or to subsidiaries that perform similar automated functions or, to a more limited extent, to an organization with widely varying data processing needs. The advantage of DAM is that all dispersed sites can participate in the decisions and implementation of a developing network, while management can monitor all changes.

If a company chooses to use the distributed automation management approach, it must establish well-defined, company wide standards for structuring all information subsystems and software modules. Modules should have clearly defined interfaces and functions. Without such definitions, system integration and testing will be extremely difficult and time consuming. If each subsystem and module is relatively self-contained, several development teams located at various distributed sites can work on different components, which should prevent major integration problems as the project nears completion.

The distribution of the program development process has more risks than a centralized effort. For example, the distributed modules will usually require more complex interfaces and, therefore, are subject to a higher probability of error. Moreover, DAM entails additional overhead for project coordination among the sites, so the travel budget will most likely increase. Nonetheless, if the network is properly controlled, it will be more efficient, resulting in a more satisfied staff.

Figure 10-19 depicts the elements and structure of the architecture. At its most general level, DAM contains information on project management, data/data base administration, software inventory, equipment inventory, and sources and uses of the company's reports. These responsibilities are broken down into functions, which are handled by the individual elements and subelements (see Table 10-2).

An uppermost element in the DAM architecture, system-life-cycle management (SLC), is similar to the conventional centralized approach in which systems development is performed in phases such as analysis, design, testing, and cutover. The primary difference between SLC and the centralized method is that the SLC process involves all the distributed departments and sites. SLC includes milestones, due dates, exception reports, descriptions of end products, commitments for equipment installation, testing plans, and other project-management information. It may also contain relationships between the development-cycle phases and the other architecture components.

In addition, SLC identifies the acceptance sites for the software development projects; that is, it assigns an impartial group with testing new or revised programs. Since acceptance-site personnel are unbiased toward the program, test data is usually better structured and more fully exercises the program logic.

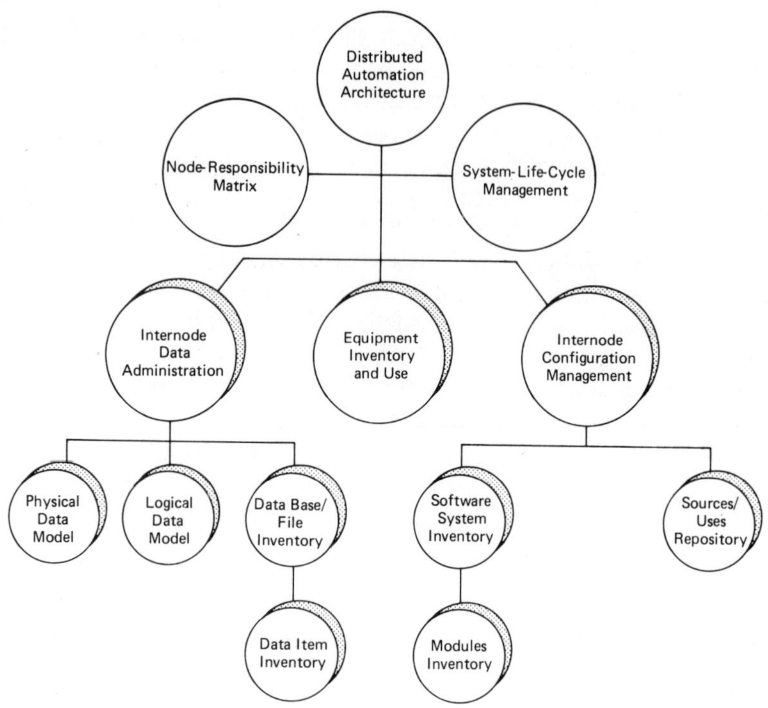

FIGURE 10-19. DAM Architecture.

The node-responsibility matrix (NRM), closely related to system-life-cycle management, identifies the sites' ongoing responsibility for the distributed programs, equipment, and data files. In certain instances, an organization may assign specific custodianship for key data bases or programs and define an authorization procedure for altering these automated elements. The node-responsibility matrix is a logical tool for this function, since it contains rules for the changes and identifies the sites that are to receive the altered components. Subsets of the information in the system-life-cycle are moved to the node-responsibility matrix as network nodes are placed in production mode.

NRM also contains technical audit information—an important component of the DAM architecture—to assist review teams in assessing node design and project control. It identifies the dates, participants, and status of the audits of the distributed components, as well as the personnel responsible for implementing the auditors' findings and recommendations.

A company can place security-related entries in the node-responsibility matrix. For example, certain network data, software, or hardware may require limited-access restrictions. Although an operating system and a data base-management system usually provide for passwords and other user identification measures, NRM offers a higher level of security—perhaps with additional identifier codes or some form of encryption access for each node.

Internode Data Administration

Internode data administration (IDA) can assist management in organizing the company's data files and data items across all distributed sites and in determining both the effects of data changes in the network's replicated data bases, where the data should reside, and how to control the partitioning and replication of data bases. IDA and its subelements, the physical data model, the logical data model, and the data base file and data item inventories, include data descriptions, data definitions, types of data bases, plus the home location of data—where the data resides and which sites have access to it. IDA also provides identifier schemes and naming conventions for data items and data bases. Though this portion of the architecture is similar to a centralized data administration model, it includes the additional vital information pertaining to the distributed sites' use of the data.

One IDA subelement, the physical data model (PDM), provides the data's actual physical storage characteristics: its location and format, the physical record layouts, the data-item relationships (as implemented by "pointers"), and the access methods. These aspects of the model closely resemble portions of a data base management system (DBMS).

The physical data model also contains important data-usage statistics, one of which—the hit ratio—is vital for managing distributed data bases. The hit ratio, calculated as total-data-obtained-at-a-node divided by total-data-items-requested-at-that-node, can be used to determine the appropriate data locations. For example, the prevalence of low hit ratios may indicate that the data is mislocated or should be replicated at those sites experiencing low hits. Be aware that replicating the data usually leads to decreased line costs. (Hit ratios are discussed further in Chapter 11.)

Since data directories often mask the problem of mislocated data in the network, the physical data model's usage statistics are critical to determining data base replication/partitioning alternatives and data use of network communications lines. Yet usage

TABLE 10-2. DAM Elements.

Elements and Subelements	Functions and Stored Information
System-Life-Cycle Management (SLC)	Identification of project phases Designation of site responsibility for project deliverables Descriptions of end products Milestones, due dates for deliverables Exception-reporting procedures Description of post-implementation review milestones Designation of site for acceptance testing
Node-Responsibility Matrix (NRM)	Designation of site responsibility for ongoing operations Identification of sites' security responsibilities Description of rules for change to common components Designation of audit sites and milestones for audits
Data Base/File and Data Item Inventories (DFII)	Description of files and data bases for distributed sharing Description of data groupings Relationships among data items over time Description of unique data characteristics Relationships between data items and data bases and files
Equipment Inventory and Use (EIU)	Description of network hardware components Repair history of components Vendor information Maintenance statistics and dates Financial information
Internode Configuration Management (ICM)	
Software Systems and Modules Inventories (SSMI)	Information on available programs General description of program

Internode Data Administration (IDA)	Information on organization's data files, databases, and data items across all distributed sites Description of data Node awareness of and access to data Description of identifier schemes and naming conventions Data relationships between local and remote sites		functions Identification of which sites use program Detailed description of each module's functions Descriptions of hardware architecture, compilers, and job control language used to run software Listing of parameters for module use
Physical Data Model (PDM)	Physical location of data Physical device view of data Internal representation of data Access methods to data Data-usage statistics	Sources/Uses Repository (SUR)	Description of input forms (sources) Description of output reports (uses) Relationships between data sources and uses and the distributed software and data base components
Logical Data Model (LDM)	Application program view of data Description of program interface to physical data model		

statistics should be gathered at selected intervals—not continuously since this is too costly.

The logical data model (LDM) provides an end user (subschema, see Chapter 4) view of the data and the interfaces of the software programs to the physical data model. It also enables virtual access to data from local or remote sites.

For example, if an application issues a request for data, the program need not be aware of where the data is located. The logical data model passes the request to the physical data model, which examines the home locator information and initiates requests to network protocols for the data. The required data is located, passed to the logical model, formated to the application's view, and then given to the application. A change in the physical location of the data will not affect the application-program logic or the logical model. Rather, the physical model can be modified to reflect the data file movement.[9] The approach closely resembles that of layered network protocols and has similar advantages.

The data base/file and data item inventories are closely related subelements that contain the relationships between data items and other data items, data bases, data files, and the distributed nodes. Their descriptions of the data elements can be used during the initial stages of a system life cycle to determine if the company has either the needed data or data that is very similar to the required data. Recent experience has shown that organizations can decrease data management costs by establishing a data inventory and passing all new data requirements through it prior to initiating expensive data-acquisition and data-reporting endeavors. This approach can be especially important if data is distributed and managed (perhaps created) at various locations in the organization.

The data item inventory contains standardized names for the many data elements in the distributed network. A distributed network whose components use common data element names offers several advantages: communicating and disseminating data changes to all sites is simplified, distributed development teams can more easily read and comprehend code and, since learning curves are reduced significantly by this approach, code is more easily maintained.

Naming standards also facilitate the use of end-to-end protocols by allowing user programs to exchange data that is commonly understood. Moreover, layered protocols sometimes examine data elements emanating from other layers to invoke

[9]These ideas are discussed further in Chapter 11.

certain editing or routing algorithms. Common names allow the various layers to more easily interface with each other. What is more, variable length records from the application program or the data base are more easily implemented if standard names precede the data content.

If an organization has limited resources and cannot afford to implement all aspects of DAM, then, at a minimum, it should establish the data item inventory by defining its data elements and establishing company data names for them. The benefits will be significant.

The item inventory of the DAM architecture can handle derived data and time-dependent data. Derived data reflects the manipulation or calculations of other data items. It is often aggregated from individual data items. For example, a company's liquidity position, sometimes called *current ratio,* is calculated, or derived, from current assets and current liabilities. As the company grows and as financial instruments (such as investments and loans) change over time, the elements that constitute the ratio may also change.

The automated systems' data inventory can aid management in comparing the company's liquidity positions in current and previous planning periods by calculating the ratios as economic factors change and by storing the time-dependent relationships. It can also help a firm's distributed subsidiaries track and use the meaning of financial indicators across disparate systems by storing time-dependent and derived data.

Equipment Inventory

Another element in DAM, equipment inventory and use (EIU), describes the network's hardware components and holds information on vendor names, model numbers, service dates, unique characteristics, and repair history. Some organizations are embedding pricing and depreciation data in this component and linking this data to their computerized financial systems.

EIU should allow for temporary storage of data that reflects network planning changes, which can serve as input to financial planning models. Then, the network configuration alternatives, such as acquisition of multiplexers, can be evaluated in parallel with the cost of hardware changes. The linkage of this component to the financial systems can be useful to the network manager and the company's comptroller.

The equipment information inventory may also contain usage statistics for analyzing peak periods use, saturation, equipment loading, and hardware sharing. The usage statistics can be collected from other components in the architecture (for example, the hit ratios in the physical data model) and serve as input to line loading and network configuration models, which in turn may provide input to the financial systems.

Internode Configuration Management

Internode configuration management (ICM) is a vital part of the architecture, because it links several DAM components together—primarily the software resources at the distributed sites. Properly implemented, it provides an overview of the organization's software resources and allows for companywide coordination of network development and maintenance.

Two of ICM's subelements, the software systems and modules inventories, describe the network's available programs, including their major functions. They also show which sites use what software—a vital point for coordination of distributed-network maintenance.

The modules inventory provides a detailed description of an individual program's functions and the input and output parameters in the modules. It allows the distributed sites to access the descriptive information and browse through the functional descriptions. The sites may then choose to use a particular module instead of developing a redundant capability. As with the data inventory, new applications program requirements should pass through the software inventory in order to avoid redundant development. If the site uses the component, it can then examine the inventory to find information such as compiler language used, architecture-specific features, and job control language.

The sources/uses repository (SUR) provides the final link between the organization's reports and its distributed network. It details the input-data forms (for example, a division's balance sheet) and which applications accept them. Further, it indicates the site responsible for preparing the forms. The repository also contains an inventory of output reports (for example, a stockholders' report) and their relationships to the automated programs that produce them.

SUR also provides information on the data files and data items that are used for the input forms (sources) and the output reports (uses). The data element names that are maintained in the data inventory identify specific sources and uses of the company's data flows.

As a simple example, an inventory data base at a warehouse in St. Louis might use form ABC containing a data element named WIDGET.6A.B. If an order-entry system in Chicago uses the same data for output report XYZ, it will likewise use the same WIDGET. 6A.B. The DAM identifies that St. Louis and Chicago use this data item respectively in report ABC, the inventory data base, and in report XYZ, the order-entry file. Consequently, the DAM facility would alert management that a possible alteration to form ABC used in St. Louis might affect some of the automated operations and management reports in Chicago. The changes could then be coordinated with full knowledge of the effects. Many companies have little idea of the impact or cost of making even such a simple change.

Using DAM

The complete architecture could function as follows. A company decides it needs a new automated system. The functional requirements, data needs, and output report requests are reviewed by a coordinating group, which examines DAM to determine if the new requirements can be partially or completely fulfilled by existing automated resources.

The data base/file inventory determines if some of the data is already available. The software system inventory provides the planner with the option of selecting existing code for use or modification to support the new system. The sources/uses repository determines which of the company's many reports will be affected.

DAM provides information on all forms, reports, files, software modules, hardware components, and sites that will be altered by the new system. If the change affects hardware performance, DAM contains statistics and information that help management determine how to reconfigure the equipment and perhaps shift computing to other resources.

Consequently, DAM helps management make cost benefit judgments in solving the problem, thereby controlling undesirable ripple effects into the distributed components. It also provides a means to integrate the revised and new components, determine the sites that are responsible for the implementation, and manage the progress of the effort.

Substantial resources are required to maintain the many elements in the distributed information automation management facility, especially for a large organization that has widely dispersed components. Therefore, an organization may choose to

TABLE 10-3. Division of Responsibilities.

Centralized		Distributed
• Definition of central and local responsibilities	⇨	• Actively participates in definition, but policies established at headquarters
• Selection and design of applications to serve multiple locations	⇨	• Participation in selection and design of applications to serve multiple locations
• Maintenance of corporate data dictionary. Review of local sites input to dictionary	⇦	• Input of local data to corporate dictionary
• Choice of network standards, data description language and data base software	⇨	• Participation in selection of standards, data description language and data base software
• Maintenance of master data bases	⇦	• Input to local copies of data bases, and upline loading to master data bases
• Selection of applications for intersite use	⇨	• Development of intersite applications
• Review of intersite application for adherence to standards	⇨	• Follows up on review findings
• Review of local application development trends	⇦	• Local application development, within standards
• Review of local data base trends	⇦	• Design of local data bases, within standards
• Review of local subschemas	⇦	• Design of subschemas for system-wide data bases
• Review of equipment selection trends	⇦	• Selection of locally used equipment, within standards
• Consulting services	⇨	• Receives consulting services
• System security policies	⇨	• Participation in development of security policies
• Auditing policies and procedures	⇨	• Participation in development of auditing procedures
• Coordination of operational reviews	⇨	• Conducting operational reviews
• Coordination of secondary site testing	⇨	• Secondary site testing

TABLE 10-3. (Continued).

Centralized		Distributed
• Review of trends	⇦	• Use of site-specific exists from multiple-site applications
• Global data base administration with design review authority	⇨	• Local data base administration functions

implement only those aspects of the architecture that appear practicable and cost-effective to its environment.

DIVISION OF RESPONSIBILITIES

This chapter has emphasized the need for unified and central direction if an organization is to use distributed processing successfully. Of equal importance is the realization that the distributed sites must be given responsibility and control over many of their resources. The managers at the distributed sites must feel a sense of participation with the undertaking. Simply stated, a successful distributed environment has strong centralized direction with considerable well-defined delegation to the distributed departments. Table 10-3 provides a list of responsibilities that are assigned either to the distributed sites or central headquarters. The arrows point to the staff that is affected by the initiating site—that is, the site that begins or initiates the actions or tasks.

11

DESIGN CONSIDERATIONS

INTRODUCTION

This chapter discusses the basics of communications systems design. The design problems and solutions are relatively simple and straightforward. Consequently, the material should be used as a vehicle for developing a general understanding of the process. A portion of the chapter uses a case study to explain certain concepts. The chapter is organized into three major sections: (a) communications line loading and network configuration, (b) software design; (c) network data base design.

Several calculations are included and the case study is structured around the calculations. The reader may choose not to examine the calculations in detail, but it is recommended they be reviewed in order to better understand the accompanying explanations. The calculations can be very useful; it is the author's experience that organizations who use a disciplined approach (such as the one described herein) find themselves more knowledgeable of the many factors that determine the network's traffic and, therefore, are in a better position to discard or accept the implications of the calculations. Their greatest benefit is to provide an inexpensive, rapid, and manual method to estimate an approximate size and number of the network components.

The calculations are described in four parts: (a) a formula, (b) a brief explanation of the formula, (c) a description of the application data, and (d) the actual calculation. Some of the

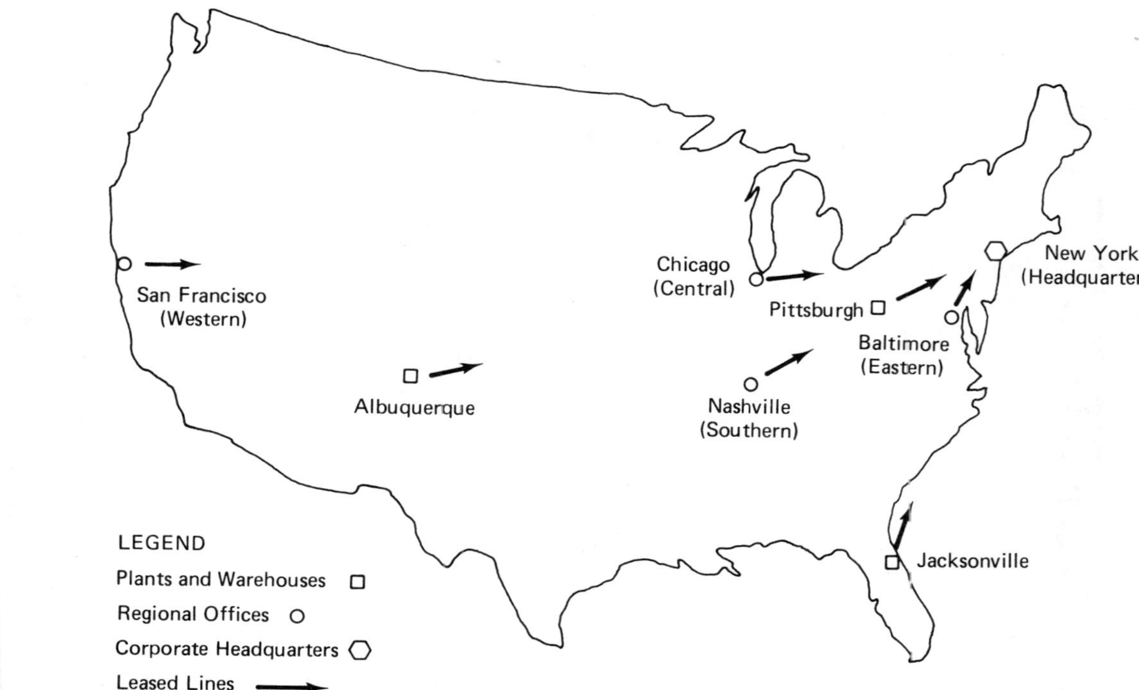

FIGURE 11-1. Current Network Structure.

calculations do not pertain directly to the case study and do not contain parts (c) and (d).

COMMUNICATIONS LINE LOADING

Case Study Background

A manufacturing company has decided to offload some of its centralized automated functions to its plants, warehouses, and sales offices throughout the United States. In the past, the company has relied on either centralized processing at its corporate site in New York or timesharing services from various vendors. Presently, leased lines and dial-up lines are used to connect remote points to the central New York computer facility where several large main frame computers provide support to over 40 sites in the country.

Figure 11-1 depicts the current network structure of this company. The four regions have headquarters located in relatively large cities. Intradistrict distribution points are also located at the regional headquarters. The sales offices within the regions use dial-up and leased lines to access the New York computer. The central facility processes inventory queries and sales transactions and sends pertinent data to the sales offices, plants/warehouses, and regional offices.

Figure 11-2 shows the Southern region, which will be used for this study. The Southern headquarters in Nashville, Tennessee, is responsible for 13 sites. The sites for this region consist of sales offices and local distribution facilities. The company's policy is to keep frequently ordered components near the customer due to the time-critical nature of the customer's needs for the material.

The company intends to install computers at the four regional headquarters. These computers will provide support to the offices in the respective regions. This impetus comes from the fact that local personnel are tasked with much of the data entry and validation and the company's personnel need to be near the sales offices and customer locations. In addition, the central host computers in New York have become bottlenecks during the periods of heavy traffic and communications costs and response time delays have increased. An analysis (discussed in the data base design section of this chapter) reveals that the company should place computers and data bases away from the New York site.

FIGURE 11-2. Southern Region.

During the business day, the sales representatives and inventory personnel at the local offices will be using the regional computer to query parts inventories, order components, and keep the regional data bases current. During the evenings, selected data will be transmitted to company headquarters in New York to update the corporate files. Transactions will also be transmitted to the company's manufacturing plants and major warehouses in Jacksonville, Albuquerque, and Pittsburgh in order to replenish the regional distribution points.

Approach to the Analysis

The decision of how to configure a network can be accomplished in part by a typical analysis of user needs and a design to support the user applications (requirements identification, specification, software and data base design, etc.). Additional refinements necessary for the network will be explained in this section.

One of the more difficult aspects of designing a network to support applications is determining the actual requirements of the applications. For example, each application will send/receive a certain volume of data across the network and will process this data under response time constraints. The data and response time requirements will provide a traffic pattern from which the communications lines, terminals, and computers can be configured. However, these data are often not readily available and may be subject to some "best guessing." Users may not know how they will use the network until they have gained some actual experience.

If the applications (such as inventory and sales in this study) are not yet automated, the design team must work closely with the end user to develop the requirements. The task is made easier if a system is already automated. The designers can then analyze the activity of the system and develop statistics for use in configuring the network according to the traffic pattern of the applications and growth pattern of the company. One should not rely on statistics alone, however; there is no substitute for close work with the user.

Traffic Pattern. User requirements should be gathered into a traffic pattern showing the number of messages and amount of data that must be transmitted and received. The traffic pattern is derived from a detailed analysis of: (a) each application's input and output traffic, (b) at each site, (c) during the peak period of activity. For example, in a sales order entry system, a sales clerk may be able to handle 15 customer calls per hour. If a typical call generates eight inputs and eight outputs to/from the computer (queries and updates), then during a peak period of one hour, the clerk will generate 240 pieces of activity (transactions) onto the network (1 hour × 15 calls per hour × 16 input/outputs per call = 240 transactions). The network must be able to service this amount of activity, or management must accept degraded service in order to reduce costs.

It may be difficult to determine detail to this level. Nonetheless, the failure to develop some kind of traffic pattern may lead to an "underdesign" of the network resulting in poor performance or an "overdesign" with unnecessary high costs and unused capacity.

As stated earlier, the data from which to develop a traffic pattern simply may not be available. The users and the designers may not know what the traffic will be. Since the baseline for a design is the amount of applications data to be passed through the network, the designers and the users must knowingly establish boundaries pertaining to (a) the most likely amount of data, (b) the most likely peak period, and (c) an expected response time. Management should review these boundaries in order to make cost-performance trade-off decisions. The goal of the organization should be to meet the user requirements with a minimum cost design.

Cost-performance trade-off analysis can be very valuable. For example, network costs can often be reduced substantially by offering a longer response time for selected messages and applications. Extended response time can decrease the costs of communications lines, processing computers, and the network software. In addition, low priority traffic can sometimes be delayed until the nonpeak traffic hours. This will also reduce costs, since the traffic can be "smoothed" over a 24-hour period and costs to implement a high capacity, peak-hours network can be reduced.

The number of input and output messages is dependent on the amount of data needed by the user and the format of the messages or transactions that flow into and out of the application. Due to efficiency and buffering considerations, it may be necessary to create more than one message from an input/output activity (that is, the 16 transactions in the sales order entry example could be broken into smaller units that move on the communications path). The total traffic is stated in volume of characters. The volume is determined by the number of messages and the number of characters in each message.

Protocol Overhead. The length of the messages will also be determined by line and protocol control characters (such as End of Transmission indicator (EOT), synchronization characters (SYN), etc.) and these characters must be included in the total amount of traffic during the peak period. The design team must know the data link controls (DLCs) that are to be used and they must be able to calculate the ratio of overhead control characters to the applications data.

The data link control method and modem will also influence the ratio applications usage of the network to that of protocol and overhead usage. For example, if a half-duplex protocol is used, the line will not be available while the receiving modem is "turning around" and resynchronizing itself for transmission in the other direction. On an aggregate basis, these modem turnaround times can become significant. Moreover, the manner in which the protocol controls the message flow will determine the ratio of applications data to overhead data. As discussed in Chapter 6, an acknowledgment of every message (IBM's Bisync) will require more overhead than inclusively acknowledging every one to seven messages (IBM's SDLC).

The number of errors that occur is likely to affect total traffic throughput since messages that are received in error will probably be retransmitted. The design team must factor probable error rates into the total traffic load. The manner in which errored messages are retransmitted requires that the design team understand how the data link control handles the errors. For example, a Go-Back-N DLC will give a different traffic throughput figure than a Selective Repeat DLC.

Performance. The performance of the system will be of paramount importance to the design team. Performance analysis entails the balancing of response time (or delay) and throughput. Response time to individual users must be weighed against total throughput for all users. Fast response time requires that minimum delay be encountered in moving the message through the network. Small delays rely on relatively short messages in order to reduce the time required to receive and check all bits of the message. Fast response time also benefits from short message queues since the shorter queues will decrease the aggregate waiting time for message processing.

On the other hand, high throughput requires longer messages in order to reduce the ratio of overhead characters to applications-specific characters. High throughput also benefits from long queues in order to smooth the traffic load. During periods of peak network activity, the lower priority messages can be placed on queues and processed later when the traffic decreases. In this manner, throughput improves overall line utilization.

It should be recognized that excessive optimization of response time to each user reaches a point of diminishing returns and overall response time to all users will eventually suffer. Obviously, an increased line speed and powerful computers will improve both response time and throughput, but the design team must still make trade-offs between these two important elements.

Finally, the network support functions must be considered carefully. Certain functions may extend response time (security provisions, code translation, providing for different terminal types, protocol conversion), while others may reduce throughput (adding department, data base, terminal or user identifiers to the message header). Yet, these functions may be essential in meeting the user requirements. Without question, reliability is a major function of any network and certain reliability functions will decrease response time. For example, redundant data bases and audit trail logging will increase delays to the network users.

Consequently, the designers will find themselves facing trade-offs between the factors of throughput, response time, and support functions (sometimes called the DDT or designer's dilemma triangle, see Figure 11-3.) It is usually not possible to provide optimum results at one point of the triangle without degrading results at the other two points. There are exceptions (the support function of data compression will increase throughput), but the DDT triangle should be uppermost in the minds of the design team.

The managers and users of the network should participate in decisions that affect the location of the point on the DDT triangle. It

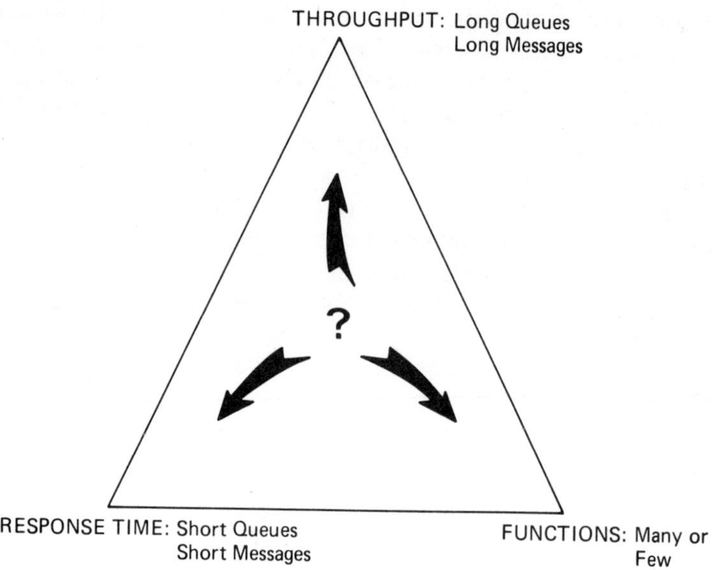

FIGURE 11-3. Designer's Dilemma Triangle.

is the responsibility of the design team to act as a catalyst for the decisions; it is *not* the function of the team to make these decsions.

Applications Requirements. The company's technical personnel have performed an extensive analysis of each site within the four regions. The analysis has lead to the following requirements for the Southern region:

The peak traffic load occurs once a day for 60 minutes from 10:15 to 11:15 a.m. Since timeliness is an important requirement, the network must accommodate to the peak period and the network must be designed to handle the traffic during the one-hour peak period.

The analysis team conducted studies at each sales office and tabulated the transactions for the peak hour traffic load (see Table 11-1). A *transaction* is defined for this study as consisting of one application input to the host and one application response from the host. This transaction activity during the peak period provides valuable input to the traffic pattern calculations.

The offices use sales and marketing information during the peak period. (Generally, other traffic is not allowed during this time—personnel, administrative, etc.) The marketing representatives use terminals for entering sales orders and querying of orders.

Inventory personnel use the system to provide component-availability information to company personnel and to customers. It is anticipated that the terminal operators will key in transactions at about .5 characters per second.

The terminal market was examined and the company has selected terminals with an EBCDIC eight-bit character set and a printer speed of 66 characters per second (CPS). The terminals use buffered memory and will input and output at 300 bits per second. The modems selected operate at 300 bits per second (bit/s).

The team has designed the message formats for the communications system. Figure 11-4 illustrates the format and applications data content of the messages for the sales and inventory applications. Since a transaction is defined as an input to the host and a response from the host, the sales application will require 100 characters and the inventory 130 characters for each complete transaction.

The design team estimates that the processing time at the regional computer will be 200 ms for each transaction. The decision to acquire faster or slower machines will be based on the effect of the 200 ms delays on line throughput, response time, and computer capacity. This study will determine these effects. The design time will use the 200 ms figure for the initial calculations. A processing time estimate may be difficult to obtain with any

TABLE 11-1. Traffic During Peak Period.

Sales Offices in Southern Region	Number of Transactions During Peak Period
Lubbock, Texas	422
Austin, Texas	466
Dallas, Texas	979
Houston, Texas	1,184
Little Rock, Arkansas	507
New Orleans, Louisiana	683
Jackson, Mississippi	296
Birmingham, Alabama	386
Gainesville, Florida	392
Miami, Florida	916
Atlanta, Georgia	1,002
Charleston, South Carolina	1,155
Raleigh, North Carolina	643
Nashville, Tennessee	1,178
	10,209

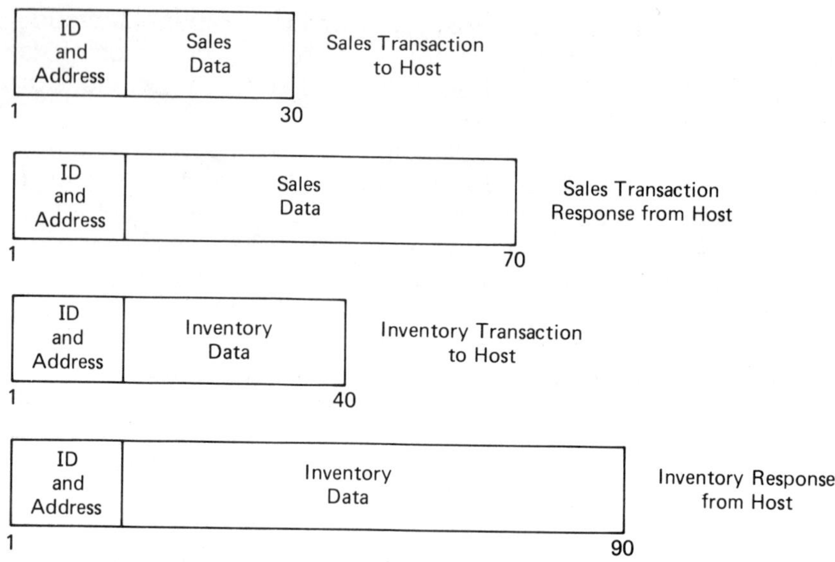

FIGURE 11-4. Message Formats.

assurance of accuracy. However, a "rough" figure can prove very useful for the first iteration of the calculations. We will cover this topic in more detail later in the case study.

The sales transactions currently comprise approximately 70% of the workload and the inventory transactions account for 30%. It is anticipated that this ratio will remain consistent for the foreseeable future.

Company management projects an annual growth rate in sales for 6% for the next five years. Obviously, this growth will affect the use of the network. Based on the estimated growth and a turnpike effect of 15%, the traffic volume is projected (Table 11-1). The *turnpike effect* is simply a "margin of safety" factor to help ensure that the system is not underdesigned. Its name is derived from the experiences that turnpike designers had several years ago. The first turnpikes were designed to carry long-haul traffic. The designers did not think a local driver would use the facility due to the trouble of getting on and off the turnpike. However, local traffic did indeed use the turnpikes and they were underdesigned.

The turnpike effect need not be just a guess. Some organizations develop scenarios that estimate a linear growth rate of traffic and then factor other contingencies such as offering new products, altering the organization's mission, etc. These contingencies would be part of a turnpike factor.

Leased vs. Dial-Up Lines. Leased lines have been selected to support the sales and inventory applications in this network. Several trade-offs exist between dial-up and leased facilities and the design team must consider both options. For example, dial-up lines provide for a flexible backup: if a line is lost, the operator simply redials into the network. Dial-up facilities are also attractive for low-volume users.

If the dial-up option is chosen, response time is lengthened due to the operator's time in dialing and connecting to the computer (perhaps 8 to 13 seconds). An alternative is for the computer to dial the terminals and communicate with the terminals' interface logic (or a transmission control unit) and *not* the human operator. Computer-generated dial-ups are, in effect, a form of polling. They offer three significant advantages to operator generated dial-ups:

1. Delay is not dependent upon the manual keying in of the data. The data is entered into a buffer and, upon a dial-up connection, transmitted onto the communications line at a much faster rate than the operator typing speed. The data is ready for transmission when the connection is made (or the dial-up is terminated by a "hang-up" and disconnect).
2. The operator does not experience delays in performing the dialing.
3. Since the operator does not dial, he/she does not get a blockage (i.e., a busy signal).

Leased lines are cost-effective for the transmission of larger volumes of data for extended periods (for example, more than three hours a day). Moreover, since leased line arrangements permit the permanent connection of the communication path within the telephone system, the lines can be monitored and "conditioned" for better service than is available on switched, dial-up facilities. Some applications require leased facilities if the delay involved in dialing numbers presents a problem. Our case study shows a great amount of activity during a critical peak hour and, therefore, warrants the leased lines. (One is now faced with what to do with the expensive lines during the non-peak hours.)

The costs of leased lines are based on (a) line distance measured in air miles, (b) rate schedules, (c) local loop costs (termination charges), and (d) installation charges. AT&T uses a Cartesian grid coordinate system for this purpose (see Figure 11-5). The grid establishes air mile distances between sites. The cost for

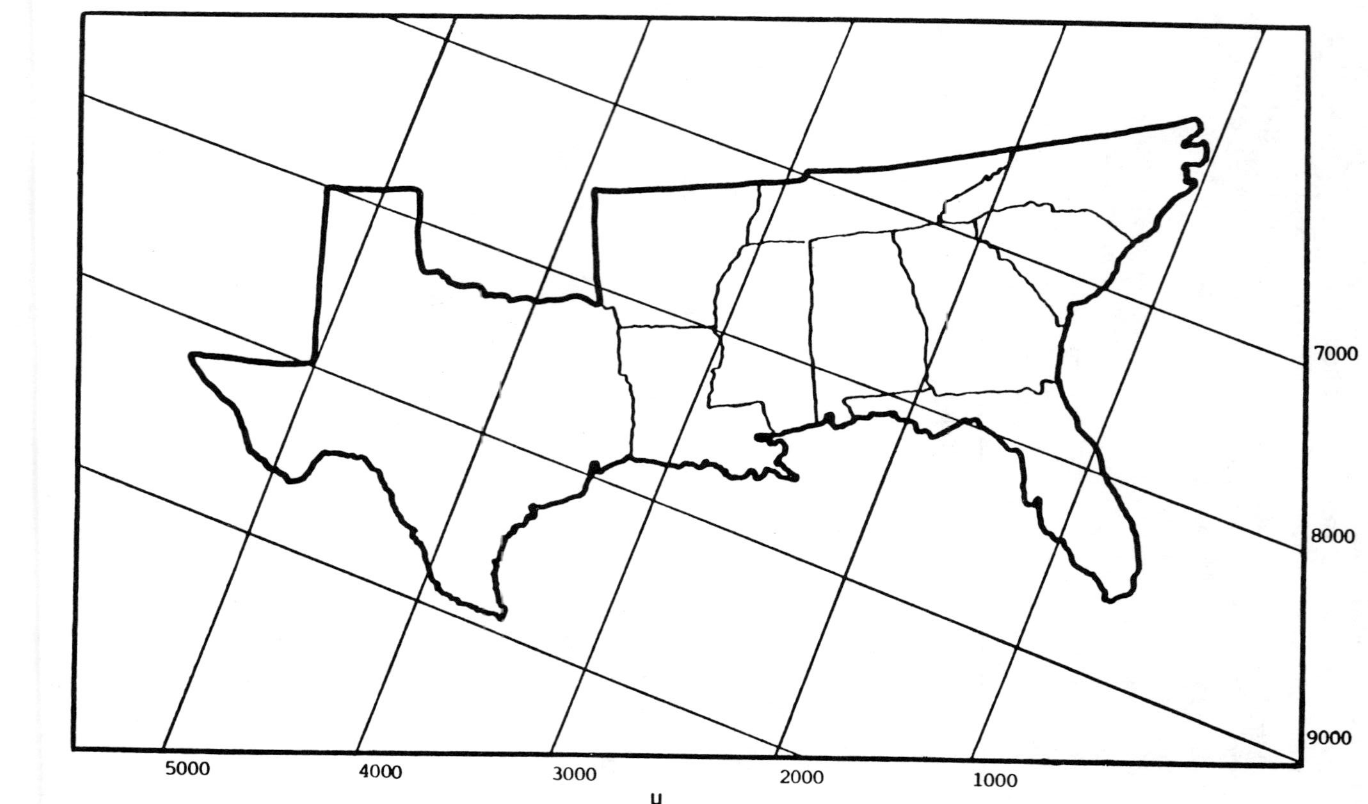

FIGURE 11-5. V-H Grid Map of the Southern Region.

line distance, installation, and termination stations are readily available from the carriers. However, be aware that varying line rates from city to city will affect network layout decisions. Some cities' interconnections will cost more than others.

Line Data Link Control (or Protocol). The line protocol will be half duplex with line turnaround time (TAT) of 110 ms. The line turnaround allows for the sending modem to resynchronize itself to receive and the receiving machine to "turn around" to send. The rationale for half duplex is due to the relatively simple and low capacity requirements of the system. Moreover, an organization usually has existing communications facilities with certain types of protocols. In many cases, the selection of a different protocol (e.g., full-duplex) may require extensive software and hardware changes. Consequently, a company may choose to use an existing protocol in order to reduce development conversion costs.

The protocol is a conventional polling/selection, stop-and-wait, asynchronous type and handles a terminal conversation in the following manner:

1. Poll to terminal from host.
2. Message sent by terminal.
3. Acknowledgment by host.
4. End of Transmission (EOT) from terminal; process the transaction.
5. Address the terminal from host.
6. Acknowledgment by terminal.
7. Message sent by host.
8. Acknowledgment by terminal.
9. End of Transmission (EOT) from host.

These nine elements will require additional characters for line control (polling, selection, EOT) and line and device synchronization. As stated earlier, the design team must add the overhead characters to the application characters in order to determine a total message length. Figure 11-6 provides an illustration of the relationship of the nine elements and the overhead and applications data. The reader may wish to review Chapter 6, which is devoted to line protocols or data link controls (DLC).

The nine elements, as well as the number of characters (bytes) associated with each element, are shown in Figure 11-6. The sales application will require 65 overhead characters and 100 data characters; the inventory applications will require 66 overhead characters and 130 data characters. (The longer inventory message requires an extra synchronization character.) Consequently, the total data sent through the network to complete one full transaction are 165 characters for sales and 196 characters for inventory.

ANSI Protocol Description. The American National Standards Institute (ANSI) describes the conversation between a terminal and the host as consisting of five phases. The complete sequence, which includes the nine elements in Figure 11-6, consists of five major phases:

Connection establishment phase. If switched lines are used, this phase represents the time to physically or electrically connect the two ends of the circuit. The activities in this phase are (a) dialing and (b) activating several RS232-C circuits (CC, CD, CA). When the CB circuit comes up, this phase is completed. Private lines do not go through this phase.

Line establishment phase. Phase two consists of (a) the poll, (b) the line turnaround (on half-duplex), and (c) a negative response to the poll or the initiation of the sending of data.

Information transfer phase. Phase three begins with data being over the RS232-C BA circuit. A protocol control character such as SYN or SOH signifies the beginning of this phase. The messages are then sent and acknowledged until the sending terminal has no more data and so indicates with a control character such as ETX.

Link termination phase. The sending station signifies the end of a transmission with a protocol control character such as EOT. Afterward, the RS232-C circuit CD goes off and the phase is completed.

Connection clearing phase. During phase five, the physical or electrical link is disconnected.

Line Loading Calculations

Block error rate. As previously stated, the design will account for some messages containing errors. These messages must be retransmitted and will impact total message throughput. The

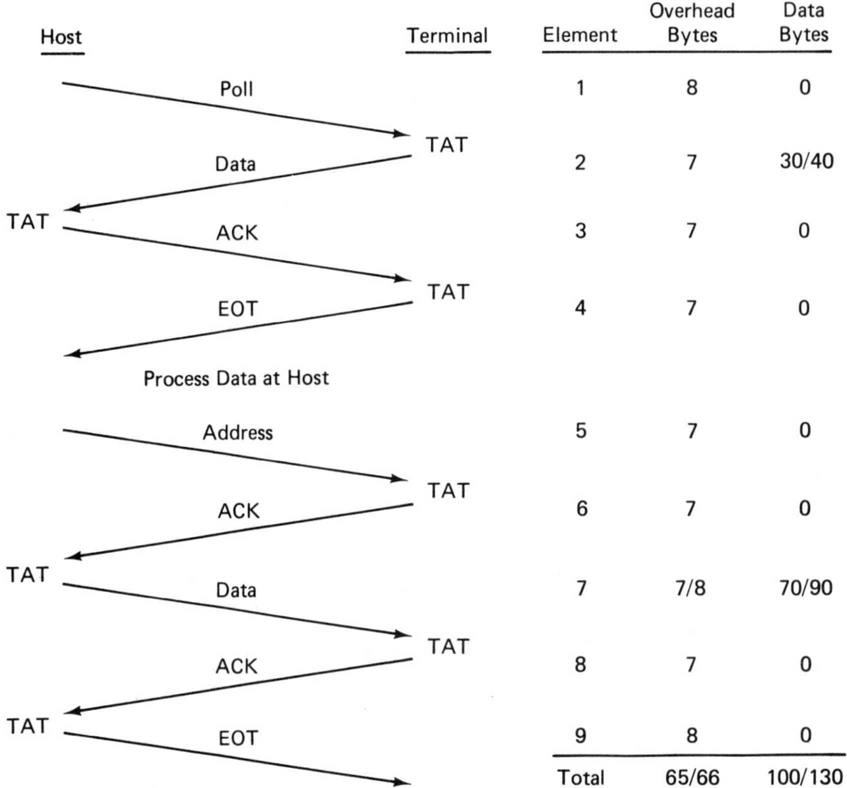

Element	Overhead Bytes	Data Bytes
1	8	0
2	7	30/40
3	7	0
4	7	0
5	7	0
6	7	0
7	7/8	70/90
8	7	0
9	8	0
Total	65/66	100/130

FIGURE 11-6. Nine Elements of the Transaction.

communications line can be tested and monitored to determine a probable error rate. The common carrier can also provide probable error statistics based on the type of line and the level of conditioning. The engineers can provide data on bit test patterns, from which the design team can calculate a probable block error rate. Chapter 5 discusses bit error rates (BER) and block error rates (BLER).

The size of the message block is an important design consideration. For example, during noisy periods on the line, longer messages are more likely to be retransmitted since bit errors occur more often in a longer bit stream. Yet, shorter messages decrease overall throughput due to the additional control headers and trailers required for each message block.

For purposes of simplicity, assume the tests have been performed on bit patterns and on the actual messages block lengths that were established by the design team.

Calculation 1: Block Error Rate

Formula: $$BLER = \frac{TBE}{TBR}$$

Where: BLER = Block error rate
TBE = Total blocks received in error
TBR = Total blocks received

Application data: Technical team conducted random tests onlines and ascertained the following:

TBE: 1000
TBR: 100,000

Calculation: $$BLER = \frac{1000}{100{,}000} = .01$$

Throughput. Throughput is a very important design consideration. The term refers to the amount of data that can be sent over a communications channel during a given time period. Throughput is described in bits per second (bps) and is usually calculated on a one-way transmission. There are a number of methods to calculate throughput. The following approach is commonly used and is based on American National Standards Institute (ANSI) Standard X3.44.

Calculation 2: Throughput

Formula: $$TP = \frac{M(1 - P)}{M/R + T}$$

Where: TP: Throughput
M: Applications message length (in bits)

P: Probability of one or more bit errors in block (see Calculation 1)
R: Line speed in bits per second (bps)
T: Time between blocks in seconds

Application data: M = 920 Avg.

Sales:	$100 \times 8 =$	800
Inv:	$130 \times 8 =$	1040
		1840

$1840/2 = 920$

$P = .01$
$R = 300$
$T = .110$

Calculation:

$$TP = \frac{920\ (1 - .01)}{920/300 + .112} = 287$$

The parameter M is derived by calculating an average message length of the sales and inventory messages. Be aware that an average might skew the throughput figure if one application's message is significantly longer than others and has considerably more traffic on the line. For purposes of line loading, tuning, and simplicity, fixed-message lengths for all applications should be considered. This approach requires that a logical user message be segmented into pieces (packets) and transmitted separately, but it greatly simplifies the design, implementation, and operation of the system.

The formula illustrates the importance of the message block length on network performance. For example, very long block lengths will increase the significance of probability P because longer messages increase the probability of receiving an erroneous bit in a message. The parameter T will be less a factor with longer messages because the time between message blocks decreases as messages increase in length. On the other hand, T assumes more significance with short messages since the relative amount of time between blocks increases.

The parameter T depends to a great extent on the data link control (DLC) logic that manages the flow of messages between the host and secondary sites. For example, the delay between blocks rests on the efficiency of the polling/selection routines and how well they keep the line "busy." Conversely, the DLC delays are sometimes masked on a half-duplex line since the modem introduces a delay in its turnaround. (This means the Request to Send (RTS)/Clear to Send (CTS) signals are reversed between blocks.) The modem and DLC vendors should be consulted before computing the T factor. This calculation assumes a delay between the blocks; later calculations will include modem turnaround time.

Utilization factor. The utilization factor is simply a number indicating the applications' actual use of the line's stated throughput capacity. In this study, the 300 bps line would not allow 300 bits per second of sales and inventory data to be passed over the line due to the overhead characters, line turnaround, errors, and polling. In some applications, such as a tape file transfer or a remote job entry (RJE), response time is not an important component and the line utilization factor is used for capacity planning. Response time calculations (see Calculations 14 and 15) must also be used if responsiveness is a user requirement. Interactive applications must be given careful consideration in relation to response time.

In simple systems, the utilization factor is not calculated but is derived from subjective evaluations and the experiences of the designers. The actual utilization usually ranges from 40% to 95%. The wide variation results from factors such as switching technology, queueing delays, multipoint complexities, and the need for low utilization to enhance response time.

The utilization factor can also be used to account for the overhead of unsuccessful polls, in which a terminal is polled and has nothing to send to the host. The design team must account for unsuccessful polls, call set-ups, and call clearings for some protocols since this type of overhead may use a significant amount of the available line capacity. Calculation 3 is a simple illustration of determining the utilization factor. Be aware that many elements can influence the actual line utilization.

Calculation 3: Utilization Factor

Formula: $$UF = \frac{TP}{R}$$

Where:
- UF: Utilization factor
- TP: Throughput (see Calculation 2)
- R: Line speed in bps (theoretical maximum capacity)

Application data: TP = 287 (see Calculation 2)
R = 300

Calculation: $UF = \frac{287}{300} = .956$

Message line time. The reader should note that Calculation 4 includes the applications characters *and* the overhead characters. Returning back to Calculation 2, it can be seen that overhead data was not used in the throughput computation. Several methods exist to factor in overhead data; the important point is not to overlook it. The 37.5 characters per second (CPS) is derived from the 300 bit/s line using 8 bit characters (300/8 = 37.5).

Calculation 4: Message Line Time

Formula: $MLT = TC \times CLT$

Where: MLT: Message line time
TC: Total characters transmitted in complete poll and address (to complete a full transaction)
CLT: Character line time (how long it takes for one character to move through the line)

Application data: TC = Sales: 165 characters (see page 360)
Inv: 196 characters (see page 360)
CLT = 26.6 ms per character (1000 ms/37.5 CPS)
Note: 1000 is used because 1000 ms = 1 second

Calculation: MLT (Sales) = $165 \times 26.6 = 4{,}389$ ms.
MLT (Inv) = $196 \times 26.6 = 5{,}213$ ms.

Network transaction time. The next calculation provides the time required to process a complete transaction. This includes the computer processing time, line turnaround time, and the time required to move the complete transaction (i.e., the nine elements previously discussed) from start to finish.

The processing time parameter (PT) will not be a factor in line loading if the line is released for other transactions while the computer is processing the ongoining transaction. This "released-line" discipline is commonly found in the larger systems. Typically, a front end processor handles the held-line or released-line tasks. Page 122 contains additional information on this topic. Also, pages 378 through 385 discuss the processing time parameter in more detail.

Calculation 5: Network Transaction Time

Formula: NTT = MLT + PT + TAT Time

Where: NTT: Network transaction time
MLT: Message line time (see Calculation 4)
PT: Processing time
TAT: Line turnaround time

Application data: MLT = Sales: 4,389 ms.
Inv: 5,213 ms.
PT = 200 ms. for both applications
TAT = 770 ms. (7 TATs × 110 ms. each, see page 361)

Calculation: NTT (Sales) = 4,389 + 200 + 770 = 5,359 ms.
NTT (Inv.) = 5,213 + 200 + 770 = 6,183 ms.

Weighted average transaction time. The sales and inventory applications will not use the network equally since sales will comprise 70% of the traffic and inventory the remaining 30%. Therefore, weighting factors must be applied to compensate for this uneven use.

Calculation 6: Weighted Average Transaction Time

Formula: $$\text{WATT} = \sum_{N=1}^{K} (\text{TPCT}(n) \times \text{NTT}(n))$$

Where: WATT: Weighted Average Transaction Time
TPCT: Total percentage of network traffic for this application
NTT: Network Transaction Time (see Calculation 5)
K: Number of applications

Application data: TPCT = Sales: 70% of traffic
Inv: 30% of traffic
NTT = Sales: 5,359 ms
Inv: 6,183 ms

Calculation: WATT (Sales) = .7 × 5,359 ms = 3,751 ms
WATT (Inv.) = .3 × 6,183 ms = 1,854 ms

3,751
1,854
5,605 ms or 5.605 seconds

Line capacity for applications. The next calculation determines how many transactions can be accommodated on a 300 bps line during the peak period. Since previous calculations have shown how long a full transaction occupies the line, it remains now to divide the applications line time into peak period time to determine the line capacity for the one hour.

The utilization factor is also applied. This may appear to be doing "double accounting" for overhead (since overhead was part of Calculations 4 and 5). However, as stated earlier, multipoint lines with queuing delays, polling delays, and unsuccessful polls introduce additional overhead. The utilization factor is one method to include these factors.

Unsuccessful polls occur when the host scans a polling table, obtains an address of a terminal, and issues a poll without obtaining a response because the terminal has nothing to send. The terminal operator may be keying in data or simply thinking. Whatever the case, unsuccessful polls can account for a substantial amount of the total network traffic. In some instances, the network overhead also consumes a substantial portion of the computer capacity. It should be noted that an actual situation may require the substitution of the utilization factor calculations with data derived from the queueing formula and polling models discussed on pages 381 through 385.

Calculation 7: Line Capacity for Applications

Formula: $$LC = \frac{LENPP}{WATT} \times UF$$

Where:
- LC: Capacity of a line to handle traffic
- LENPP: Length of peak period in seconds
- WATT Weighted Average Transaction Time (see Calculation 6)
- UF: Utilization Factor (see Calculation 3)

Calculation: $$LC = \frac{3600}{5.605} \times .956 = 614$$

Therefore, 614 transactions can be accommodated per line per peak period.

Total number of lines. We now have sufficient information to determine how many lines will be required at each branch office in the Southern region. Calculation 8 shows that the two applications will require 14.70 communications lines. It is very unlikely that 13,209 transactions an hour can be handled by a small scale (read mini) computer. The design team must develop the loading profile (discussed on page 378) and response times (page 384) in order to size the proper computer(s) in the workload.

The design team must also make evaluations on where to place the computers within the Southern Region. (This study has assumed that the host is situated at regional headquarters). The section on distributed system design will discuss this issue in more detail.

Calculation 8: Total Number of Lines

Formula: $$TL = \frac{TV}{LC}$$

Where: TL: Total lines required
TV: Total volume
LC: Capacity of a line to handle traffic (see Calculation 7)

Application data and calculations:

Site	TV /	LC =	TL
Lubbock	422	614	0.69
Austin	466	614	0.76
Dallas	979	614	1.59
Houston	1,184	614	1.93
Little Rock	507	614	0.82
New Orleans	683	614	1.11
Jackson	296	614	0.48
Birmingham	386	614	0.63
Gainesville	392	614	0.64
Miami	916	614	1.49
Atlanta	1,002	614	1.63
Charleston	1,155	614	1.88
Raleigh	643	614	1.05
Nashville	1,178	614	1.92
	10,209	614	16.62
		(Nashville)	−1.92
			14.70

The total volume of 10,209 transactions for the Southern region represents a sizeable workload for the region computer. The reader may have surmized by now that an average processing time of 200 ms for each transaction means the 10,209 peak hour traffic cannot be effectively accommodated on a computer of this capacity unless it is completely dedicated to the two applications. Theoretically, the 200 ms window allows 18,000 messages per hour to be processed if the messages arrive at a fixed rate (200 ms: 5 transactions a second × 60 seconds × 60 minutes = 18,000 transactions per hour). However, the 18,000 messages represent a 100% utilization of the computer's CPU cycles which *cannot be obtained* due to the asymptotic nature of large workloads and response time. Moreover, the 200 ms processing time assumes all resources function without further delay and without error, which is rarely the case.

In other words, the hypothetical machine capacity of 200 ms may not suffice. As noted earlier, the design team used the figure for preliminary calculations and, in order for the line sizing data to make sense, the computer capacity (processing time) should be evaluated with response time calculations (see Calculations 14 and 15) to permit the full effect of the traffic load to be analyzed.

Practically speaking, an organization faced with this type of problem has several options:

- Exmaine the market for machines of greater capacity.
- Consider installing additional machines, with a front-end processor to route the traffic among the computers.
- Offload work from the regional site by locating small computers at some of the sales offices.
- Downgrade the system response time by allowing fewer transactions during the peak hour (for example, only critical inquiries) and transmitting less important traffic at a later time.

Whatever the options chosen, the decision will affect line sizing and will require another iteration of Calculations 1 through 8.

Choice of network topology. The line layout for the Southern region is illustrated in Figure 11-7. The line layout shows several locations configured with multiple lines. For example, the traffic volume from Houston will require 2 (1.93 lines, rounded to 2) lines to satisfy the peak load throughput requirements.

The design team's job is not yet complete, however, because the present arrangement may not present the most cost-effective approach. The lines may now be configured to give a minimum cost network topology that satisfies the performance objectives. For example, the New Orleans and Jackson branches will need 1.11 and .48 lines respectively to satisfy their requirements. Full line capacity must be given to these branches since it is not possible to lease a fraction of a line. Consequently, the two sites have unused line capacity.

The lines can be reconfigured to provide the same service with fewer lines. This is accomplished by placing a line from New Orleans to Jackson and moving .11 of the New Orleans traffic to the partially used line of .48 at Jackson (see Figure 11-8). The route of the altered data flow means leasing a line of shorter distance. As indicated earlier in this study, line costs are based on the air mile distance of the line, and the line from New Orleans to Jackson will cost less than from New Orleans to Nashville.

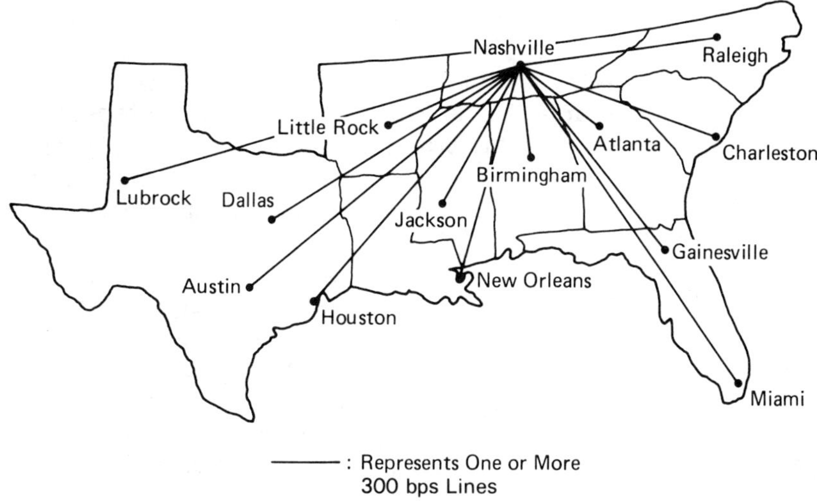

FIGURE 11-7. Network Lines for the Southern Region.

The design team could perform the network configuration analysis by hand but, with 13 sites involved, the task would involve hundreds of combinations. Computer models are available to perform programmatically the configuration. The common carriers provide this service as part of their product; consulting firms have the capability. IBM and other companies also have network

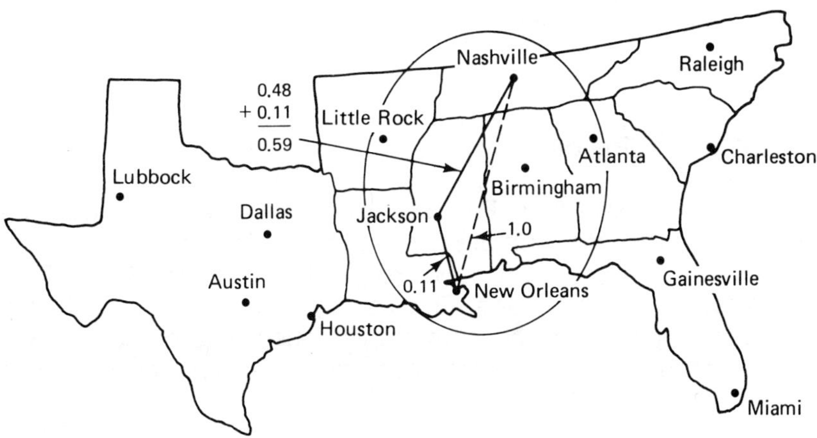

FIGURE 11-8. Alternate Topology Example.

configuration models. The IBM model is called the Communications Network Design Program (CNDP). This program accepts input parameters of site-to-site distances and rates. Network paths are "grown" from the farthest note toward the host. CNDP repetitively adds a node to a link, reevaluates the change in relation to previous configurations, and compares costs. The detailed explanations of Esau-Williams, the Prim approach, and others are available from vendors.

The Network Analysis Corporation (NAC) uses a product called MIND (Modular Interactive Network Designer) for line configuration, cost analysis and topology layout.[1] The MIND user inputs parameters describing each node (ID, telephone company location (area) codes, input and output traffic, number of terminals at each node, certain node termination costs, and other parameters.) MIND then provides the following information:

- Network costs based on current tariffs.
- Assignment of terminals.
- Effect of possible use of multiplexers.
- Analysis of multipoint vs. point-to-point.
- Line load.
- Analysis of effect of increasing/decreasing line speed and/or traffic and/or terminals.
- Reliability indicators.
- Waiting and response time reports.
- Simulation of different data link controls.

Other Considerations

Data link protocols. Protocols are available other than the half-duplex, stop-and-wait method that was selected by the design team. Two full-duplex protocols frequently used are GO-BACK-N and Selective Repeat (see Chapter 6). The use of full-duplex protocols requires that the communications line provide for two-way, simultaneous transmission. This capability is provided by two

[1]Network Analysis Corporation, 130 Steamboat Road, Great Neck, New York, 11024.

separate lines attached to full-duplex modems. Another popular alternative is using one line and dividing the channel bandwidth into subchannels. The subchannels operate at slower speeds but may be sufficient to handle the traffic. The two subchannels are connected by the secondary pin connections on the RS232-C interface.

The reader will note that these calculations eliminate the T factor (see Calculation 2) because a full-duplex protocol does not require a line turnaround. Also, the Selective Repeat formula does not contain the N factor (probable number of retransmissions of correct messages) that is in the GO-BACK-N calculation.

Calculation 9: Throughput With GO-BACK-N

Formula: $$TPGB = \frac{M \times (1 - (N \times P)}{(M/R)}$$

Where:
TPGB: Throughput
M: Applications message length (see Calculation 2)
P: Probability of one or more bit errors in block (see Calculation 1)
R: Lines speed in bits per second (bps)
T: Time between blocks in seconds
N: Probable number of retransmitted messages

Application data:
$M = 920$
$P = .01$
$R = 300$
$T = .110$
$N = 3$ (assumption for purposes of this example)

Calculation: $$TPGB = \frac{920 \times (1 - (3 \times .01))}{920/300} = 291$$

Calculation 10: Throughput with Selective Repeat

Formula: $$TPSR = \frac{M \times (1 - P)}{M/R}$$

Where: Same as above

Calculation: $$TPSR = \frac{920\ (1 - .01)}{920/300} = 297$$

The improvement in throughput for Selective Repeat is generally not much better than GO-BACK-N if line conditions are reasonably good and the protocol does not allow a great number of messages to be outstanding before a NAK or ACK is required. However, the throughput rates of 291 bps and 297 bps are better than the rate of 287 bps that was calculated for the half-duplex protocol. The differences become more evident on higher speed lines.

Modem upgrade. The study has focused on a network that will service the applications' peak period traffic needs. However, the design team must also consider the cost-effective use of the network during times other than the peak period. For example, the Dallas site is configured to accommodate 979 transactions during the one-hour peak period and requires more than one 300 bit/s line. However, what happens to the two lines when traffic is low?

During nonpeak periods it is probable that the lines will not be fully utilized and the organization may incur unnecessary line costs. An alternative approach is to upgrade the line speed with faster modems, which would allow placing all traffic on one line at Dallas. In this case, the higher cost of the faster modems could be offset by the decreased line costs.

The reader might wonder why the design team would choose to use lower speed equipment. It might first appear to be more reasonable to select higher bit rate modems, such as 4800 bps or 9600 bps, and place more traffic on one line. First, the high speed modems are considerably more expensive than lower speed alternatives. Moreover, as discussed earlier in the book, the higher bit rates necessitate a higher signal rate per second on the line. For example, a signal rate of 300 baud (cycles per second) means that the signal occupies a line for .003 seconds (1 sec/300 cycles per second). On the other hand, a signal rate of 9600 baud has each discrete signal on the line for .0001 seconds (1 sec/9600 cycles per second). The short line time of higher speed transmissions present more problems in synchronization and are more subject to line errors (voltage spikes, noise). In any event, the 300 bps rate is probably more than adequate for use in operator-to-computer applications because of the manual aspect of data entry.

The modem vendors offer equipment that provides multiple terminal attachment to a modem. These modems are called *multiple access units* (MAU) or *piggy-back modems*. Typically, an MAU accommodates one to four modems and will allow cascading terminals onto a line as was done in the line configuration in Figure 11-7.

A component in the network cannot be altered as if it were independent from other components. Beware of the ripple effect. The modem upgrade could improve response time, thereby possibly encouraging increased use of the network by the users, which in turn might lead to bottleneck problems at the host. Worse yet, a speed or capacity upgrade could lead to computer saturation if the ripple effects are not anticipated and controlled.

Consideration for multiplexers. Multiplexing is widely used and has proven quite effective in reducing network communications line costs. Chapter 3 provides a discussion of multiplexers. Multiplexing allows the high capacity line to be shared with other terminal devices. In this way, several separate transmissions are sent over the same line and the line efficiency ratio can be significantly improved.

Multiplexer capacity. Let us suppose the Lubbock line is tied into the Dallas site and the two lines at Dallas are tied into a multiplexer with an output rate of 4800 bps. This configuration would have three input lines at 300 bps each (one from Lubbock and two from Dallas). The calculation reveals that the 4800 bit/s multiplexer could handle this traffic load.

Calculation 11: Multiplexer Capacity

Formula: $A \times .97: \sum_{I=1}^{K} (N(I) \times C(I) \times L(I))$

Where: A: Aggregate capacity in bps of output side

N: Number of input channels at this speed and code
C: Speed of input channels in characters per second
L: Length of character set (in bits)
K: Total number of input channels

Application data:	A:	4800 bps
	Lubbock:	(.69 line, rounded to 1)
	N:	1
	C:	37.5 cps
	L:	8
	Dallas:	(1.59 lines, rounded to 2)
	N:	2
	C:	37.5 cps
	L:	8

Calculation: $4800 \times .97: (1 \times 37.5 \times 8) + (2 \times 37.5 \times 8)$
4,656 : 900. Thus, the output side is of greater capacity than input side

The FDM and TDM use fixed allocations of frequency and time, respectively. This approach can lead to wasted line capacity when the terminals are idle (for example, an operator pausing for think time). The line capacity is still reserved for the terminal regardless of whether the terminal is using the line.

The use of Statistical Time Division Multiplexers (STDM) can increase the actual use of the line. The STDM does not provide fixed allotments of time for each station. Rather, the time slots are dynamically allocated to active terminals only. In a sense, the STDM plays the odds that all terminals will not be operating at the same time, because the sum of all input rates to the STDM are greater than the output rate.

The reader is cautioned that, while STDMs are valuable devices, their use requires a more detailed analysis in order to obtain proper performance and sizing. For example, the design team must examine the following areas for inclusion into STDM configuration planning:

Call rate per terminal: Frequency of terminals' use of the communications system.

Hold time per terminal: Once terminal establishes a session, the length of time the terminal is active.

Message arrival distribution: Frequency of receiving and transmitting messages at each terminal and the amount of data in each message.

Acceptable blocking probability: Possibility that STDM must queue up messages.

STDM techniques add overhead with each additional active terminal in order to properly identify data from those terminals on the line. Be aware that the addition of overhead causes a nonlinear (exponential) effect on the line.

Satellite links. Satellite links are available from a number of specialized common carriers and the small-volume user can purchase voice-grade lines at a reasonable price. However, satellite circuits entail additional problems for the designer. A primary consideration results from propagation delay. This delay is the time required to move the message from the sending terminal to the host. A typical propagation delay from the San Francisco office to the New York site would be around 20 to 30 milliseconds using a land circuit. Satellite transmission must traverse a greater distance since the satellites are usually located 22,300 miles above the equator. A San Francisco-New York one-way satellite transmission would require about 240 to 270 milliseconds, plus additional time for delays at earth stations and the user's site.

The propagation delay poses a number of problems for the design time. First, the data link protocol may be affected. Typically, a polling/selection protocol polls a station and waits a given time period for a reply. If the polled terminal does not respond within the time period, the protocol logic usually repolls and eventually executes some error routines. The delay on a satellite link would result in protocols spending much of their time in the timeout and error logic because many protocols are designed for the faster terrestrial links.

The second problem relates to response time. The longer delays on a satellite link may not be acceptable for some applications (for example, real time applications) and the design team must be aware of the user's response time requirements. However, a propagation delay of less than one-half second is not a problem for many applications and several satellite carriers have developed techniques for dealing with the delay problem.

Calculation 12: Satellite Propagation Delay

Formula: $T = TFP + TRR + TRP + TRS$

Where:
- T: Propagation delay time
- TFP: Forward delay time
- TRR: Reaction time at receiving node
- TRP: Reverse delay time
- TRS: Reaction time at sending node

The object is to determine the number of messages that must be present (i.e., N, as described in Calculations 9 and 10). N can be described as the number of messages of M characters in length that can be sent within total delay time T, given the channel speed R.

Calculation 13: Resending Messages on Satellite Links

Formula:

$$N = \frac{T}{M/R}$$

Where:
- N: Messages transmitted
- T: Total time
- M: Message length
- R: Channel speed

Thus, Calculations 16 and 17 can be used to determine the N parameter for Calculations 13 and 14.

Computer Capacity Considerations. Calculation 5 includes a parameter for the computer time required to process a transaction. The computer processing time is often quite difficult to calculate precisely, especially if the host is multiprogrammed, multiprocessed, or coupled to other hosts. It is possible to determine the sequence of operations (and instruction set) of the applications programs that will be used to process a transaction and this data can be very useful for performance tuning and analysis of the possible impact of increased traffic. However, the processing time is dependent on factors that vary from period to period, such as transaction load and job mix. Moreover, the operating systems and data base management systems vendors may not provide the instruction sequences for those packages. The designer should be aware of these factors.

The processing time parameter is often influenced by disk I/O channel capacity. Many smaller machines have a single I/O bus that does not permit overlapping I/O transfer. Other systems use I/O for fetching instructions on disk and, again, this approach can add many variables to the situation.

The software can contribute to the wide range in the processing parameter. For example, an operating system that performs time slicing on an applications transaction may build up large execution queues. On the other hand, an operating system that allows tasks to go to completion could also present queueing problems, but of a different nature.

It is naive to think the vendor's specifications will provide a definitive guideline. The machines' power measure in (a) memory fetch time, (b) instruction time, (c) millions of instructions per second (MIPS), or (d) thousands of operations (ADD, MOVE) per second (KOPS) are useful figures but measure internal processing speed and must be evaluated in view of many other considerations.

One approach to sizing a workload for a transaction-based system is depicted in Figure 11-9. With this technique, the various processes to service a transaction are measured. The factors (such as number of instructions, number of data base I/Os) are translated into times and computer workloads. The designers can then determine approximate delays at the serving processes and calculate a processing time (PT) for the transaction. As previously stated, this task is complicated by lack of knowledge of O/S and DBMS instruction sets, multiprogramming, and time sharing operating systems, but the exercise can be helpful in understanding more about the traffic workload. Moreover, the simpler micro- and minicomputer is being used increasingly in data communications systems, such as distributed systems and local area networks. These machines, often tailored to operate with specialized software and limited instruction sets, permit a more exact calculation of the sizing factors. In any event, sizing is an absolute necessity for systems requiring fast response time and efficient execution.

Standard work unit. An additional tool for designing an on-line, transaction-oriented environment is the Standard Work Unit (SWU). The SWU is used to regulate and predict the resources consumed by each on-line transaction in the network. The SWU concept requires that each transaction consume a limited amount of computer and data resources. For example, the SWU could define (a) a maximum number of data base CALLS, (b) a maximum of machine cycle executions for each transaction, or (c) a maximum amount of CPU time.

LOADING PROFILE

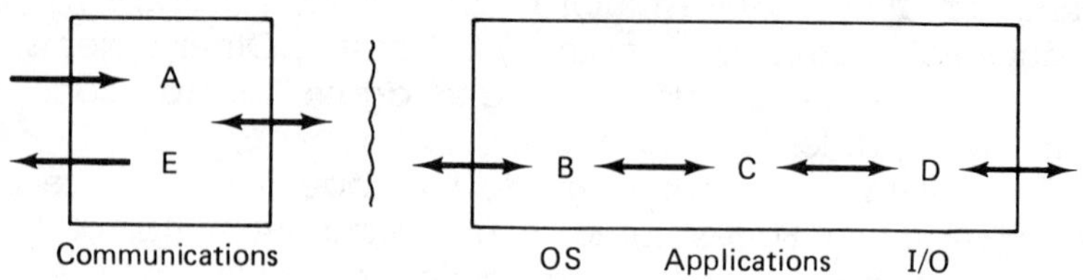

SERVERS	TASKS	SIZING FACTORS
A	Network Input	Message Size; Line Speed; Delay
B	OS Process	Number of Instructions
C	Applications Process	Number of Instructions
B	OS Accepts Request	Number of Instructions
D	I/O Access (Directory)	Peripheral I/O Rate; Block Size
D	I/O Access (Data)	Peripheral I/O Rate; Block Size
B	OS Accepts I/O	Number of Transactions
C	Applications Process	Number of Instructions
E	Network Output	Message Size; Line Speed; Network Delay

FIGURE 11-9. Workload Sizing.

The idea behind SWU is illustrated in Figure 11-10. The Nonstandard Work Unit imposes unpredictable loads on the computer, resulting in unanticipated peak load saturation and long-range computer resource planning. In contrast, the SWU provides for a more stable operating environment and a better ability to predict the implication of adding systems to the network. The SWU is particularly useful in view of the nonlinear relationship between increased workload and response time.

Benchmarking and modeling. The use of benchmarking or modeling is a practical approach in determining an appropriate computer architecture. These techniques allow the designers to examine a number of alternatives before committing to a configuration to support the network traffic.

Modeling has long been a popular method of computer capacity sizing. It attempts to represent the performance of a computer through the use of programs that accept and process known or assumed patterns of behavior on workload. Modeling can be expensive, and the programs can become large and complex. Nonetheless, the technique is a powerful tool and should be considered by the designers.

Use of Queueing Models. Many data communications systems designs are based on providing fast response time, in contrast to obtaining consistent, high throughput or providing "just enough" capacity to meet the traffic demands. Regardless of the design objectives, it is quite important to examine response time as part of the analysis and design process.

Queueing formulas are quite helpful in this analysis and have seen increasing use for designing transaction-based data com-

STANDARD WORK UNIT CONCEPT

Purpose: Attempt to Regulate Resources Consumed by Each Execution of on-line Transactions.

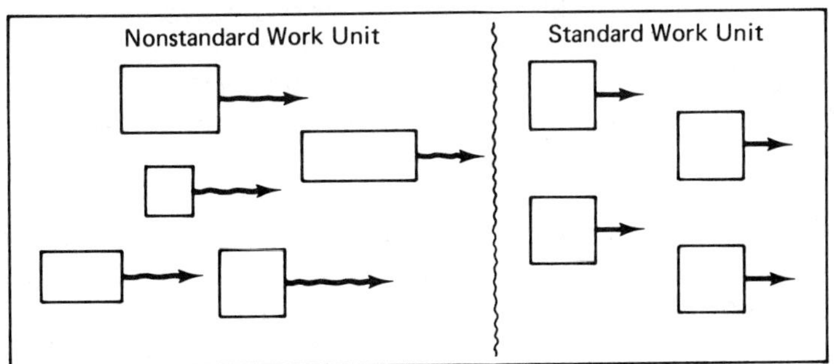

CAPACITY PLANNING

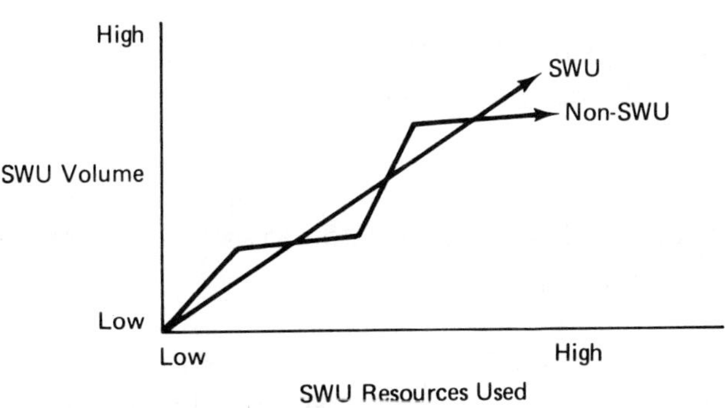

FIGURE 11-10. Standard Work Unit.

munications systems. The formulas are integrated into computer programs to provide models for sizing response time.

Queueing analysis is quite important because of the nonlinear relationship between workload and response time. This occurs because response time is a function of processor time (PT) and the waiting time in queues. The waiting queues exist at several points in the process. For example, in Figure 11-9, servers A and E usually have significant queues and the other servers may also have waits associated with them.

The important point to remember about this discussion is that the time a message spends in a queue depends on the arrival rate of messages into that queue and/or the service time for that queue. Consider the following response time formula:

$$R = E(ts)/(1 - E(n)E(ts))$$

Where: R = Response time
$E(n)$ = Message arrival rate
$E(ts)$ = Service time at server

Since $E(n)$ is a denominator in the formula, $E(ts)$ is not a linear function of $E(n)$. Therefore, as service time or message arrival rate increases, response time increases nonlinearly, as shown in Figure 11-11.

Queueing models do not obviate the systems analysis and user requirements study. Indeed, the formulas are only as valid as their input parameters, which are based on the applications' traffic pattern. The following user data are the primary input parameters to the response time calculations:

- Average arrival rate of applications messages at computer site.
- Time between arrival of messages.
- Number of queues (the messages' waiting lines).
- Time in the queue.
- Number of messages in queue.
- Number of servers (e.g., computers to service the message).
- Average service time.

Given these data, the formulas can then be used to calculate an expected response time. A design should use and substitute various parameters and repeat the formula calculations. In this manner, sizing and cost trade-offs can be compared against the

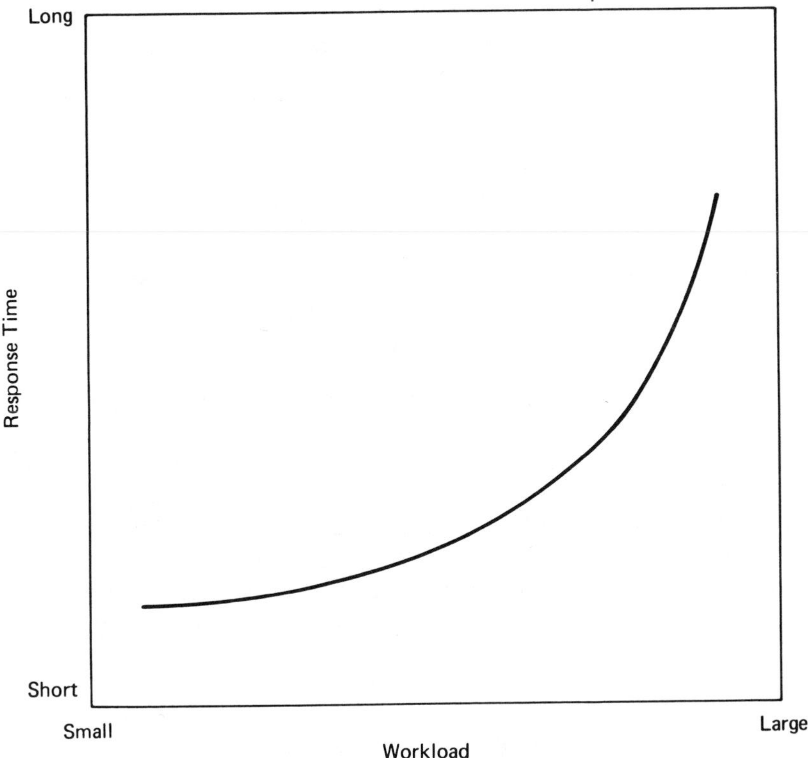

FIGURE 11-11. Workload and Response Time.

response time requirements and adjustments can be made as dictated by user needs and the company's budget.

We know from previous calculations that the 200 ms processing time is inadequate to service the peak period traffic. Therefore, let us assume the design team sizes a more powerful machine for the response time calculations. In this situation, the projected computer will use 100 ms for each of the 10,209 transactions—a significant upgrade over the initial 200 ms processing time.

Calculation 14: Utilization of Server

Formula: $$U = \frac{E(n) \times E(ts)}{M}$$

Where: U: Utilization of server
E(n): Average arrival rate of messages

E(ts): Average service time
M: Number of servers during test period

Application data: E(n): 170 per minute
10,209/60 min. = 170
E(ts): 100 ms or .1 second
M: One computer

Calculation:

$$U = \frac{170 \times .1 = .283}{1 \times 60 \text{ sec. per min.}}$$

The facility utilization of .283 says that the computer will be occupied 28.3% of the time during this peak period. This formula can also be used to determine the utilization of a communications line (which was calculated in calculations 1, 2, and 3), but the reader is encouraged to examine queueing calculations in more detail before applying the formula. In its simplest form, utilization is the time the server is occupied divided by total time available. This approach yields slightly different results than do Calculations 1, 2 and 3.

Calculation 15: Response Time

Formula:

$$E(tq) = E(ts) + \frac{U \times E(ts) \times A}{(1-U)}$$

Where: E(tq): Average queue and service time (response time)
U: Utilization of server (Calculation 15)
E(ts): Average service time
A: Adjustment factor, to allow for estimating errors

Application data: U: .283
E(ts): .1 second
A: 1 (A can also be calculated; see references)

Calculation:

$$E(tq) = .1 + \frac{.283 \times .1 \times 1}{1 - .283} = .14 \text{ seconds}$$

The .14 second response time is an excellent response time for manually operated terminal applications. If the design team chooses to size with a slower computer, the response time will increase. However, the response time is deceptive. For example, a computer yielding a 300 ms processing time will probably not service the large traffic volume since Calculation 14 will yield a utilization factor of greater than 85%. The 100 ms machine is a fairly close fit and is likely near saturation. The exponential nature of queueing will result in markedly degraded performance as the traffic load increases.

Summary

The reader should now have an awareness of the basic elements in data communications line loading and topology design. This section introduces the subject; it is a discipline and profession unto itself. Nonetheless, you should have a sufficient background to understand the process. The next two sections on network software design and data base design will round out the material in the first section.

PARTITIONING AND ALLOCATION

One common approach to network software design is the Partitioning/Allocation Method (PAM). The object of PAM is to divide and distribute the automated resources to the proper nodes in the network. PAM can be used to achieve more consistent load leveling, higher throughput, and faster response time at sites in the network. For example, as depicted in Figure 11-12, (and Figure 10-1) code can be partitioned (divided) into parts so that each part can be executed in parallel. The subcomponents can then be loaded to separate processors for simultaneous or parallel processing, thus decreasing the overall processing wait time that would be encountered by using one computer. By placing the program on four processors, the time required to execute the code is reduced from T + 4 to T + 1.

Network software partitioning requires more computers, which of course is an added expense to the system. However, the rapidly declining cost in hardware has provided the impetus to develop parallel processing systems. The trend is toward distributed parallel processing using machines and software tailored for specific functions.

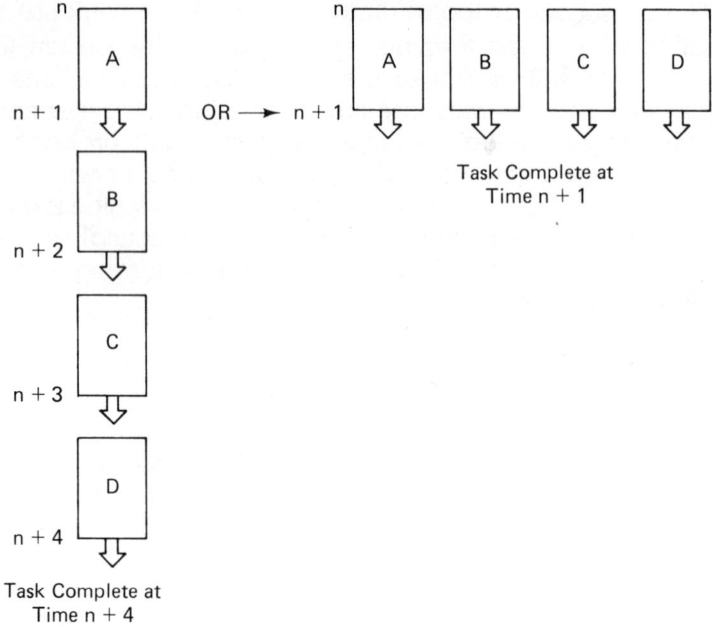

FIGURE 11-12. Parallelism.

Partitioning involves the initial description of the design at a very general and narrative level. The design moves to more detail as the requirements are refined and described in greater detail. Eventually, the partitions are realized in the form of source code (see Figure 11-13).

Since the partitions may be allocated to separate or remote machines, the partitioning design is an important part of the design process. The partitions must be as independent and self-contained as possible, yet contribute to the overall system goals as an integral part of the whole. The partitions should have few connections to each other in order to minimize complexity and error propagation. Each module should have only one entrance into its code and one exit and should connect (couple) only to modules directly above and below it. The ideas are illustrated in Figure 11-14. The multiple connections in Figure 11-14(b) greatly complicate the system and lead to excessive traffic across the communications lines. For example, if modules C and E were loaded into computers in Nashville and Dallas, respectively, execution of code in module C could affect module E. This kind of connection results in transac-

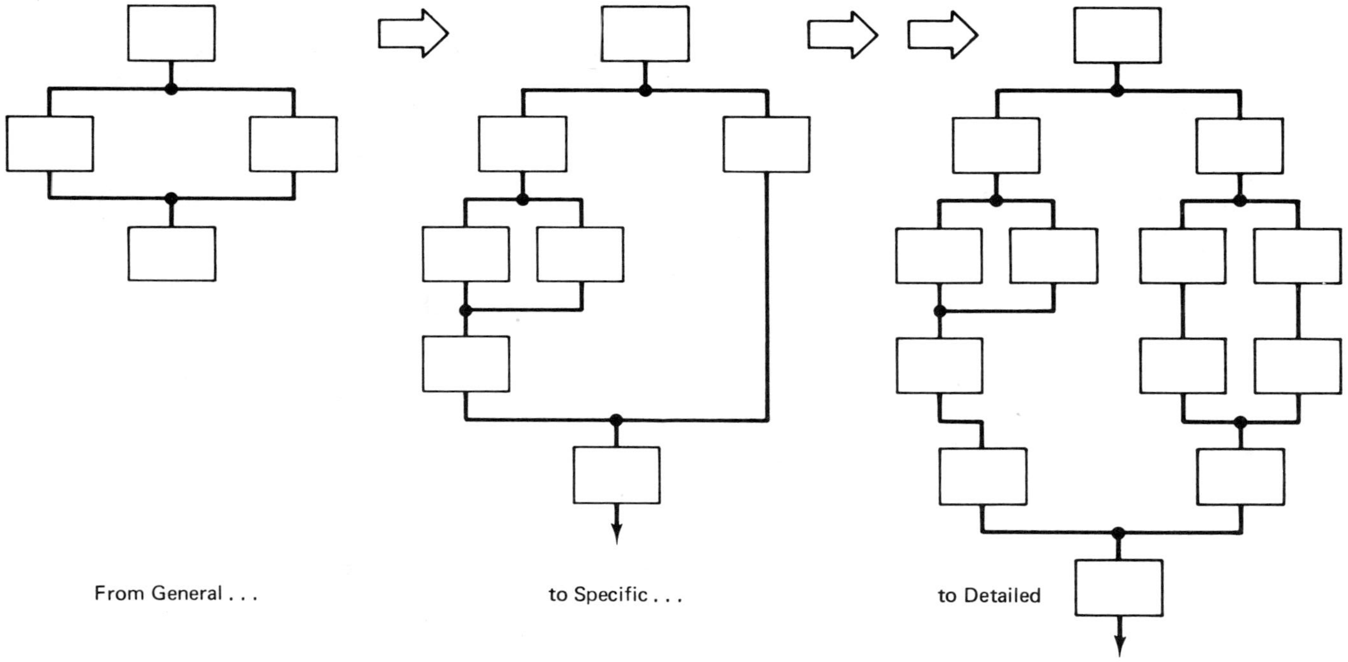

From General . . .

to Specific . . .

to Detailed

FIGURE 11-13. Partitioning.

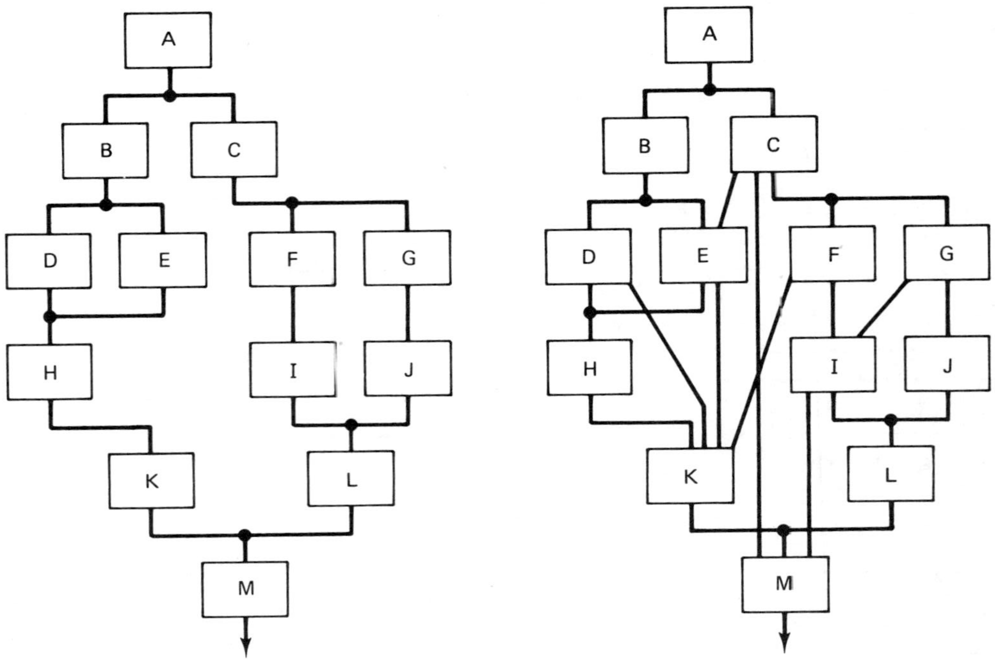

(a) Desirable Connections

(b) Undesirable Connections

FIGURE 11-14. Connecting Network Partitions.

tions and messages being transmitted between the two sites. Obviously, messge flow among computers is inevitable, even desirable. But the code must be organized into modules to avoid internodal dependency.

To achieve simple connections, the partitions should exhibit strong logical cohesiveness. The goal of cohesiveness is achieved when the elements (or code) within the partition perform one function or a set of closely related functions. Once the instructions in the partition are executed, the process *should not require* other sites to complete the task. The objective is to partition programs into efficient and self-sufficient modules.

Systems that exhibit strong cohesiveness within modules are said to have tight logical binding. Tightly bound modules provide the added effect of loosely coupled modules. This means that the software modules executing in the network computers do not depend on other modules since the functional code is located locally.

Figure 11-14 also illustrates another critical aspect of PAM: keeping the scope of effect within the scope of control. This simply means that the execution of an instruction should affect only its own module and the modules connected directly below it. Figure 11-14(a) achieves this goal; Figure 11-14(b) does not. The danger of the connection in Figure 11-14(b) is the initiation of undesirable and unpredictable ripple effects throughout the network.

Partitioning and allocation work best when adhering to the principles of atomic actions.[2]

- An action is atomic if process X is not aware of the existence of other active processes and the other processes are not aware of process X during the time process X is performing the action.
- An action is atomic if the process performing the action does not communicate with other processes while the action is being performed.
- An action is atomic if the process performing the action can detect no changes except those performed by itself and if it does not reveal its state changes until the action is complete.

Upon the completion of the partitioning, resource allocation can be accomplished. Figure 11-15 depicts how the system modules

[2] B. Randall, *Operation Systems: An Advanced Course.* New York: Springer-Verlag, 1979, pp. 282–391.

in Figure 11-13 are allocated to sites A, B, and C. In this example, site A initiates the action by accepting the message and routing it to the proper site for processing. It could go to sites B and C for parallel execution. The results of the processing at C and B are transmitted back to A for final execution.

Data Flow Systems

To further illustrate partitioning and allocation, we will examine an actual example of software.[3] The following code is a series of FORTRAN arithmetic assignment statements. FORTRAN code is read as follows: The value of the variable on the left side of the equals sign is replaced by the calculation on the values of the variables on the right. That is, in statement 1, A is replaced by the division of B by C:

Statement Number	Program Instruction
1	A = B / C
2	D = A + E
3	F = P − Q
4	R = G • S
5	T = W • D
6	V = R / T
7	X = D + F

The code need not be executed sequentially from statements 1 through 7. Several different sequences of execution are possible, one being: 1; 2 and 3 simultaneously; 7; 4 and 5 simultaneously; 6. Another: 1 and 4 simultaneously; 2 and 3 simultaneously; 5, 6, and 7 simultaneously. The alternate sequences permit the program to be processed on multiple execution units.

Most large-scale mainframes provide this function in the form of look ahead decoding, pipelining, and parallel processing. The important point to note here is that applications, telecommunications, and data base code can be designed to execute within a network on different processing elements. The technique should be

[3] The partitioning of individual instructions within a software module is a complex undertaking. It is explained here briefly to give the reader an appreciation of the process. A user application would rarely partition down to this level, but today's operating systems are evolving toward this process.

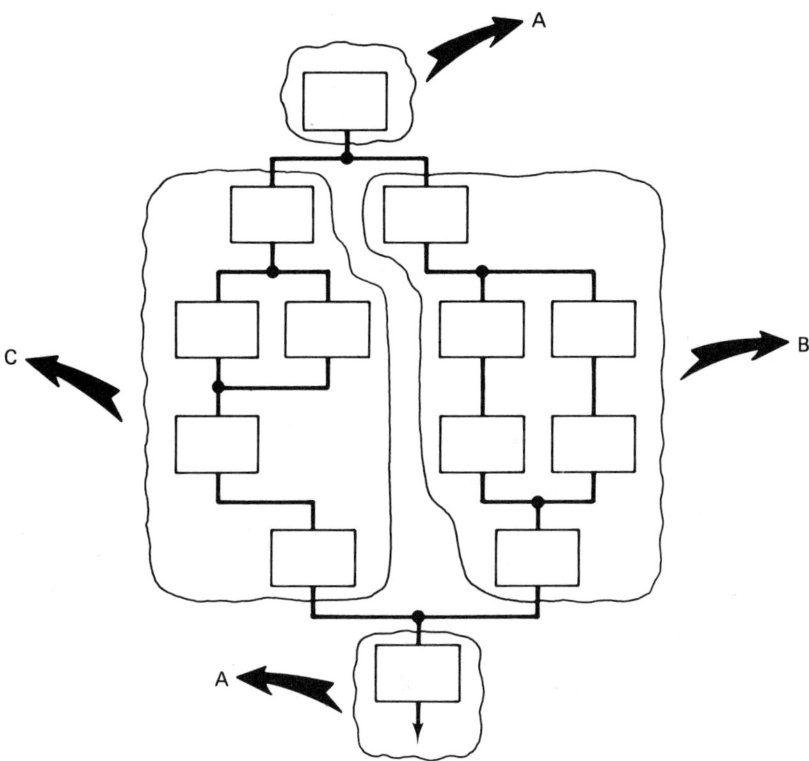

FIGURE 11-15. Allocation.

examined in view of cheaper processors and the resulting benefits of load leveling and decreased delay.

These ideas require a different design mentality since many people in the applications programming area have become accustomed to thinking in terms of sequential code execution, sequential module execution, and sequential job dependencies. Consequently, the PAM technique will require training for the software staff.

The FORTRAN program might be partitioned and allocated as shown in Figure 11-16. The first option is a conventional sequential processing (Figure 16(a)). As an alternative, in Figure 16(b) statements 2 and 3, 4 and 5 are executed simultaneously to cut down overall delay from T + 7 to T + 5. In Figure 16(c), additional parallelism gives a response of T + 4.

The FORTRAN coding example illustrates the concept of data flow systems. In contrast to the classical Von Newmann approach of

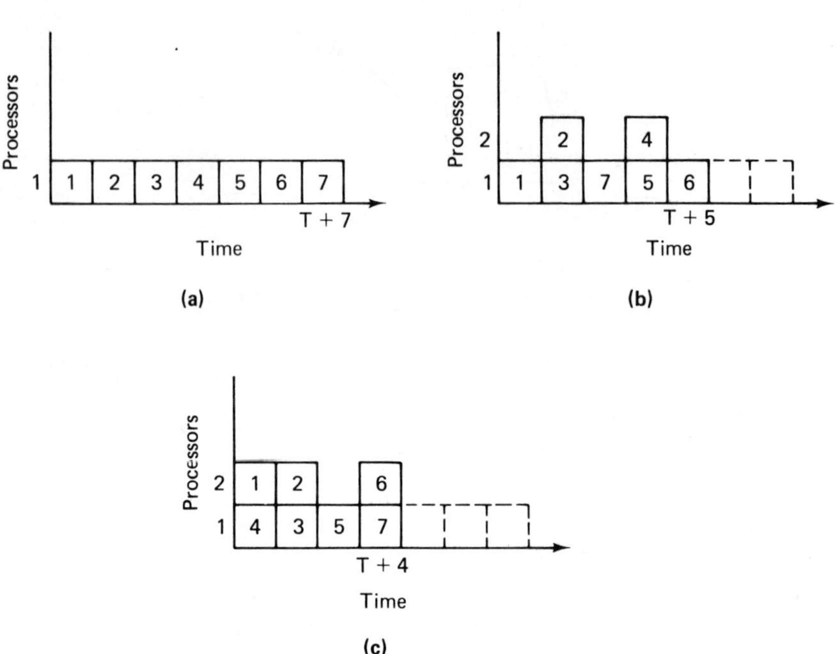

FIGURE 11-16. Partitioning for Parallelism.

a single program counter and memory addresses, data flow systems execute instructions *only* if all required values to the instruction have been computed and are available. The instructions compute a value and the value is passed to subsequent instructions for further processing. All sequencing is based only on data dependencies.

Data flow ideas are attractive for several reasons: (a) partitioning to achieve parallelism is simplified; (b) data flow programs are very modular; (c) the software achieves tight binding and loose coupling; and (d) the scope of effect of an instruction or module execution is quite limited. All these points are important to a communications network containing computers that share resources such as data bases and CPU processing.

The code on page 310 presents data conflicts because the sequencing of its operations are (a) not based on data dependencies and (b) use duplicate work spaces for the same data value. Data flow concepts could improve the code by ordering instruction execution by data availability. Obviously, this would require rethinking and redesigning computer systems that are sequentially bound and allow independent transaction execution.

The system would be required to provide for complete integration of all transactions running on the computer. Data flow systems for multiple transactions on partitioned/replicated data bases would be very complex but very powerful. We will eventually see commercial implementation of the concepts. Recent research on distributed data base systems is pointing toward increased use of these ideas.

The PAM concepts must be approached carefully (and data flow systems have not seen extended commercial application). Excessive partitioning will consume inordinate delay and overhead in transporting the data between the processing elements. It should be remembered that a common carrier path operates in a milliseconds world and the local area network path in microseconds; whereas, the processors operate in nano and pico seconds. The relatively slow paths between the processors should be used to shift workload and move transactions only if the rules of binding, coupling, atomic actions, and scope of effect/control are followed. Otherwise, the network will spend needless time in managing excessive traffic and the resulting delays.

Dimension Analysis

Dimension analysis is used to determine how many processors should be used for the allocation. It is also used to determine the load on each processor. The concept is shown in Figure 11-17. The system is described by its logic width and logic depth.

The depth of the logic determines the time required to execute a process and depends on the number of modules that are executed sequentially and on the number of instructions that are executed in each module. The depth of the logic is also dependent upon the data depth (amount of data) since recurring instances of data may result in the interactive execution of instructions. Obviously, varying data depths from transaction to transaction creates ambiguity and problems in obtaining an accurate logic depth assessment. Consequently, an accurate dimension analysis should use the ideas of a Standard Work Unit discussed on page 379.

The width of the logic determines the number of concurrent executions of modules that can be executed simultaneously. Logic width is an especially important consideration for local area networks in which multiple computers share a task, or in applications where parallelism and equal work distribution on computers is important.

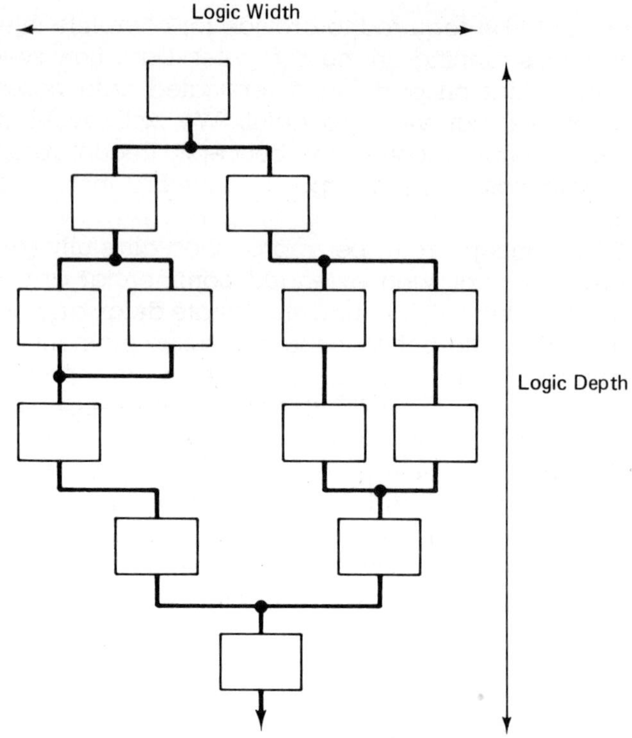

FIGURE 11-17. Dimension Analysis.

Partitioning Resources in the Southern Region

Once the software is partitioned according to the concepts of binding, coupling, scopes of effect and control, dimension analysis, and atomic actions it can then be allocated effectively to the computers in the network. To illustrate this point, let us assume in the case study that the Houston and Dallas volumes grow to the extent that the computer in Nashville cannot handle all the traffic of the region. For purposes of response time and communications costs, the decision is made to add a computer in Dallas to process sales and inventory data for the Houston-Dallas offices. If the system has been designed according to PAM conventions, the tasks of further partitioning and allocation are greatly simplified. Figure 11-18 shows why.

The software system to process the sales and inventory transactions is allocated to the computers in Nashville and Dallas as shown in Figure 11-18 and as follows:

Module	Location	Functions
A	Nashville, Dallas	• Accepts input transactions from either sales or inventory. • Edits transactions. • Sends data to B1 or B.
B1	Dallas	• Modification to B to satisfy Dallas' local environment.
B	Nashville	• Performs housekeeping chores. • Obtains specific customer profile from data bases. • Passes data to D and E.
D, F	Nashville, Dallas	• Performs functions for sales application.
E	Nashville, Dallas	• Performs functions for inventory application.
G	Nashville	• Performs tasks to send component ordering data to warehouses and plants in Jacksonville, Albuquerque, and Pittsburgh.
H	Nashville, Dallas	• Posts updates to files • Formats data for I.
I	Nashville	• Performs regional summaries. • Formats data for N.Y. headquarters. • Transmits data to plants and warehouses.

It should be noted that the adherence to PAM conventions allows module G to be used only at the regional computer. Since G deals only with the warehouse/plant function, the Dallas site is not concerned with this function. This code is *not* embedded into other modules. Moreover, modules A, D, E, F, and H are shared because they are designed to perform specific tasks relating to customers' sales and inventory functions. Typically, modules must be modified for local needs, so module B was changed to accommodate Dallas' unique requirements, giving module B1. As the network continues to grow, dimension analysis may reveal other opportunities for further PAM distribution. For example, at the Nashville office, modules D-F and E-G are candidates for additional partitioning.

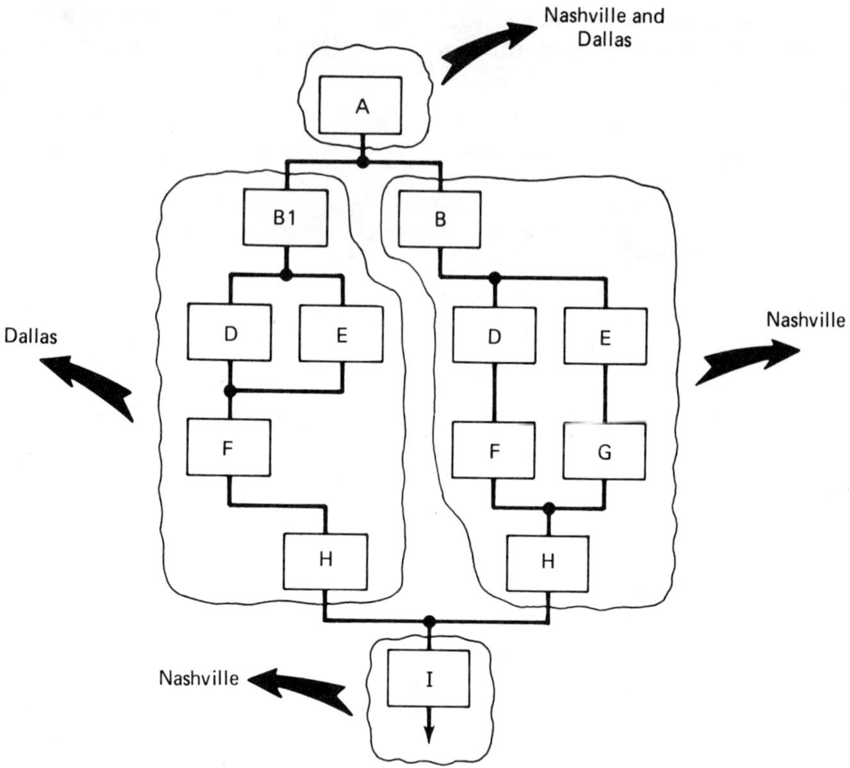

FIGURE 11-18. Southern Region Allocation.

Our case study does not exhibit the complex and sophisticated parallelism discussed earlier. However, it does exhibit load leveling, resource sharing, and distributed processing. It exhibits another characteristic of utmost importance: it is a simple solution to the problem.

Finally, the ability to downline load program modules to the Dallas site depends upon the compatability of Nashville's and Dallas' information systems. The nodes must have the same hardware architecture, communications protocols, DBMS, compilers, and operating systems.

NETWORK DATA BASE DESIGN CONSIDERATIONS

Chapter 4 contains introductory information on data bases and their role in communications networks. Chapter 10 expands the discussion, focusing on distributed data bases. This section continues the discussion and provides additional guidelines for designing data bases for a distributed network.

Ratio Analysis

Ratio analysis is often used to configure data bases for a network. It is based on an examination of the frequency of use of the data by each user at each site in the system. Data base decisions regarding centralization, partitioning, and replication are based on the analysis. The process is based on five ratios:

- Interest Ratio
- Miss Ratio
- Update/Retrieval Ratio
- Site Update/Total Update Ratio
- Precision Ratio

Interest Ratio. This is a useful pre-implementation tool to determine where data bases should reside. The interest ratio is derived as follows:

$$IR = \frac{TNI}{ANI}$$

Where IR: Interest ratio
TNI: This node's interest in the data
ANI: All nodes' interest in the data

TNI represents a node's probable frequency of use of a data base or a segment of the data base. A typical user/customer requirements analysis is used to obtain the interest statistics. TNI is divided by ANI to obtain an indicator of one node's interest relative to all node's interest. Ratios heavily weighted toward one node are an obvious sign favoring the location of the data at that site. Thus, the interest ratio provides a "home location" indicator.

The ratio is only as valid as the analysis that produce the TNI and ANI parameters. It is often impossible to predict data usage, and the formula's parameters may be "best guesses." On the other hand, if the data is already automated and stored in a data base, interest and usage statistics can be obtained fairly easily.

In most instances, it is not practical to perform a data interest analysis to the level of an individual data item. Rather, the design team usually groups similar or common items together for the interest ratio calculation. This approach, called *clustering*, is widely used for general design decisions. The grouping can also serve to evaluate the clustering of data at the distributed sites, and to relate the clustering to the users subschemas.

The interest ratio can also aid the team in making decisions on the distribution of software and should be used to complement the partition allocation (PAM) methodology. For example, if the ratio reveals that a site (e.g., one of the four regions) will execute a software program frequently, it makes sense to downline load a copy of the program at that site instead of using a remote computer (i.e., New York). Moreover, the ratio will reveal the site or sites that have the primary or "vested interest" in the software. Primary interest should be a major factor in determining which site is responsible for software development, maintenance, and downline loading control to other sites. A key in a successful distributed system is giving the responsibility of software/data management to the site(s) with the vested interest. Those sites are motivated to properly manage the resources because, to a large extent, their performance depends on the information emanating from the software or data bases under their purview.

Figure 11-19 provides an example of the data interest worksheet. The data obtained from the analysis and recorded in this sheet was used to justify the earlier decision in our case study to place computers and data bases in the four regions and three plants. For example, the Nashville analysis indicates a total of 10,209 sales and inventory transactions against the data bases during the peak period. This data is used operationally at Nashville more often than at the New York headquarters site. Figure 11-19 shows 3,063 transactions against the sales system (30% of 10,209) and 7,146 against the inventory system (70% of 10,209). In marked contrast, the New York site has 62 and 75 transactions against these two data bases. The interest ratio shows:

$$IR = \frac{TNI}{ANI} \quad \text{or:}$$

$$74 = \frac{10{,}029}{137} \quad \text{(only New York is interested)}$$

The Nashville site has a probable use (or interest) in this data by a factor of 74, an obvious reason for home locating the data in Nashville.

This illustration, while a bit simple, serves the purpose of showing how the interest ratio is used. Its value is greater in

	Data / Site	New York	Pittsburgh	Albuquerque	Jacksonville	Nashville	Chicago	Baltimore	San Francisco
*	Price	U: 2 R: 60				U: 979 R: 2084			
*	Inventory Location	U: 5				U: 4573			
	Stock Deletion Date	R: 70				R: 2573			
	Personnel								
	Payroll								
	≈	≈	≈	≈	≈	≈	≈	≈	≈
	Market Trend Analysis								
	Corporate Sales MIS								
	Corporate Inventory MIS								
**	Southern Sales MIS	U: 0 R: 4050	U: 680 R: 1490	U: 100 R: 130	U: 900 R: 1750	U: 1600 R: 3800			
	Southern Inventory MIS								
	Central Sales MIS								

NOTE: Blank Entries in Worksheet are Not Required for Examples.

U = Frequency of Update Transactions
R = Frequency of Retrieval Transactions
* = Accessed During Peak Period Hour
**= Accessed During 24-hour Period

FIGURE 11-19. Data Interest Worksheet.

circumstances where several sites have interest in the data, and the usage magnitude differences may be less obvious. For instance, after the Nashville site has processed the peak period workload, it transmits sales and inventory data to the three plants in Pittsburgh, Albuquerque, and Jacksonville and to the headquarters computers in New York. The data is aggregated with accounting information (accounts receivable, accounts payable), parts invoices, and other data to form the Southern Region Inventory MIS (Management Information System). Five sites have need of the Inventory MIS base. As depicted in Figure 11-19, the interest ratios reveal the following:

New York: 28%

$$.28 = \frac{4050}{14500}$$

Pittsburgh: 14%

$$.14 = \frac{2{,}170}{14500}$$

Albuquerque: 1%

$$.01 = \frac{230}{14500}$$

Jacksonville: 18%

$$.18 = \frac{2650}{14500}$$

Nashville: 37%

$$.37 = \frac{5400}{14500}$$

Nashville has greater usage of the data at 37%; New York follows with a 28% usage ratio; next Jacksonville (18%) and then Pittsburgh (14%). Albuquerque does little support work with the Southern region (1% ratio).

The designers now have some data to use for rational design decisions. Several options are available. First, the Southern Region Inventory MIS could remain centralized at the Nashville site. However, this means 63% of the transactions would have to be transmitted across the expensive communications path to the Nashville computer. Second, all sites could be loaded with a replicate of the data base. However, this approach would be wasteful in view of the fact that Albuquerque and Pittsburgh do not have much interest in the data. Third, replicated copies could be placed at the Nashville and New York sites, giving 65% of all transactions a local access capability. The three plants would then transmit transactions to Nashville and New York to fulfill the remaining 35% of the total traffic. As we shall see when the update/retrieval ratio is examined, the third alternative is particularly attractive from the standpoint of simplicity.

The interest indicators are useful analysis tools. Once a system has been implemented, the data and software used at each site must be reexamined frequently in order to determine if the distributions still make sense. Data usage does not ordinarily remain stable in growing and dynamic organizations. Data flow changes as the company introduces new products, changes its mission, and acquires or divests itself of subsidiaries. Consequently, the interest ratios should be examined periodically to determine if the initial analysis is still valid.

Miss Ratio. The miss ratio is useful for this purpose. The ratio can be used to relocate software and data bases—in essence, to determine if the home locations are still appropriate. The ratio indicates how frequently data is requested at a local site but obtained elsewhere.

$$MR = \frac{TDNO}{TDR}$$

Where: MR: miss ratio (at the local site)
TDNO: total data not obtained (at the local site)
TDR: total data requested at the local site

Obtaining data for the TDNO and TDR parameters will probably require the writing of software that "traps" these statistics. For example, all outgoing transactions must be logged, with information on where the transaction is being transmitted. The statistics should be gathered periodically, not continuously. The overhead to collect the data can become quite expensive.

Interest and miss ratios should be evaluated with other factors. They should not be used to provide the "final answer," but as a tool to be used with other considerations:

- *Frequency of use vs. need for timeliness:* A site may not use data often, but may need it immediately upon its infrequent request. In such a situation, it may make sense to keep redundant data at the sites that use it regularly *and* at the sites that need it rapidly.

- *Frequency of use vs. vested interest:* A site that accesses data frequently may not necessarily be the site that considers the data most vital to its operations. Corporate headquarters in New York could have vested interest in certain data, yet actually use it infrequently.
- *Update vs. retrieval:* The design team must make decisions on how to provide optimum performance for users who retrieve information and for users who update the data bases. It is difficult to provide ideal performance for both types of users, especially if the data is dispersed to multiple sites. We examine this problem with the next ratio.

Update/Retrieval Ratio. The decision to replicate a data base in the network depends to a great degree on the frequency of updates versus retrievals that occur. The case study worksheet in Figure 11-19 provides an example.

Chapter 10 (pages 315 through 318) explains the complexity and potential problems of updating replicated data—especially in a real-time mode. As a general rule, real-time updates should be confined to centralized or partitioned data. The Southern Region Inventory MIS update/retrieval ratio is calculated as follows:

$$URR = U/R$$

Where URR: Update/Retrieval ratio
U: Number of updates against a data base
R: Number of retrievals against a data base

The formula reveals that the sites have these URR ratios: New York = 0%; Pittsburgh = 45%; Albuquerque = 76%; Jacksonville = 51%; Nashville = 42%. We examine the significance of these ratios after examining the next ratio.

Site Update/Total Update Ratio: In many situations, the total number of updates created at each site relative to the other sites is another important consideration. Consequently, the site update/total update ratio is also used:

$$UUR = TUTS/TUAS$$

Where UUR: Update/Update ratio
TUTS: Total number of updates at this site
TUAS: Total number of updates of all sites

The IR, UUR formula gives the following results: New York = 0%; Pittsburgh = 20%; Albuquerque = 3%; Jacksonville = 27%; Nashville = 49%.

The URR and UUR data provide a basis to make the following conclusions:

- New York presents no update complexity problem since it has no update traffic. Moreover, the interest ratio reveals New York has 28% of the total transaction traffic. These two factors support the decision to replicate a copy of the data at the New York site.
- The Pittsburgh, Albuquerque, and Jacksonville plant sites have 45%, 70%, and 51% update/retrieval, respectively. At first glance, perhaps the high incidence of updates warrants replicated data at these sites. *However*, the total update traffic revealed by the update/update ratio is much smaller for the plants: 20%, 3%, and 27% compared to Nashville's 50%. Nashville has primary update interest, a factor favoring the location of data at that site.
- Moreover, and most important, since the three plants do generate update traffic, replicating the data at those sites may create the timing and synchronization problems discussed in Chapter 10.

Consequently, the decision to replicate the data at New York and Nashville keeps all updates at the one data base location in Nashville. The three plants will transmit their relatively small number of updates and retrievals to Nashville. (Pittsburgh could reasonably use the New York copy for its retrieval as well.) At some point in the 24-hour processing cycle, Nashville will downline load a fresh copy of the data to New York (practically speaking, perhaps using air express). Thereafter, New York's 4,050 average load will be done locally.

Precision Ratio. We return to the problem of physical layout on the disk (discussed on pages 139 through 141). Retrieved requests from remote sites in the network should create responses containing the exact data needed—nothing more and nothing less. For example, let us assume the Inventory MIS has 200 separate items or fields stored in a physical block on the data base. The five sites' subschemas vary; some need certain fields within the record and not others. Moreover, the sites have different subschemas depending on a specific retrieval requirement.

The precision ratio is an ongoing analysis tool used to ensure that the physical data base storage accommodates the total user community. This is especially important if an organization's system transmits and receives disk blocks without reformating into smaller more precise messages. The precision ratio is calculated as follows:

$$PR = \frac{RIR}{TIR}$$

Where PR: Precision ratio
RIR: Relevant items retrieved
TIR: Total items retrieved

The ratio is used to analyze each user logical view (subschema) and determine how concisely the data base is structured to satisfy the request. For example, a user at the New York site has a subschema to view items 1-30, 78, 100-175 of the 200 item block, the precision ratio is 53% (.53 = 106/200). The 53% figure means that 47% of the data in the disk block was read, sent across a channel into a computer buffer (and perhaps even transmitted across a communications link) and not used. Is this a reasonable ratio? The answer is that it depends upon the ratios of *all other* users. If a precision ratio aggregate reveals that users at several sites in the network are receiving low ratios, it is quite likely that the physical data bases need to be reformated and reorganized. Precision ratio analysis is especially important for data bases with a large community of users in the network. They must be as precise as possible for the entire user community, or individual users will "spin-off" data and create their own data structures. Such actions can create wasteful redundancies and data conflicts in the organization.

Data Base Design Decision Trees

In addition to the ratios, decision trees are useful design aids. The graphic nature of their structure can be used as a briefing tool design aid. Figure 11-20 shows a simple decision tree based on two factors: amount of data and real-time update frequency. Each organization should determine the specific parameters for the two variables. The decisions to centralize, replicate, or partition are general guidelines and certainly subject to change as individual circumstances warrant. Figure 11-21 shows a more involved tree. The update/retrieval ratio has been added to the structure. Finally, Figure 11-22 is an example of a tree developed for post-installation

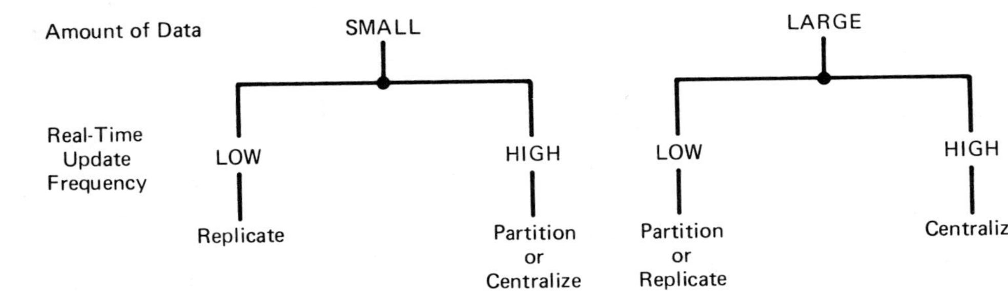

FIGURE 11-20. Decision Tree.

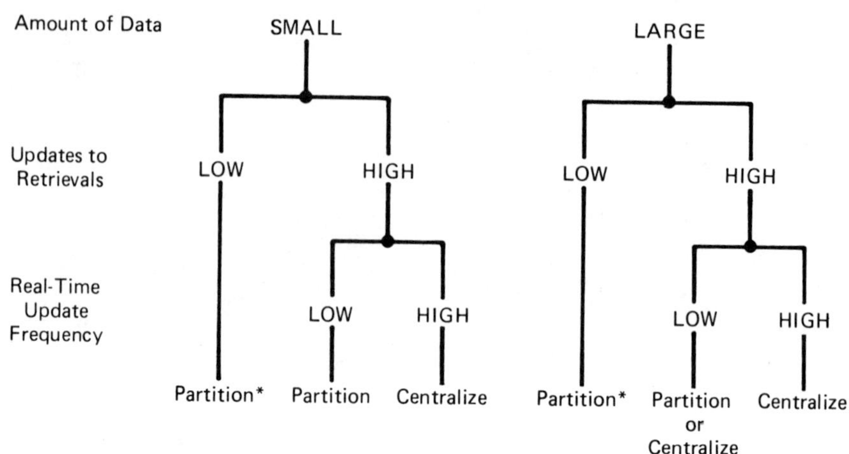

FIGURE 11-21. Decision Tree.

analysis based on the amount of data, the miss ratio, and the real-time update frequency. The miss ratio might indicate a restructuring of the network data bases. While not illustrated here, the precision ratio can also be used with the decision tree. The reader is encouraged to draw a tree using the precision ratio as it relates to his or her organization.

Each organization should design the trees and the final recommended approaches based on its specific requirement. Since the three trees presented here are approaches that have been substantiated by the experiences of network and data base designers, the reader can use them as the initial working assumptions.

Synchronization of Network Databases

In many instances, ratio analysis and the decision tree will lead to the decision to partition and replicate data bases. This section examines methods to keep the segmented and copied data concurrent and consistent across the nodes.[4] The subject is covered

[4]The approach discussed is based on research conducted by the Computer Corporation of America (CCA), 575 Technology Square, Cambridge MA 02139. Numerous papers and articles on the SDD-1 (A system for Distributed Databases) model are available from this company. SDD-1 is based on relational data bases and CCA claims it to be the first operational and correct distributed data base system in existence.

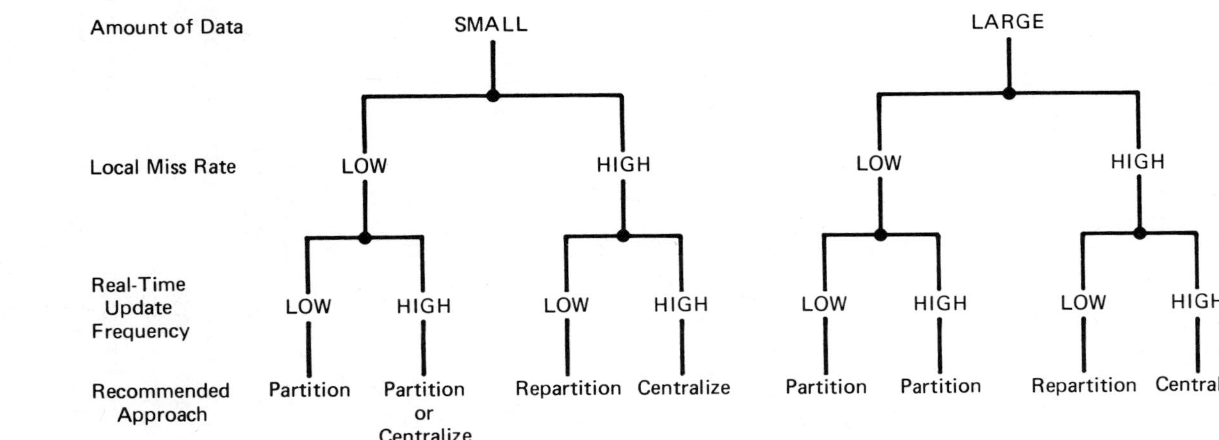

FIGURE 11-22. Decision Tree.

with four discussions: (a) system architecture, (b) serialized scheduling, (c) conflict analysis, and (d) network data management.

System Architecture. The network data base system is organized around the transaction manager or module (TM) and the data manager or module (DM). Each site has a computer running the TM and DM software (see Figure 11-23). The TM supervises transactions and *all* network action to satisfy the user's transaction request. The DM manages the data bases and the DBMS. In a sense, the DM is a backend data base processor and it is conceivable that DM functions could be placed in a separate backend computer. The architecture functions through four operations at the transaction—TM interface:

1. BEGIN: TM sets up work space for the user transaction (T). It provides temporary buffers for data moving into and out of the data bases.
2. READ(X): TM looks for a copy of data X in T's work space. It returns it to T. If the data is not in the work space, it issues a dm-read(**x**) command to one of the network DMs, which accesses the data and returns it to T's work space.
3. WRITE(X): TM looks for a copy of X in T's work space. If found, it updates the "old" copy with the current value of X. If the data is not in the workspace, a copy of the current value is placed into it.

   Notice that no changes have yet been made to the network data bases. That is, no dm-write(X) commands have been issued by the TM to the DMs. The system uses a *two-phase commit* procedure to assure (a) restart/recovery integrity, (b) adherence to atomic commitment, and (c) resiliency across nodes.

   - *Restart/recovery:* The DBMS can restart a transaction at any time before a dm-write(**x**) is executed.
   - *Atomic commitment:* Two-phase commit keeps all nodes' data base actions related to one transaction isolated from each other.
   - *Resiliency:* Failure of a component during a Write does not lock up or bring down other components in the network. In other words (and using another term), a component failure does not produce "sympathetic failures" of other components executing the affected transaction.

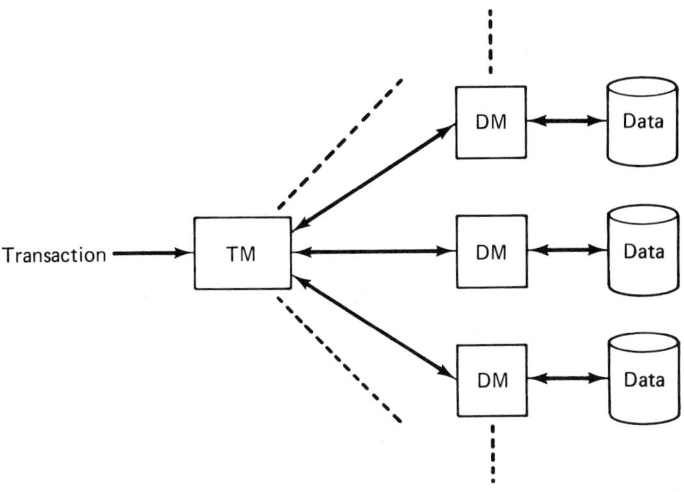

FIGURE 11-23. TM/DM Architecture.

4. END: For the first phase of the two-phase commit, TM issues a *pre*Write(x) to each DM in the network. The DM copies X from T's workspace onto some form of secure storage. A DBMS failure at this juncture does no harm. After all DMs have executed the preWrite, the second phase of the commit begins by the TM issuing a dm-write(x) to the DMs. The DM then updates the data base from T's secure storage.

The prewrite commands specify all DMs that are involved in the two-phase commitment. Consequently, if a TM fails during the second phase, the DMs time-out, and check with other DMs to determine if a dm-write(x) had been received. If so, all other DMs use it as if it were issued by the TM. Moreover, in the event of a DBMS failure during the Write, secure storage is used to recover data.

Serialized Scheduling. Network data bases using partitioned or replicated copies are increasingly using the concept of serialized scheduling, wherein the effect of executing multiple, interleaved transaction executions is the same as running the transactions serially. Figure 11-24 shows the effect of serialized scheduling of three transactions executing across three sites. The activity logs of the three DMs reveal the following order of operations (R_i (X_j)) means transaction i evokes a READ of X database at DM_j):[5]

[5]Philip A. Bernstein and Nathan Goodman, "Concurrency Control in Distributed Database Systems," *ACM Computing Surveys*, Vol. 13, No. 2, June 1981, p. 196.

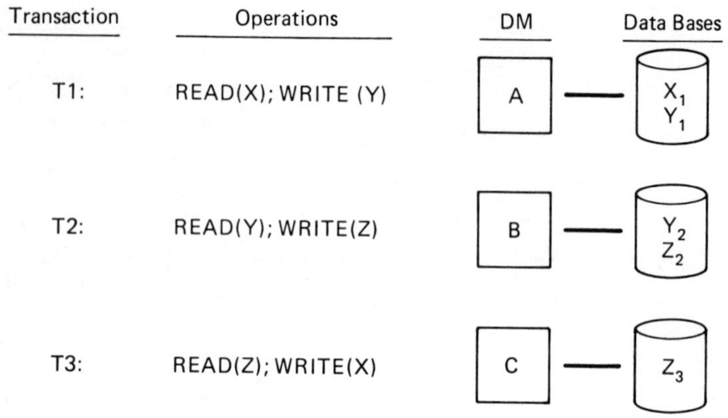

Transaction	Operations	DM	Data Bases
T1:	READ(X); WRITE (Y)	A	X_1 Y_1
T2:	READ(Y); WRITE(Z)	B	Y_2 Z_2
T3:	READ(Z); WRITE(X)	C	Z_3

FIGURE 11-24. Serialized Scheduling. (From "Concurrency Control in Distributed Database Systems" by Philip A. Bernstein and Nathan Goodman. *ACM Computing Surveys*, Vol. 13, No. 2, June 1981, p 196.)

DM A: $R_1(X_1)$ $W_1(Y_1)$ $R_2(Y_1)$ $W_3(X_1)$
DM B: $W_1(Y_2)$ $W_2(Z_2)$
DM C: $W_2(Z_3)$ $R_3(Z_3)$

The scheduling establishes transaction 1 to precede transaction 2, which precedes transaction 3. The schedule has the same effect of running each transaction serially. The DMs process all of transaction 1 actions before executing transaction 2, and the transaction 2 Write is executed at DM C before transaction 3.

Serialized scheduling is accomplished through the use of timestamps and message classes. Timestamps are placed on each transaction in accordance with a systemwide logical clock. The universal clock provides rules for keeping all sites at approximately the same logical time. The timestamp is made unique by the placement of a node identification within the timestamp. The transactions are divided into classes. The class is predetermined, based on the network data the transaction reads and writes. Since some transactions access the same data, their executions can possibly interfere and cause an inconsistent state in the data base. Consequently, transactions arriving at a node with conflicting classes must be processed in timestamp sequence. Since the transaction classes are predefined, all nodes know of potential conflicts.

Conflict Analysis. The analysis of class conflicts using timestamps involves two types of conflicts: Write/Write and Read/Write. Each type is based on class scheduling, wherein nodes will not process a Read with a conflicting class until all affected earlier Writes have been processed. Figure 11-25 provides an example of Write/Write conflict analysis.

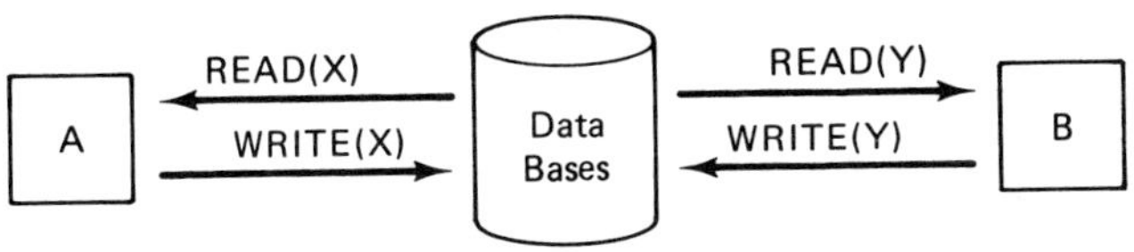

FIGURE 11-25. Conflict Analysis.

Site A transmits a Read transaction for a data base. Let us assume that the Read(X) transaction has a timestamp of t6. The node DBMS knows that classes X and Y may conflict. Since no class Y transaction with a t equal to or greater than t6 has arrived, site A's transaction is held.

Site B transmits a Read to the data base with a timestamp of t4. Since t4 < T6, B could possibly send a Write (say, at t5), *and* since B's timestamp is earlier than A, site B's Read is executed.

Site B obtains data, and then issues a Write (Y) with a timestamp of t5. Its Write is executed since site A's timestamp is t6. Site B then sends a timing message with a timestamp of t6. Since A's t is equal (or less than B's t), its Read and subsequent Write can now be executed.

Network Data Management. A network data base must do much more than scheduling and conflict analysis. SDD's TM provides many other functions, some of which are illustrated in Figure 11-26.[6] The relational data bases are stored at the nodes in pieces called *fragments*. (Ratio analysis would be a good tool for deciding how to fragment the data.) The data fragments are accessed and provided to the end user through the following sequence of events:

1. The transaction is analyzed by the Request Parser to determine if the requested data is available and the user

[6] James A. Martin, *Design and Strategy for Distributed Data Processing.* Englewood Cliffs NJ: Prentice-Hall, 1981, p. 337.

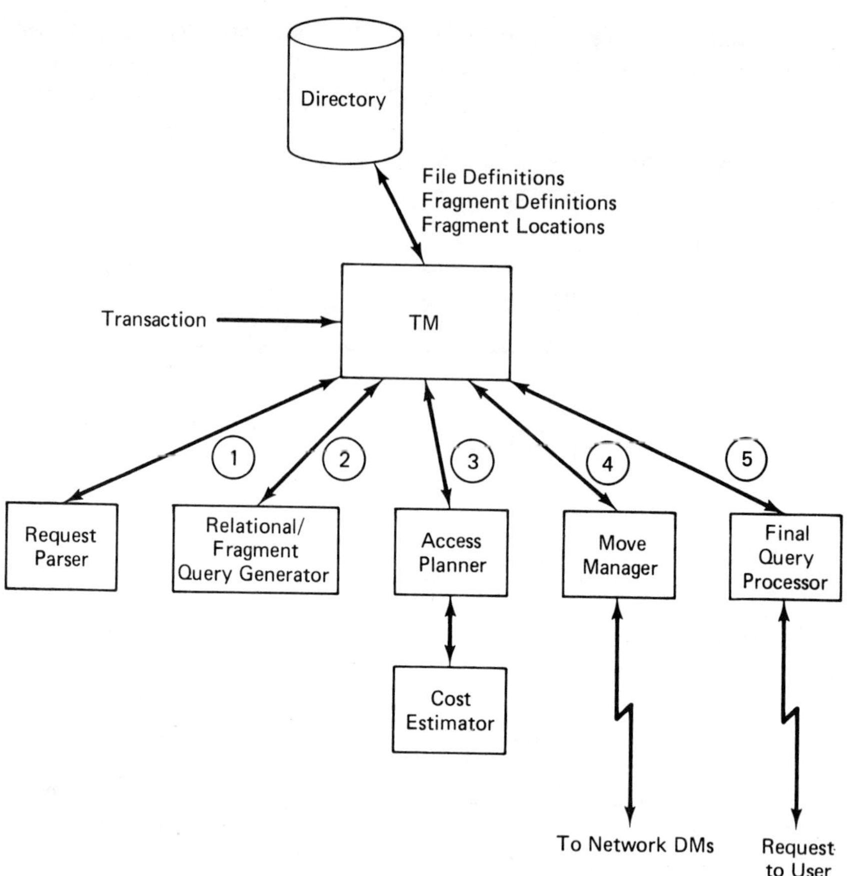

FIGURE 11-26. Network Data Management. (From *Design and Strategy for Distributed Data Processing* by James Martin. Englewood Cliffs NJ: Prentice-Hall, 1981, p. 337. Reprinted with permission.)

has authority to access. The Parser obtains this information from a systemwide Directory.

2. The request is translated into relational access language (see Chapter 4). The Directory is used to obtain fragment definitions and the Fragment Query Generator transforms the access language to match the fragments.
3. The Access Planner obtains the site locations of the fragments from the Directory. The Cost Estimator establishes how much data is to be moved to satisfy the user request. The Planner develops an access plan for obtaining the data. This process is critical; the plan must provide for

efficient and fast data movement. "Roundup" overhead (see Chapter 10) must be kept as low as possible.

4. The Move Manager executes the access plan and generates requests to the affected DMs in the network.
5. Data is moved to one site, assembled by the Final Query Processor, and presented to the user. During this process, other activities such as serialized scheduling and conflict analysis are also taking place.

Network Data Base Considerations: A Summary

Network data bases can be cumbersome to use and complex to update. As discussed previously, if multiple copies of one file exist at several sites, an update requires rather extensive coordination among the sites to keep all copies concurrent and identical.

The problems with updating multiple copies are reduced substantially by batching updates and processing them during a nonpeak period; the software complexity is reduced and the possibility of a site's "down" data bases does not pose as big a danger during an off-hour. This is a popular and well-proven approach.

Of course, the advantage of multiple copies for local retrieval may outweigh the problems with the updates. Using local queries, the transaction is processed on-site, thereby providing opportunities to reduce communications costs. The potential bottleneck problems are relieved by having multiple service points and the backup copies provide for failure recovery.

The use of partitioned data bases presents different considerations. In this case, the single copy is divided (partitioned) and placed at more than one location. This approach provides a simpler environment for updates; the update request is sent to the proper site, the data base is locked out, updated, and released. The one copy eliminates many of the overhead and delay problems associated with multiple copies.

A possible disadvantage of partitioned data bases called *roundup overhead* arises when a data retrieval request requires data from several of the data base partitions. The data request must be transmitted to all sites where the data resides, sent to the requestor and assimilated into a logical request, and presented to the user. The data retrieval could experience inordinate delay and overhead. Nonetheless, if the data is partitioned based on valid home location indicators, multiple retrievals to the distributed sites can be reduced.

The segmentation of the data files will affect line usage since the duplication of data could decrease the transaction rates during the peak period. For example, as was suggested in the case study, the larger offices in the Southern region might acquire their own machines and data files due to the findings of the distributed systems analysis. In a practical sense, several offices in the Southern region could likely benefit from local computing capacity. This means fewer transactions at the Nashville regional headquarters. Consequently, an organization must periodically and frequently reevaluate the traffic flow because the network topology will be affected.

The ripple effect must be evaluated with *each change* to the network. If duplicate data is created, the replicated data bases will require additional transactions during the peak period in order to keep the data consistent across sites, thus creating additional overhead. Response time and throughput must then be reevaluated. Again, the ripple effect must be uppermost in the minds of the design team.

The distribution of data presents significant opportunities, yet must be approached carefully. The environment should be kept as simple as possible. As an organization moves to multiple copies of data, fast response time requirements, and larger data volumes, the distributed systems become more complex and subject to synchronization problems.

12

CONCLUSION

INTRODUCTION

The communications industry and communications common carriers have been highly regulated in most countries because of the vital role they play in nation's economic, social, and security programs. In the United States, the communications common carriers grew up as regulated monopolies—as a compromise to America's free enterprise system. Other countries have treated communications as part of the government structure in the form of Postal, Telephone, and Telegraph (PTT) ministries.

The 1934 Communications Act gave the Federal Communications Commission (FCC) the responsibility to oversee the communications trade. Prior to the Act, the Interstate Commerce Commission (ICC) held jurisdiction due to the location of the majority of telephone and telegraph facilities on railroad rights-of-way. The Act's primary objective was to provide. . . ." so far as possible, to all the people in the United States a rapid, efficient, nationwide and worldwide wire and radio communications service with adequate facilities at reasonable charges. . . .[1]

AT&T was incorporated in 1885 and grew to become one of the largest companies in the world. It controls 80% of the nation's telephones and most of the vital telephone and communications

[1]The Communications Act of 1934, Title 1, Section 1, U.S. Government Printing Office, Washington, D.C.

network. Its influence can perhaps best be appreciated by considering the following facts: (a) it has over 1% of the total U.S. workforce; (b) its 1980 revenues were about 2% of the Gross National Product (GNP); (c) its annual budget is about 7% of the total business expenditures in this country.

The 1934 Communications Act and the growth of AT&T lead to a highly centralized, one-company industry. While about 2,000 common carriers existed as late as 1970, the nation's important communications infrastructure was dominated by AT&T.

The computer industry is considerably younger and it is a highly competitive industry. It is marketing oriented and its accounting methods encourage rapid change. The common carriers, having evolved under a regulated, protected environment, have been known as "order takers" only. A high quality communications service has been available, but a customer had few options other than the local "Ma Bell." The communications industry in the U.S., as critical as it is to the nation, had not been designed for the rapid changes that resulted from the computer-based information revolution.

Events in the early 1940s, considered insignificant at that time, established a path for these industries to join forces. Several attempts were made to attach computer-related devices to communications facilities. One of the earliest recorded events was in 1940 when Bell Labs connected a Teletypewriter at Dartmouth College to an electronic calculator in New York City to demonstrate a merging of two infant technologies.

RECENT HISTORY OF THE COMMUNICATIONS INDUSTRY

Foreign Attachments

Until January 1969, the common carriers had restricted customers (subscribers) from attaching any noncommon carrier device (such as a modem) to the public dial telephone network. Their reasoning was that a quality end-to-end service required the carriers to have control over all the facilities in the network.

This part of the common carrier domain was first breached in 1956. The Hush-a-Phone Corporation marketed a rubber shield to place around the telephone mouthpiece to provide for privacy in a conversation. AT&T contested the use of the device and the case went to the courts. After considerable controversy, the U.S. Court of Appeals ruled that the tariff restrictions on this device were unduly restrictive and not detrimental to the public. AT&T was ordered to revise its tariff to allow the use of the Hush-a-Phone device. AT&T acceded, but past practice continued to discourage the attachment of "foreign" non-AT&T devices to the public telephone network.

The most significant foreign attachment ruling against AT&T occurred in the late 1960s, as a result of an antitrust suit by the Carter Electronics Corporation—the Carterfone case. This Dallas firm sold a device for connecting a private communications system (such as radio devices connected into the dial network). The product was very successful in spite of pressure from the common carriers to discontinue its use.

An antitrust suit by Carterfone went before the FCC and the courts, and AT&T lost. The common carriers were required to revise their tariffs in January 1969 to allow the attachment of non-AT&T modems. The Carterfone decision is considered to be a watershed in the foreign attachment issue. Without question, the decision gave impetus to competition in the industry and provided a path into AT&T's heretofore restricted domain.

The Data Access Arrangement

The 1969 tariff revisions did not give subscribers complete freedom of how to attach their devices. The devices had to adhere to certain electrical standards and all network signals such as dialing numbers were to be performed by AT&T. In addition, subscribers were required to place an AT&T-supplied device between their modem and the network. This device, a Data Access Arrangement (DAA), limited the power of the attached modems and protected the network against high voltage.

The independent foreign attachment vendors vigorously opposed this move. They contended their equipment was technically superior to Bell's machines and argued that the DAA should be replaced with standards.

AT&T backed away from the strict DAA policy in late 1974. The company stated a protective module could be built by an independent vendor if it were designed by AT&T. However, in 1975 the FCC ruled that non-AT&T DAA-equivalent circuitry could be placed in the foreign attachment after the device was tested and certified by the FCC. Thus, the final DAA decision provided further inroads into the AT&T domain.

The foreign attachment rulings precipitated protests from the common carriers. AT&T held that the decisions would eventually lead to a deterioration in the nation's communications system. This does not appear to have been the case. A more serious issue has developed regarding the interconnection of noncommunications equipment (i.e., nonregulated) into an industry that is highly regulated. This problem has become even more visible today as the technologies of the communications and computer industries merge together. We will return to this problem shortly.

Line Sharing and the Value Added Carrier

In the past, a user of AT&T leased lines was not allowed to share the line with other users. This meant that customers often had wasted and costly excess line capacity. The problem was exacerbated by tariffs restricting a "part-time" use of a leased line. The leased line customer got something or nothing at all, and the something was a 24-hours-a-day, 7 days-a-week voice-grade line. Excess capacity could not be shared or resold.

Certain exceptions were granted by AT&T. The sharing of the high capacity TELPAK lines was permitted for a limited number of users such as public utilities, railroads, and government agencies. Also, a customer was allowed to connect to other users to transmit data concerning the customer's business, but the other users were not allowed to transmit to each other (for example, security brokerage firm inquiry services). Another notable exception to line sharing restrictions was granted in 1937 by the FCC in giving Aeronautical Radio, Inc., (ARINC) permission to provide a communications network to the airlines industry. AT&T opposed this FCC ruling.

AT&T revised its tariffs in February 1969 to permit shared use of certain private lines. However, under existing tariffs and as late as 1968, emerging networks would have been forbidden to share and resell their facilities. It is difficult for a newcomer to relate to this environment, but it serves to illustrate how much the industry has changed.

On July 16, 1976, the FCC ordered all common carriers to eliminate restrictions on the shared use and resale of their facilities. The decision gave birth to the value added carrier (VAC) industry and lead to VAC offerings that have created a much more diverse and competitive national communications system. At the same time, AT&T saw its monopoly position successfully challenged once again.

The first value added carrier was Packet Communications, Inc. It is no longer operating but several other VACs are in existence. The public packet networks such as Telenet and Tymnet are viable value added carriers. The facsimile transmission companies, such as Graphnet and Faxpax, are other examples of value added carriers.

Interconnection, Competitive Lines, and the Specialized Common Carrier (SCC)

The foreign attachment decisions clarified issues on connection of noncommon carrier instruments into the common carrier facilities. However, AT&T had not faced a serious challenge in its monopoly position of other services, such as long-haul lines and the interconnection of those lines into the local loops of a customer through the common carrier's end office.

Microwave Communications, Inc., (MCI) provided the challenge. In 1963, MCI asked the FCC for tariff approval to construct microwave common carrier services between St. Louis and Chicago. The offering was meant to constitute a service from a specialized common carrier (SCC): customer services would not be provided on an end-to-end arrangement and services would be specialized to a single type of carrier (in this case, microwave). The former point implied that existing carriers would provide the local loop service for the specialized common carrier.

The MCI request and the specialized carrier concept was vigorously opposed. AT&T held that SCCs would fragment the nation's communications system, leading to differing standards and the inevitable interface problems. Second, AT&T stated that the SCC would construct duplicate systems that were not needed to support the nation's communications requirements. Third, AT&T opposed the SCC using the argument of "cream skimming": the SCCs would only service the high-density, lucrative parts of the country and the common carriers would eventually be left in the unenviable position of servicing low revenue, rural areas.

A landmark decision by the FCC on August 13, 1969, provided for the SCC to begin operations.[2] The FCC ruled that the MCI request was in the public interest and the common carriers would provide local loop connection:

---

[2]FCC Decision regarding MCI, August 13, 1969, released August 14, 1969. Federal Communications Commission, U.S. Government Printing Office, Washington, D.C.

> . . . (Commission believes) that MCI's offering would enable such subscribers to obtain a type of service not presently available and would tend to increase the efficiency of operation of the subscribers' business
>
> . . . MCI may come directly to the Commission with a request for an order of interconnection. We have already concluded that a grant of MCI's proposal is in the public interest. We likewise conclude that, absent a significant showing that interconnection is not technically feasible, the issuance of an order requiring the existing carriers to provide loop service is in the public interest.

The ruling was viewed with dismay by AT&T, other common carriers, as well as selected segments in the industry. Opponents viewed the ruling as completely unfair to the Bell System which, as a regulated monopoly, had provided America with the most reliable and efficient communications system in the world. Nonetheless, the ruling held. MCI and other specialized common carriers came into existence and, as AT&T predicted, no one saw an SCC construct a communications link to service the residents of Muleshoe, Texas, and Prairie View, New Mexico.

Indeed, the SCCs skimmed the cream, but not without difficulty. In the 1970s, MCI contended that AT&T was denying them timely local loop connections, resulting in lost customer accounts. AT&T denied the charges, but again AT&T lost. In 1980, a Chicago court awarded MCI $1.8 billion, based on a 1974 suit of local loop denial. (The suit represented an enormous amount of money to MCI but just a few weeks of AT&T gross operating revenue.) Thus, AT&T found itself once again in a position of "lost turf."

Specialized Common Carriers in the Switched Network

The MCI decision resulted in an erosion (however slight) of AT&T's leased lines monopoly. In 1975, AT&T also lost its monopoly on switched long distance service. Its adversary, once again MCI, began marketing EXECUNET. Using Bell's local loops, a customer dials into an MCI computer, which then switches the call through its own network. Other companies followed MCI; Southern Pacific (Sprint), ITT (City-Call), and Western Union (Metro I) initiated switched offerings to the public. The SCCs offered "discount" prices, claiming a customer could save 8 to 19% on a five-minute call and 4 to 16% on a 10-minute connection. AT&T contends that the discount direct dial companies cause little concern. (They have less than 1% of the market.) However, switched calls represent a sizable revenue base ($26 billion in 1980) and SCC inroads will certainly present problems for AT&T.

The Computer Inquiries

A 1956 consent decree prohibited AT&T from entering any business other than common carrier communications. However, the definition of what was communications and what was data processing was becoming very hazy. FCC initiated its First Computer Inquiry in 1966 (ending in 1973) in order to provide guidance to the industry. The FCC established definitions of data processing and message switching and allowed all carriers except AT&T to perform both functions. Bell was prohibited because of the consent decree.

The inquiry did not settle the issues because the computer and communications industries continued to merge. Finally, in 1976, the FCC initiated the Second Computer Inquiry. Its final and very significant conclusion in 1981 established that common carriers could split their organizations into basic and enhanced services. The basic services such as telephones would be regulated. The remaining enhanced services would be unregulated.

The "Open Skies" Policy and Satellite Technology

In 1971, the FCC stated that it would license any responsible applicant to operate satellite communications systems. The statement was meant to foster competition and encourage the application of satellite technology to new services and functions. At the time of the policy announcement, AT&T was banned from owning satellites until 1979 and was required to lease channels from other carriers.

Certainly one of the most significant results of the announcement was the entry of IBM into the communications field. IBM formed the Satellite Business Systems (SBS) Corporation in partnership with COMSAT General and Aetna Life and Casualty and successfully petitioned the FCC for a satellite carrier license. In December 1975, SBS went into operation and subsequently launched a geosynchronous satellite system. In 1981, it acquired its first customer, and in 1982 used the space shuttle to launch its third satellite.

At the present time, SBS is consolidating its operations, gathering a user base, and working toward recovering its greater than $600 million investment.

The Communications Satellite Corporation (COMSAT) was created prior to the open skies policy to foster and advance communications satellite technology. COMSAT and NASA were instrumental in bringing the United States to a position of world leadership in space communications. In the 1960s and 1970s, the government provided foresight and support to the communications industry. However, NASA's role as an R&D agency has not been changed; COMSAT has a limited charter; and the open skies policy did not have many takers due to the extraordinary costs involved. Few private sector organizations can afford a satellite program.

Unfortunately, the U.S. position in future commercial satellite space uses is being eroded due to the lack of a cohesive national plan and a lack of support from the Congress and government agencies. On the other hand, the Japanese and European governments have developed active programs to support their companies in this important field. NASA and COMSAT need to have their missions expanded to help the U.S. develop and market satellite communications technology, as well as other uses at the satellites. Once again, U.S. research will likely be exploited by other countries.

THE INFORMATION SOCIETY

The Merging Technologies and the World Economy

The FCC Computer Inquiries recognized that the communications and computer technologies are merging. The once clear distinctions between the industries no longer exists. Equipment and software such as PBXs, word processors, rooftop satellite antennas, and electronic mail systems all contain computer and communications components. The problem is distinguishing what is to be regulated (if anything) and what is not subject to regulation. It makes little sense to allow an unregulated industry to enter the common carrier field and, at the same time, forbid the common carriers to compete in the computer industry. Recent court decisions and FCC rulings show very encouraging signs, but the present structure is not suited to foster the development of a sound national computer-communications infrastructure.[3] IBM and AT&T are important companies; both are assets to the country. Each must be given sufficient leeway to grow naturally and that means both companies must be allowed to compete in each other's domain.

At the same time, the other companies in these industries must be given fair opportunity to introduce products and garner a piece of the marketplace without unfair impediment from the two giants. The nation must develop a policy to (a) accommodate and nurture the merging communications-computer industry, (b) give more latitude to the industry leaders to continue their growth, and (c) provide sufficient shielding and incentives to the smaller innovative companies to allow them to contribute and prosper. Without question, all this is a sizable task. Yet, it makes a great deal of difference to our business fortunes in the U.S. if we treat this problem as a fragmented, special interest issue or whether we view it as an issue justifying a national policy and requiring acquiescence of vested special interests. The Computer Inquiry II decisions, the IBM-SBS venture, and the recent AT&T-American Bell announcement are very positive steps and should lead to other major changes in the future.

We must also rethink our traditional approaches in supporting private enterprise. As the world moves toward a more interdependent and integrated economy, it is vital that the U.S. adopt programs that *help not hinder* our private sector to compete in the world market. At the present time, three of America's greatest economic assets are its computer, communications, and space satellite industries. Our present government programs are doing little to make these industries viable in the future world market—while other industrialized nations have already implemented support programs for their high technology firms. We can no longer view competition as an intranational issue; our long-range fortunes and wealth depend on how well we fare in the world competitive arena.

---

[3]In June, 1982, AT&T made a major announcement as a result of the Computer Inquiry II decision: the creation of American Bell, a wholly owned subsidiary for data communications and enhanced services. American Bell places one of the nation's technological giants squarely into the vital communication and computer arenas. In the near future, American Bell will implement NET/1000, a large public data communications network.

Managing the Information Society

Some very thought-provoking discussions on the information society are emerging in the communications and computer industries, in Congress, and in business and academic circles. The discussions revolve around the idea that the information society has come to rely on knowledge as its principal product and commodity for economic growth. As Peter Ducker stated, knowledge (derived from information) has become the primary industry in the United States.

As discussed in Chapter 1, many people believe that the generation and use of information actually creates wealth. Information increases the Gross National Product (a) by enhancing productivity and (b) by reinforcing research and development (R&D). Productivity and R&D lower the costs of producing the society's goods.

This recognition has lead some industries to a greater commitment to increased automation and has created a degree of employee unrest as jobs are replaced or eliminated by the computer (e.g., the printers' trade). With increasing frequency, organizations are focusing on plans to achieve more output per unit of input of capital, energy, materials, and people. And, with the diminishing supply of cheap energy, our society has had to rethink the traditional idea of increasing output by a stepwise increase in the inputs. Without question, the computer and communications industries are supplying many of the answers to these problems.

At the same time, other questions are being raised about the treatment and management of information as a resource, product, and commodity. For example, information is quite different from other products (such as a building, a television set, or a car):

- It is often difficult to assign a price to information.
- It is often difficult to assess the quality of information.
- Once sold, information is not always relinquished to the buyer. The seller often retains the product. This does not fit in very neatly with some of our capitalistic notions.
- Information as a commodity does not always depreciate as do most of society's commodities. This fact presents some challenges in the future for the nation's tax structure and business tax incentives.

According to neoclassical economic theory, a society benefits from a competitive environment for the production, distribution, and marketing of its products. The "purist" holds that unrestricted competition leads to the best product and the best use of society's investments—as long as the competition does not infringe upon the competitive rights of others.[4] However it is recognized that certain areas do not benefit from competition, and that a price/demand approach does not work well for resources that are common to the society (e.g., parklands, rivers, air). This idea was part of the rationale for making the communications industry a regulated monopoly.

Certain sectors of our society have raised the idea that information (or certain aspects of information) are common to all, and competitive strategies will not function properly. Several years ago, practically everyone (except the common carriers) agreed that increased competition in the communications industry would surely lead to significantly reduced prices and costs to the consumer. Today, some of these deregulation proponents are having second thoughts.

The issues are complex, especially in view of the merging of the computer and communications technologies and in view of the increased competition to these vital industries from other countries. Our leaders are facing what appear to be contradictory choices: (a) enforce the antitrust laws, leading to the breakup and destabilization of several major U.S. firms that are quite successful in the international market or (b) provide support and encouragement of the large computer and communications firms (as Japan does for its industry giants), at the likely expense of the smaller, yet truly innovative companies. Surely, we can hope to find a middle ground, but recent experience in Congress and the Executive Branch does not seem very encouraging. The problem is summarized very well by an editorial in *The Washington Post*:[5]

> With good reason, those concerned with efficiency, low prices, and effective competition have supported vigorous enforcement of antitrust laws. . . . Critics . . . agree . . . that breaking up "stifling, gluttonous, masses" which charac-

---

[4]This leads to an interesting logical anomoly: Some economists state that the best social structure is one that permits a substantial amount of information about a commodity, so that the purchaser can make wise purchase decisions. However, if a potential buyer has substantial information about an information-type commodity, there may be no need to purchase it!

[5]Louis T. Wells and Marshall I. Goldman, "Save the Business Baroness," *Outlook, The Washington Post*, September, 1978. Since this article was written, the major government antitrust suits against IBM and AT&T were dropped.

> terize many American industries is obviously the superior solution to the antitrust problem.
>
> But while such a policy made sense in the past, it is increasingly not only shortsighted but detrimental in many cases today. Indeed, the need now in important instances may be to alter antitrust law and enforcement to encourage the concentration of economic power.
>
> Does it make sense, for example, to tie up the courts with cases seeking to restrain or breakup companies like IBM, GM or Xerox, when countries like Japan and Germany are considering how to consolidate and strengthen their computer, auto and copying industries so they can outproduce their American counterparts? Certainly it is ironic that, at a time when we need to improve our balance of trade and bolster the troubled dollar, many of our largest manufacturers and exporters find themselves in court because they have been too effective. It is wasteful, to say the least, to see a company like IBM spend several million dollars a year for almost 11 years on legal fees for its antitrust suits from Justice. It seems particularly ill-advised now that Hitachi has announced plans to introduce a computer as (direct competition to one of our most successful industries).

Most of us realize that the merging of the communications and computer technologies is inevitable—the two are a logical whole. A wise national communications policy should encourage increased integration of the computer with communications. Such a policy would provide improvements to our information society.

The issues are critical to our economic and personal well-being, for computers and the communication links between the computers will have as profound an effect in the future as did the use of railroads and roads in the past on this country's ability to exploit the industrial revolution. The cotton gin would have had limited value if roads and rivers were not available to transport the cotton to the consumers. The Clyne Report makes this point.[6]

---

[6]The Clyne Report on information technology is available from the Canadian Embassy, 1746 Massachusetts Ave., N.W., Washington, D. C. 20036 Refer to: Clyne Report *Telecommunications and Canada*, 1979.

> Rich countries of today are those that exploited the industrial revolution in the 19th century; the rich countries of the future will be those that exploit the information revolution to their own best advantage in the 20th & 21st centuries.

The transportation network for the information society consists of the nation's communications resources. These resources form the basic infrastructure for the information society, allowing the computers to effectively communicate with each other. The full benefits of the information society can only be realized if a cohesive and efficient infrastructure exists. Recent FCC rulings, court decisions, and the advent of AT&T's American Bell and IBM's SBS are steps in the right direction, but do not go far enough. A broad national policy is needed that continues to encourage a merging of the technologies, fosters their movement into the global market place, and recognizes the importance of the infrastructure to our economic growth and social well-being.

APPENDIX

STANDARDS ORGANIZATIONS

CCITT STUDY GROUP ACTIVITIES

Study Group I. Defines operational aspects of Teletex, Videotex, and facsimilie. Produced Recommendations F.200 for Teletex and F.300 for Videotex. Telematic is a term proposed to cover this area.

Study Group VII. Defines standards for public data networks. Many of the X-series recommendations have come from this group. It is also responsible for the ISO Reference model. The group consists of several Working Parties (WP) and Special Rapporteurs to address special issues.

Study Group VIII. Defines terminal equipment recommendations for Telematic services. The group is addressing the upper four layers of the OSI Reference model. The group has published recommendations on Telematics (S-series) and facsimilie transmission (T-series).

Study Group XI. Defines standards for telephony switching and signalling. The group will publish the Q.900 series.

Study Group XVII. Defines data transmission standards over a telephone network. The group developed the V series for modem interfaces.

Study Group XVIII. Defines standards for digital networks. The group will publish the I series documents.

CCITT has nine other Study Groups:

Number	Subject
II	Telephone operation
III	General tariff principles
IV	Maintenance of networks
V	Electromagnetic impairment control
VI	Cable sheaths and poles
IX	Telegraph networks
XII	Telephone performance
XV	Transmission systems
XVI	Telephone circuits

U.S. participation in CCITT is provided by the State Department and CCITT Study Groups (SG)

Number	Subject
SG A	U.S. government regulatory policies
SG B	Telegraph operations
SG C	Worldwide telephone network
SG D	Data transmission

ISO TECHNICAL COMMITTEES AND SUBCOMMITTEES

Technical Committee 97. TC 97 defines standards in the ISO Reference model for distributed systems. It consists of three subcommittees (SC) and several Working Groups (WG):

SC-6: Data Communications

WG-1: Data Link Layer
WG-2: Network Layer
WG-3: Physical Layer

SC-16: Open Systems Interconnection (OSI)

- WG-1: Reference Model
- WG-4: Application and System Development
- WG-5: Application and Presentation Layers
- WG-6: Session and Transport Layers

SC-18: Text Preparation and Interchange

- WG-1: User Requirements
- WG-2: Symbols and Terminology
- WG-3: Text Structure
- WG-4: Text Interchange
- WG-5: Text Preparation and Presentation

ANSI TECHNICAL COMMITTEES

X353—Data Communications (Works with ISO SC 6)

- X353.1: Planning
- X353.2: Vocabulary
- X353.3: Network Layer
- X353.4: Data Link Layer
- X353.5: Quality of Service
- X353.6: Signalling Speeds
- X353.7: Public Digital Network Access

X3T5—Open Systems Interconnection (OSI) (Works with ISO SC 16)

- X3T5.1: Reference Model
- X3T5.2: Planning
- X3T5.4: Application and System Management
- X3T5.5: Application and Presentation Layers
- X3T5.6: Session and Transport Layers

X3V1—Office Systems (Works with ISO SC 18)

- X3V1.1: User Requirements
- X3V1.2: Symbols and Terminology
- X3V1.3: Text Structure
- X3V1.4: Procedures for Text Interchange
- X3V1.5: Text Preparation and Presentation

The following lists contain some of the better known standards that have been developed by CCITT, ISO and ANSI.[1] These standards can be obtained by writing:

CCITT Documents:

U.S. Department of Commerce
National Technical Information Service (NTIS)
5285 Port Royal Road
Springfield VA 22161

ISO and ANSI Documents

American National Standards Institute
1430 Broadway
New York, NY 10018

CCITT Series Recommendations

Number	Title
V.1	Equivalence between binary notation symbols and the significant conditions of a two-condition code.
V.2	Power levels for data transmission over telephone lines.
V.3	International Alphabet No. 5.
V.4	General structure of signals of International Alphabet No. 5 code for data transmission over public telephone networks.
V.5	Standardization of data signaling rates for synchronous data transmission in the general switched telephone network.
V.6	Standardization of data-signaling rates for synchronous data transmission on leased telephone-type circuits.
V.7	Definitions of terms concerning data communication over the telephone network.
V.10(X.26)	Electrical characteristics for unbalanced double-current interchange circuits for general use with integrated circuit equipment in the field of data communications.
V.11(X.27)	Electrical characteristics for balanced double-current interchange circuits for general use with integrated circuit equipment in the field of data communications.
V.15	Use of acoustic coupling for data transmission.

[1]Omnicon, Inc., provides a very useful reader service on these standards. Write to Omnicon, Inc., 400 Holloway Ct. N.E., Vienna, Va. 22180. Portions of this Appendix were extracted from Omnicon, NTIS, ANSI, and the National Bureau of Standards. The author thanks Omnicon for the excellent X and V series descriptions.

CCITT Series Recommendations

Number	Title
V.16	Medical analogue data transmission modems.
V.19	Modems for parallel data transmission using telephone signaling frequencies.
V.20	Parallel data transmission modems standardized for universal use in the general switched telephone network.
V.21	300 bits per second duplex modem standardized for use in the general switched telephone network.
V.22	1200 bits per second duplex modem standardized for use on general switched telephone network and on leased circuits.
V.23	600/1200-baud modem standardized for use in the general switched telephone network.
V.24	List of definitions for interchange circuits between data terminal equipment and data circuit-terminating equipment.
V.25	Automatic calling and/or answering equipment on the general switched telephone network, including disabling of echo suppressors on manually established calls.
V.26	2400 bits per second modem standardized for use on four-wire leased circuits.
V.26bis	2400/1200 bits per second modem standardized for use in the general switched telephone network.
V.27	4800 bits per second modem with manual equalizer telephone-type circuits standardized for use on leased circuits.
V.27bis	4800/2400 bits per second modem with automatic equalizer standardized for use on leased telephone-type circuits.
V.27ter	4800/2400 bits per second modem standardized for use in the general switched telephone network.
V.28	Electrical characteristics for unbalanced double-current interchange circuits.
V.29	9600 bits per second modem standardized for use on point-to-point four-wire leased telephone type circuits.
V.31	Electrical characteristics for single-current interchange circuits controlled by contact closure.
V.35	Data transmission at 48 kilobits per second using 60-108 kHz group band circuits.
V.36	Modems for synchronous data transmission using 60-108 kHz group band circuits.
V.37	Synchronous data transmission at a data signaling rate higher than 72 kbits using 60-108 kHz group and circuits.
V.40	Error indication with electromechanical equipment.

CCITT Series Recommendations

Number	Title
V.41	Code independent error control system.
V.50	Standard limits for transmission quality of data transmission.
V.51	Organization of maintenance of international telephone type circuits used for data transmission.
V.52	Characteristics of distortion and error-rate measuring apparatus for data transmission.
V.53	Limits for the maintenance of telephone-type circuits used for data transmission.
V.54	Loop test devices for modems.
V.55	Specification for an impulse noise measuring instrument for telephone-type circuits.
V.56	Comparative tests of modems for use over telephone-type circuits.
V.57	Comprehensive data test set for high data signaling rates.
X.1	International user classes of service in public data networks.
X.2	International user services and facilities in public data networks.
X.3	Packet assembly/disassembly facility (PAD) in a public data network.
X.4	General structure of signals of International Alphabet No. 5 code for data transmission over public data networks.
X.15	Definitions of terms concerning public data networks.
X.20	Interface between data terminal equipment (DTE) and data circuit-terminating equipment (DCE) for start-stop transmission services on public data networks.
X.20bis	Use on public data networks of data terminal equipment (DTE) which is designed for interfacing to asynchronous duplex V-series modems.
X.21	Interface between data terminal equipment (DTE) and data circuit-terminating equipment (DCE) for synchronous operation on public data networks.
X.21bis	Use on public data networks of data terminal equipments which is designed for interfacing to synchronous V-series modems.
X.22	Multiplex DTE/DCE interface for user classes 3-6.
X.24	List of definitions for interchange circuits between data terminal equipment (DTE) and data circuit-terminating equipment (DCE) on public data networks.

CCITT Series Recommendations

Number	Title
X.25	Interface between data terminal equipment (DTE) and data circuit-terminating equipment (DCE) for terminals operating in the packet mode on public data networks.
X.26(V.10)	Electrical characteristics for unbalanced double-current interchange circuits for general use with integrated circuit equipment in the field of data communications.
X.27(V.11)	Electrical characteristics for balanced double-current interchange circuits for general use with integrated circuit equipment in the field of data communications.
X.28	DTE/DCE interface for a start-stop mode data terminal equipment accessing the packet assembly/disassembly facility (PAD) in a public data network situated in the same country.
X.29	Procedures for the exchange of control information and user data between a packet assembly/disassembly facility (PAD) and a packet mode DTE or another PAD.
X.40	Standardization of frequency-shift and modulated transmission systems for the provision of telegraph and data channels by frequency division of a group.
X.50	Fundamental parameters of a multiplexing scheme for the international interface between synchronous data networks.
X.50bis	Fundamental parameters of a 48 kbit/s user data signaling rate transmission scheme for the international interface between synchronous data networks.
X.51	Fundamental parameters of a multiplexing scheme for the international interface between synchronous data networks.
X.51bis	Fundamental parameters of a 48 kbit/s user data signaling rate transmission scheme for the international interface between synchronous data networks using 10-bit envelope structure.
X.52	Method of encoding anisochronous signals into a synchronous user bearer.
X.53	Numbering of channels on international multiplex links at 64 kbit/s.
X.54	Allocation of channels on international multiplex links at 64 kbit/s.
X.60	Common channel signaling for circuit switched data applications.
X.61	Signaling system No. 7—Data user part.
X.71	Decentralized terminal and transit control signaling system on international circuits between synchronous data networks.

CCITT Series Recommendations

Number	Title
X.75	Terminal and transit call control procedures and data transfer system on international circuits between packet-switched data networks.
X.80	Interworking of interexchange signaling systems for circuit switched data services.
X.87	Principles and procedures for realization of international user facilities and network utilities in public data networks.
X.92	Hypothetical reference connections for public synchronous data networks.
X.96	Call progress signals in public data networks.
X.110	Routing principles for international public data services through switched public data networks of the same type.
X.121	International numbering plan for public data networks.
X.130	Provisional objectives for call set-up and clear-down times in public synchronous data networks (circuit switching).
X.132	Provisional objectives for grade of service in international data communications over circuit switched public data networks.
X.150	DTE and DCE test loops in public data networks.
X.180	Administrative arrangements for international closed user groups (CUGs).

International Organization for Standardization

Number	Title
ISO 646	Seven-bit character set for information processing interchange—1973, confirmed 1979
ISO 1155	Information processing—Use of longitudinal parity to detect errors in information messages—1978
ISO 1177	Information processing—Character structure for start/stop and synchronous transmission—1973, revision being balloted.
ISO 1745	Information processing—Basic mode control procedures for data communications systems—1975, revision being balloted.
ISO 2022	Code extension techniques for use with ISO seven-bit coded character set—1973.
ISO 2110	Data communication—25-pin DTE/DCE interface connector and pin assignments—1980.

International Organization for Standardization

Number	Title
ISO 2111	Data communication—Basic mode control procedures—Code independent information transfer—1972, revision being balloted.
ISO 2593	Connector pin allocations for use with high speed data terminal equipment—1973, revision being balloted.
ISO 2628	Basic mode control procedures—Complements—1973, confirmed 1979.
ISO 3309	Data communication—High-level data link control procedures—frame structure—1979, revision being balloted.
ISO 4335	Data communication—High-level data link control procedures-Elements of procedures 1979—Addendum I, 1979; Addendum II, 1981
ISO 4902	Data communication—37-pin DTE/DCE interface connector and pin assignments—1980.
ISO 4903	Data communication—15-pin DTE/DCE interface connector and pin assignments—1980.
ISO 6159	Data communication—HDLC unbalanced classes of procedures —1980.
ISO 6256	Data communication—HDLC balanced class of procedures—1980.

American National Standards Institute

Number	Title
X3.1—1976	Synchronous signaling rates for data transmission.
X3.4—1977	Code for Information Interchange.
X3.15—1976	Bit sequencing of the American National Standard Code for information interchange in serial-by-bit data transmission.
X3.16—1976	Character structure and character parity sense for serial-by-bit data communication in the American National Standard Code for Information Interchange.
X3.24	Signal quality at interface between data terminal equipment and synchronous data communication equipment for serial data transmission (adopts EIA RS-334-A)—pending resolution of negative ballot.
X3.25—1976	Character structure and character parity sense for parallel-by-bit communication in the American National Standard Code for Information Interchange.

American National Standards Institute

Number	Title
X3.28—1976	Procedures for the use of communication control characters of American National Standard Code for Information Interchange in specified data communication links.
X3.36—1975	Synchronous high-speed data signaling rates between data terminal equipment and data communication equipment.
X3.41—1974	Code extension techniques for use with 7-bit coded character set of American National Standard Code for Information Interchange.
X3.44—1974	Determination of the performance of data communication systems.
X3.57—1977	Structure for formatting message headings for information interchange using the American National Standard Code for Information Interchange for data communication system control.
X3.66—1979	For advanced data communications control procedures (ADCCP).
X3.79—1981	Determination of performance of data communication systems that use bit-oriented control procedures.
X3.92—1981	Data encryption algorithm.

Several other organizations are quite active in the standards area. The European Computer Manufacturers Association (ECMA) has published some valuable documents on HDLC and local area networks (ECMA, 114 Rue du Rhone, CH-1204 Geneva, Switzerland). The Electronics Industries Association has many published standards and provides a catalog of these documents (ELA, 2001 Eye St., N.W., Washington, D.C. 20006). The National Communications System (NCS), (General Services Administration, Specification Distribution Board, Building 197, Washington Navy Yard, Washington, D.C. 20407) and the National Bureau of Standards (write to the NTIS) publish standards for government organizations.

INDEX